W9-BID-933

Comfort Heating

Second Edition

Comfort Heating

Billy C. Langley

Reston Publishing Company, Inc.
A Prentice-Hall Company
Reston, Virginia

Library of Congress Cataloging in Publication Data

Langley, Billy C.
Comfort heating

Includes Index.
1. Heating—Equipment and supplies.
2. Heating—Equipment and supplies—Maintenance and repair. I. Title.
TH223.L28 1978 697 77-28165
ISBN 0-87909-091-X

A Prentice-Hall Company
Reston, Virginia 22090

10 9 8 7 6 5 4 3 2

Printed in the United States of America

Contents

Preface

This text is designed to provide the necessary facets of theory and practice for anyone interested in the subject of comfort heating with gas, oil, electricity, coal, or solar energy.

The performance-based objectives provided at the beginning of each chapter give the student an indication of what minimum knowledge he can expect to receive from that portion of the text.

Comfort Heating provides not only the necessary theory for the student and service technician alike, but also gives examples to help reinforce that theory. Many actual manufacturers' service and/or maintenance and operating instructions are incorporated as well.

Upon completion of this study, the student will have the knowledge and confidence necessary for the proper servicing and installation of heating equipment.

Comfort Heating

1

A Brief History of Heating

The objectives of this chapter are:

- To acquaint the reader with the evolution of heating from primitive to present times.
- To cause the reader to better appreciate the advancements made in the heating industry.

The originator of home heating was probably the Stone Age Man. With his primitive ways, the Stone Age Man had no means of producing heat for himself. He had to depend on nature to provide a means of keeping warm or a method of cooking food. This came in the form of lightning which started dried leaves or wood burning. Lightning, therefore, is akin to the modern electric ignitors used on modern heating units.

Archeologists have lamps that have been traced back thousands of years. Some of these lamps were made like bowls while others were made from human skulls. These crude lamps were used by prehistoric artists in old caves where they made etchings on the cave walls. They produced light to work by and enough heat to warm the artists' hands during the cold season. The fuel for the lamps was animal fat, with wicks made from dried moss, twisted into ropelike strands. By slightly stretching the imagination, this could be called the first oil burner.

The first central heating units were put to use by the ancient Romans and Chinese. Not wanting to get their magnificent castles dirty with soot, the Romans built a furnace pit, or hypocaust, beneath the buildings. The heat would penetrate the castles through a 12 or 14 inch floor and thus heat their homes with the first man-made radiant heating systems.

The Chinese approached the matter in a more advanced way. They built ovens with air passages beneath their homes and installed warm air pipes to the rooms to be heated. In the bedrooms, the beds were placed directly over the outlets. In this manner, the Chinese heated their homes with the first central warm air systems.

Chimneys came into being in the twelfth century. With the chimney came the fireplace and the fuel was burned within the area to be heated, providing better use of the heat giving source. They were, however, very expensive at first and only a few buildings were equipped with them. The chimney and fireplace removed smoke, smells, and a good deal of heat from the home. When burning, they created such a draft that special furniture was designed to protect the occupants.

Ben Thompson helped to alleviate the draft problem some by building a restrictor, or throat, in the chimney. The restriction slowed down the burning and allowed more of the heat to be put into the room. Ben Franklin improved on the restrictor by adding a path for the smoke that made it go through the restrictor, then down behind the fireplace, below the hearth, then out the chimney, thereby making greater use of the fuel.

The early common fuels for fireplace use were wood, peat, and charcoal. Shepherds burned dried manure and in some cases dried bones, called a bone-fire, which has been changed in modern language to bonfire. Coal was recognized by Theophrastus as a fuel. However, it was usually used as a semiprecious jewel to ward off evil spirits. People were executed in England for burning coal. In the fourteenth century, the forests in England were diminishing so rapidly that Parliament asked the king to change the law governing the burning of coal and allow its use as a fuel. However, it was Queen Elizabeth I who actually passed the act to conserve the forest and allow coal to be put to use.

Louis Savot, a Paris physician, designed and installed the first heat-circulating fireplace. It was installed in Louvre Palace around 1600 A.D. The bottom and back were made of metal. Air circulated through the fireplace by entering at the bottom, below the hearth, rose to the top by convection, and exited under the mantle, very much resembling heat exchangers in modern forced-air furnaces. The same principle is used in today's fireplaces.

The slower-burning coal fires allowed the grate to be used. The coal to be burned was placed on the grate, allowing greater aeration of the fuel and more complete burning. Also, the ashes would fall beneath the grate and could be removed more easily. Outlets were placed in the first floor ceiling and allowed the heat to enter the upstairs rooms and, after a fashion, heat them. Thus, they brought central heating one step closer. Because of the grate, the Franklin stove and its descendents caused the fireplace to become more decorative than useful.

Gas, the most popular of the modern fuels, was first used in the western world by balloonists. Pilatre de Rozier, the leading balloonist of

France at the time, attempted to cross the English Channel in a combination airship that included a hot-air fire balloon connected with a gas balloon. In June 1785, high over the French coast, the airship exploded. All passengers were sent to a fiery death. Because of this incident, gas became known as a mysterious and powerful source of energy. Man did not devise a means for controlling gas until many years later.

Fuel oil began to be recognized about 1860 as a plentiful source of energy for economical heating. In 1861, Werner, a mechanic, developed the first oil burner. In his invention, oil was trickled over preheated plates where it turned into a vapor and could be readily ignited. This method is not too different from modern pot-type oil burners. In 1863, the first pressure-type spray oil burner was introduced by Brydges Adams.

Fuel oil, however, was not advancing alone. Coal was also making great strides in comfort heating. Living room stoves were being moved to the basement. The home was heated with ducts, pipes and radiators. All of these devices, together with the advancements which made it cleaner to use coal, helped to bring about practical central heating in the nineteenth century. The man of the household shook the grate and stoked the furnace in the morning and again in the evening. However, the pipes, ducts, and registers were an eyesore in the decor of the home. Because the design of the ducts was difficult and the home could not be evenly heated, steam boilers and radiators became more popular. However, the noise accompanying these early radiators made them undesirable.

Through highly sophisticated engineering, completely automatic oil heat emerged in about 1920. These units were controlled by a thermostat. There was no mess as with coal. Oil heat was accompanied by carefree operation and the equipment was very durable. The heating equipment had automatic safety controls, which was a strong factor in oil replacing coal as a heating fuel. By about 1928 there were more than one-half million oil burners in use.

In the 1950s, pipelines brought natural gas to most of the larger cities and to the homes of most users. This convenience soon made natural gas the most popular home heating fuel. About this same time public utilities operating with the security of government franchises forced their way into the comfort heating markets. Each of these utilities claimed to be the most economical and dependable.

In the 1960s, electric utilities emerged on the heating market. At this time sophisticated graphs and charts were used to disguise the operating costs, which were more expensive than most people cared to pay.

Both gas and electricity claimed to be the only truly modern fuel. However, when we search back in history we find that the Chinese used gas for heating almost 3,000 years ago. Also, electric heat was being installed and put to use as far back as 1889 in Minneapolis, Minnesota.

2

Heat Sources and Combustion

The objectives of this chapter are:

- To introduce to you the different sources of heat for comfort heating.
- To provide you with the basic characteristics of heating fuels.
- To acquaint you with the distribution systems used for different sources.
- To demonstrate to you the need for knowledge of proper combustion.
- To bring to your attention the requirements of a safe, economical heating system.

HEAT SOURCES

Strictly speaking, fuel is any substance that releases heat when mixed with the proper amount of oxygen. Only those materials that ignite at relatively low temperatures, burn rapidly, and are easily obtained in large quantities at relatively low prices are considered good fuels.

The value of a fuel is derived from the amount of heat released when it is burned and the heat of combustion is measured. This value is obtained when a given amount of a fuel is burned under controlled conditions. The apparatus used for this purpose is called a *calorimeter*. The released heat is absorbed by a definite volume of water and the rise in temperature of the water is measured. The common ratings are given in British thermal units (Btu/lb) or cubic foot (ft^3) of the fuel burned.

When a fuel contains hydrogen, the heat given off during combustion will depend on the state of the water vapor (H_2O) formed when the hydrogen is burned. A heating value known as the higher, or gross, heating value is obtained when this water vapor is condensed and the latent heat of condensation is salvaged. If this water vapor is not condensed, the latent heat of vaporization is lost and this is known as the lower, or net, heating value.

The heating values of solid fuels are given in Btu/lb. Liquid fuels may be rated in Btu/lb or Btu/gal of the fuel. The Btu rating of gaseous fuels is given per ft^3. The gas industry uses the standard conditions of a temperature of 60° F, 30 inches of mercury pressure with a saturated condition with water vapor to determine these values.

The function of a heating system depends on some source of energy for proper operation. There are four basic sources of energy used to accomplish this tremendous job. They are (1) *solids,* (2) *liquids,* (3) *gases,* and (4) *electricity.*

Even with modern advancements and techniques, *coal* is still the most popular solid fuel used for heating. Coal is a mixture of carbonaceous material, minerals, and water. Coal was formed from vegetation that had gone through the decaying processes over a long period of time. This decayed matter was then covered with layer upon layer of earth and rock. The heat caused by the decaying vegetation and the pressure of the layers of earth caused the composition to turn into several ranks of coal.

Coal is obtained by one of two methods: open-pit and deep mining. When the open-pit method is used, the overlying earth is removed with earth moving equipment and then the coal is recovered. When the deep mining method is used a tunnel is sunk into the coal shaft proper. The coal is then removed by using explosives and loading it into cars or on conveyors that take it to the surface and then through the refining process.

Coal is divided into four classifications in decreasing order of rank. The classifications are anthracite, bituminous, subbituminous, and lignite. The higher ranking coals are classified according to their carbon content when they are dry. The lower ranking coals are classified according to their calorific value.

The two most popular coals used for heating are anthracite, or hard, and bituminous, or soft. Anthracite coal will release from 13,000–14,000 Btu/lb when burned with 9.6 pounds of air. Bituminous coal will release from 12,000–15,000 Btu/lb when burned with 10.3 pounds of air. Dry air at about 70° F has a volume of 13.3 ft^3/lb.

Coal is burned by the use of several methods. The oldest method used is hand-firing. In this method, coal is thrown onto the grates through the opening where the primary air is drawn. The fuel bed con-

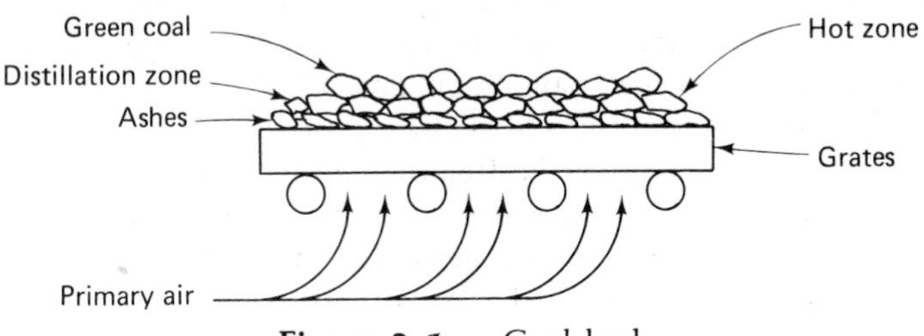

Figure 2–1. Coal bed.

sists of a layer of ashes lying directly on the grates, a hot zone where the combustion occurs, a cooler, or distillation, zone where the gases are driven from the coal, and then a layer of green coal (Figure 2–1). It is necessary to admit secondary air over the fire to obtain complete combustion for two reasons. First, hot carbon and carbon dioxide react to form combustible carbon monoxide. Second, there is distillation of gases from the distillation zone.

When hand-firing was used, one method was to push the hot coals to the back of the fire box and leave the green coal toward the front. The secondary air entering through the door would carry the gases over the hot coals, causing them to burn more completely than if the green coal were placed on the top.

Mechanical firing was the next method to appear for the use of coal as a heating fuel. This method used the principle of feeding coal and air simultaneously into the fire box. The coal is pushed into the fire box by means of a rotating worm or screw. The hotter part of the fire was either pushed to the rear or to the top of the fire box, depending on whether the front feed or the bottom feed was used.

Coal is not as popular as a heating fuel as it was at the turn of the century. Some of the factors contributing to its lack of use are inconvenience, storage problems, and pollution.

Fuel oil is the most common liquid used in the heating industry today. It is a petroleum product and is comprised of a mixture of liquid hydrocarbons, produced as a byproduct of the refining of petroleum. Petroleum has been known for thousands of years. Seepages of oil and gas around the Caspian Sea and Black Sea were known and used for heating and cooking before the birth of Christ. The Chinese drilled for oil long before the Christian era. They used percussion bits, bamboo piping, and much manpower. The Chinese discovered gas and oil while drilling for salt.

Fuel oil is a stiff competitor of natural gas in the home heating industry. Fuel oil is graded and classified according to the range of distillation. The grades range from 1 to 6, omitting number 3. Grades 1, 2, and 4 are used for heating, with 1 and 2 the most popular for domestic use. Number 4 is generally used in light industrial furnaces. The lower graded numbers are more expensive than the higher numbered oils because of impurities, such as asphalt. Grades of 5 and 6 are too thick to be used in domestic equipment and require preheating to insure a steady flow to the burner. The specifications governing fuel oils set forth by the U.S. Department of Commerce conform to American Society for Testing Materials (ASTM) specifications for fuel oils.

The proper combustion of fuel oils can be obtained only when the oil is properly atomized and mixed with air. The heat emitted by burning fuel oil ranges from 137,000–151,000 Btu/gal, depending on the grade. The heat content of number 1 is 137,000 Btu/gal and of number 2, the most popular domestic fuel oil, is 140,000 Btu/gal. (See Table 2–1.) The flash points of fuel oils will vary considerably because of the refining methods.

Table 2–1
Btu ratings of fuel oil

Grade	Heat/Gal
No. 1	132,900–137,000
No. 2	135,800–141,800
No. 4	143,100–148,100
No. 5	146,800–150,000
No. 6	151,300–155,900

The storage of fuel oil is a contributing factor in the efficient operation of the system. In cold climates, the oil should be stored inside or in some type of heating device, such as steam or hot water pipes around the tank or electric immersion heaters used to preheat it before entering the combustion area. The fuel oil may be stored outside in milder climates without it becoming excessively thick, thereby reducing the burner efficiency.

The use of fuel oils for domestic purposes is a result of the convenience and cleanliness of oil compared to coal. Despite the competition of natural gas, fuel oil is still a leader in home heating equipment, especially in the northeastern sections of the United States.

There are two major classifications of gaseous fuels used for comfort heating. The most familiar is *natural gas.* The second one is known as *LP gas.* The principal components of natural gas are methane, about 85%; ethane, about 12%; the other 3% is made up of propane and butane. They have no carbon monoxide, oxygen, olefins or acetylene in their composition; however, some have large quantities of carbon dioxide, nitrogen, and hydrogen sulphide.

Natural gases are the lightest of all petroleum products. They are usually found where oil is found, but in some cases they are found elsewhere. Theorists have long argued about the exact origin of natural gas. However, most agree that natural gas was formed during the decomposition of plant and animal remains that were buried in prehistoric times. Because these plants and animals lived during the same period as those that are presently found as fossils, natural gas is sometimes called a *fossil fuel*.

Both natural gas and petroleum are mixtures of hydrocarbons. Both are considered fossil fuels and are composed of various chemicals obtained from the hydrogen and carbon contained in the prehistoric plants and animals.

The gas industry may be broken down into the various areas of exploration, production, transmission, and distribution.

The exploration section of the industry performs the function that its name implies. The people who are employed in exploration simply explore new areas, determine the location of the gas or petroleum field, and make the necessary reports, purchases, and other essential duties prior to the actual drilling.

When all the exploration functions are completed the production department accomplishes the actual drilling of the well. The gas, or crude oil, is brought to the earth's surface and is blocked at this point. The well remains in this state until the gas or oil is needed.

When the need arises the transmission department receives the gas from the well at pressures from 500 to 3,000 pounds per square inch gauge (psig). Even with these high pressures, the resistance of the pipe and the distance covered, booster pumps are used to transfer the gas from the well to the refinery. At the refinery, the gas passes through a drying process that removes moisture, propane, and butane. During this process, most of the odor is also removed from the raw gas and an odorant is added to aid in leak detection.

After the refining processes are completed, the distribution department receives the gas through measuring gates where the number of ft^3 of gas is recorded. The gas is then passed through a series of regulators that reduce the pressure in steps. The steps are necessary to prevent the regulators from freezing and becoming inoperative.

The gas pressure is reduced to correspond with the requirements of one of two distribution systems, either the intermediate or the low pressure system. In most cases, the low pressure system is taken from the intermediate system. The intermediate distribution system maintains pressures from 18–20 psig while the low pressure system has a pressure of approximately 8 ounces. The low pressure system is generally employed where cast iron pipe is used for distribution. Cast iron pipe does not seem to hold the gas as well at higher pressures as does the copper or polyethylene pipe used in the intermediate systems.

When the intermediate system is used, the gas pressure is reduced from 18–20 psig to 4½ ounces where it enters the house. Both systems use gas meters at this point, but the low pressure system does not require a regulator. Even though the 8 ounces of pressure in the low pressure distribution system is higher than that required in the meter loop (the meter loop is all components from the main line through the meter and regulator, if used), no regulator is needed as a result of the pressure drop. This pressure drop occurs because gas does not flow in a straight line but rolls instead, reducing the flow of gas about 10 ft^3 for each turn. This rolling action also brings about the need for straightening vanes in the line directly ahead of the gas meter. If these vanes were left out the meter would not measure the flow of gas correctly.

On leaving the meter loop, the gas flows into the house piping. Most of the appliances used are manufactured to operate on 4½ ounces of gas pressure; however, natural gas furnaces are built to operate on 3½ inches of water column of gas pressure in the furnace manifold. This requires an additional regulator at the furnace.

Natural gas has a specific gravity of .65, an ignition temperature of 1,100° F and a burning temperature of 3,500° F. One cubic foot of natural gas will emit from 900–1,400 Btu/ft^3, with the greater amount of natural gas used as a heating fuel emitting approximately 1,100 Btu/ft^3. The Btu content of natural gas will vary from area to area. The local gas company should be consulted when the exact Btu content is desired.

Natural gas is made up of 55–98% methane (CH_4), .1–14% ethane (C_3H_8), and .5% carbon dioxide (CO_2). It requires 15 ft^3 of air per ft^3 of gas for proper combustion. It is lighter than air. Because methane and ethane have such low boiling points, methane −258.7° F and ethane −127.5° F, natural gas remains a gas under the pressures and temperatures encountered during its distribution. Because of the varying amounts of methane and ethane, the boiling point of natural gas will vary according to the mixture.

The second classification of gaseous fuels is *liquefied petroleum (LP)* gas. Liquefied petroleum is both butane and propane, and in some cases a mixture of the two. These two fuels are refined natural gases and were developed for use in rural areas. They are transported by truck and stored in containers specifically made for LP gas installations.

Liquefied petroleum is a liquid until the vapor above it is drawn off. When these fuels are extracted from raw gas at the refinery, it is in the liquid state, under pressure, and remains in this state during storage and transportation. Only after the pressure is reduced, does liquefied petroleum become a gaseous fuel.

LP gas has at least one definite advantage in that it is stored in the liquid state and thus the heat content is concentrated. This concentration of heat makes it economically feasible to provide service anywhere that portable cylinders may be used. The fact that 1 gallon of liquid propane

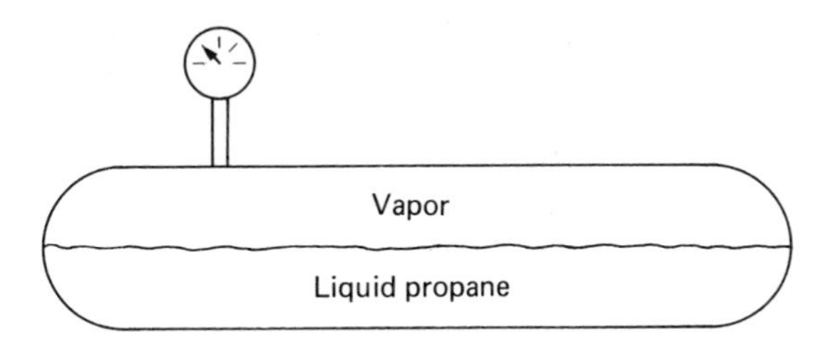

Figure 2–2. LP gas storage tank.

turns to 36.31 ft^3 of gas when it is evaporated illustrates its feasibility.

When LP gas is stored in the container, it is in both the liquid and gaseous state (Figure 2–2). To make the action of LP gas more easily understood, let's review briefly the boiling point and pressures of water (Figure 2–3). When the pressure cooker is first filled with water with no heat applied and the top remaining off, there is no pressure on the surface of the water [Figure 2–3(a)]. However, when the top is put securely in place and heat is applied, as in Figure 2–3(b), the pressure will begin to rise after the boiling point of the water is reached. As more heat is applied, more pressure will be created above the water by the evaporating liquid. When a constant temperature is maintained, a corresponding pressure will also be maintained. Likewise, when the pressure is reduced, the boiling point is reduced.

If we apply this principle to LP gas in a storage tank, it can readily be seen that as vapor is withdrawn from the tank more liquid will evaporate to replace that which was withdrawn. We must remember that each liquid has its own boiling point and pressure.

Butane (C_4H_{10}), like propane (C_3H_8) and natural gas, is placed in the hydrocarbon series of gaseous fuels, because they are composed of hydrogen and carbon. Butane has a boiling point of 31.1° F, a specific gravity of 2, a heating value of 3,267 Btu/ft^3 of vapor. It requires 30.97 ft^3 of

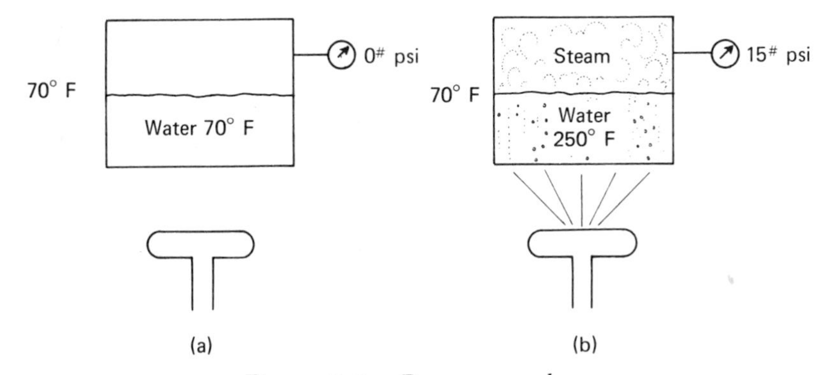

Figure 2–3. Pressure cooker.

air per ft^3 of vapor for proper combustion. At sea level, it has a gauge pressure of 36.9 pounds at 100° F. Butane expands to 31.75 ft^3 of vapor per gallon of liquid. It is heavier than air. The ignition temperature is approximately 1,100° F and the burning temperature is 3,300° F.

Propane has a boiling point of −43.8° F, a specific gravity of 1.52, a heating value of 2,521 Btu/ft^3 of vapor. It requires 23.82 ft^3 of air per ft^3 of vapor for proper combustion. At sea level, it has a gauge pressure of 175.3 pounds at 100° F. Propane expands to 36.35 ft^3 of vapor per gallon of liquid. It also is heavier than air. It has an ignition temperature of 1,100° F and a burning temperature of 2,975° F.

When we study the physical properties of LP gases we can see that each has both good and bad properties. These properties should be given a great deal of consideration when determining which fuel is to be used for any given application. The two characteristics deserving the most consideration are the Btu content and the vapor pressure. See Table 2–2 for a comparison of vapor pressures. When considering these fuels, the pressure is the major limiting factor, especially in colder climates. As we study the table, we can see that when the temperature of liquid butane reaches 30° F or lower, there is no pressure in the tank. Therefore, butane would not be suitable as a fuel at these lower temperatures without some source of heat for the storage tank. This source of heat may be steam or hot water pipes around the tank, electrical heaters around the tank, or even having the tank buried in the ground. However, these all add to the initial cost of the equipment.

If we look at the pressures of propane, we see that they are suitable throughout a wide range of temperatures. Therefore, from the pressure standpoint, propane would be the ideal fuel. On the other hand, the lower Btu rating makes it less desirable than butane. To overcome this dilemma, the two fuels may be mixed to obtain some of the desirable characteristics of each gas. An example of this may be a mixture of 60% butane and 40% propane. At a temperature of 30° F, it will have a vapor

Table 2–2
LP gas vapor pressures

Temp. °F	Propane	Butane	Temp. °F	Propane	Butane
0	38.2 psig		70	124 psig	16.9 psig
10	46		80	142.8	22.9
20	55.5		90	164	29.8
30	66.3		100	187	37.5
40	78	3 psig	110	212	46.1
50	91.8	6.9	120	240	56.1
60	107.1	11.6	130	272	66.1

pressure of approximately 24 psig and a heat content of 2,950 Btu/ft^3. Since these fuels are usually mixed before delivery to the local distributor, it is difficult to know exactly what the tank pressure and Btu/ft^3 are. As long as there is enough pressure in the storage tank to allow 11 inches water column of pressure to enter the house piping, there is little or nothing a service technician can do.

Before getting involved too deeply in work involving LP gases, the state and local authorities should be consulted. Some states maintain strict control over the personnel working with these fuels.

The use of *electricity* is becoming more important and more frequently used in comfort heating. Electricity is not a new phenomenon; man has known of its existence for centuries. The early applications of the heat producing ability of electricity were limited. It was used only in a few specialized areas, mainly industrial processes and as portable heaters to help supplement inefficient heating systems. Today, electric heating is adaptable to almost any type of construction and in any climate, and at a cost that most people can afford.

Electrical power is generated at the utility company's generating station. As it leaves the generating station, it passes through a bank of transformers to increase the voltage. The voltage is increased to aid in the transmission of the electricity. From the transformer bank, the electricity is distributed to user substations. The voltage is reduced at the substation by another bank of transformers to voltages that can be used by commercial manufacturing plants. The voltage is again reduced by the building's current transformer and is carried through the meter loop to the disconnect switch. From the disconnect switch, the electricity is distributed through the house wiring to the various appliances and electric heating units (Figure 2-4).

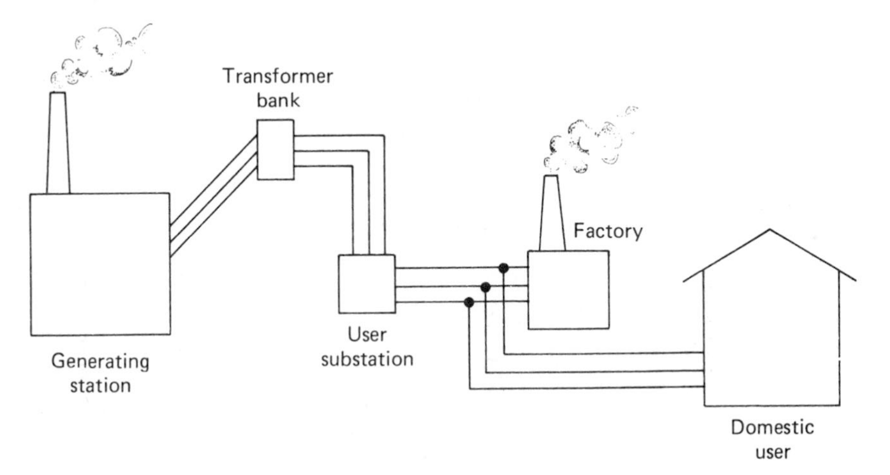

Figure 2-4. Electrical distribution.

The major methods of using electricity for comfort heating are: resistance heating, heat pump, and a combination of the two. There are three types of resistance elements used for electric heating: (1) the open wire, (2) the open ribbon, and (3) the tubular cased wire (Figure 2-5). An electric resistance heating element may be defined as: An assembly consisting of a resistance wire, insulated supports, and terminals for connecting the electrical supply wire to the resistance wire. Resistance heating will convert electrical energy to heat energy at the rate of 3,410 Btu/kW (1,000 watts). Theoretically, electrical heating is 100% efficient; that is, for each Btu input to the heating equipment, 1 Btu in usable heat is recovered.

The open wire heating elements are usually made of nichrome wire, which is wire made from nickel and chromium, but without iron, wound in a spring-like shape and mounted in ceramic insulators to prevent electrical shorting to the metal frame. The open wire elements have a longer life than the others because they operate cooler from releasing all the heat directly into the air stream. They also have a lower air pressure drop as compared to the other types.

The open ribbon elements are also made of nichrome wire and are insulated in the same manner as the open wire elements. The flat design allows more intimate contact between the wire and the air because it has more surface area. Thereby, its efficiency is increased over the open wire

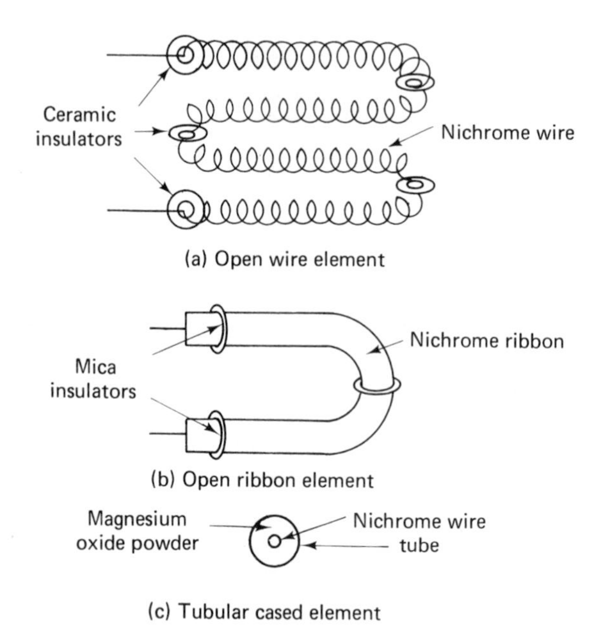

Figure 2-5. Resistance elements.

element. However, due to the increase in manufacturing costs, the open wire element is more popular.

The tubular-cased heating element is the same type as those used in cook stoves. The nichrome wire is placed inside a tube and insulated from it by magnesium oxide powder. Thus, tubular-cased heating elements do not require external insulation as do the other two resistance elements. It is also less efficient because of the energy loss caused by the extra material that the heat must pass through before reaching the moving air. It is, however, safer to use because of the interior insulation used in its manufacturing process. The tubular-cased elements have a shorter life than either of the other two elements because of their higher operating temperature. The control of these elements is more difficult than the others because of the extra material involved. Also, the air must be circulated sufficiently long to insure the proper cooling of these elements.

The *heat pump* is the most efficient method of electric heating. A heat pump is a refrigeration unit that reverses the flow of refrigerant according to the different seasons of the year. By use of a series of valves, it cools the home in the summer and provides heat for it in the winter. Electricity is used to drive the compressor and if the system design conditions—usually 72° F and 50% RH indoor conditions—are maintained, a heat pump will release three to four times as much heat as could be obtained from resistance elements. Therefore, with an input of 1 kW (3,410 Btu) approximately three times the input or 10,230 Btu are released by the heat pump.

A heat pump is less efficient at lower outdoor temperatures because the evaporator temperature must be lower than the ambient temperature in order to absorb enough heat to evaporate the refrigerant. This lower temperature is accompanied by a lower suction pressure that reduces the compressor capacity. Also, at the lower temperatures, there is not enough heat absorbed to replace the heat loss through the walls of the home.

A heat pump is classified by its heat source. These heat sources are: Air to air; water to air; water to water; ground to air, and solar assist. The first of each of these combinations refers to the heat source for winter operation. However, the air to air unit is the most common due to the installation costs.

When these units are installed in cold climates, additional heat is usually required to satisfactorily heat the home. Because of the reduced efficiency, resistance elements are used in conjunction with these systems.

If the cost of equipment per hour of operation is considered, a heat pump is more economical to own than a cooling system with a separate heating unit. The reason for this is that when one part, either heating or cooling, is used, the other part is not operating. Therefore, the cost per hour of operation is more.

COMBUSTION

The available energy contained in a fuel is converted to heat energy by a process known as *combustion.* Combustion may be defined as the chemical action of a substance with oxygen resulting in the evolution of heat and some light.

There are three basic requirements for combustion: Sufficiently high temperatures; oxygen; and fuel (Figure 2-6). When the air-fuel mixture is admitted to the combustion chamber some means must be provided to bring the mixture to its flash point. This is usually done by a pilot light. If, for any reason, the temperature of the mixture is reduced below its flash point, the flame will automatically be extinguished. For example, if the temperature of a mixture of natural gas and air is reduced below its flash point of 1,100° F, there will be no flame.

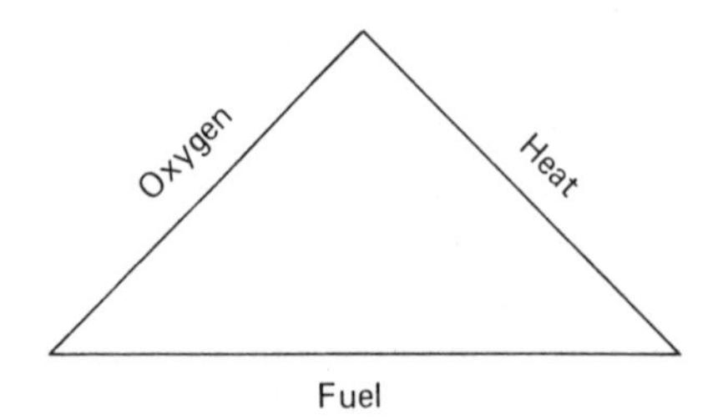

Figure 2-6. Basic combustion requirements.

Also, an ample supply of properly distributed oxygen must be supplied. The oxygen requirements governing the combustion process will vary with each different fuel. They also depend on whether or not the fuel and air are properly mixed in the correct proportions.

The third requirement for combustion is the fuel. Properties of fuel were dealt with earlier in this chapter. The physical properties of each fuel must be considered when determining its requirements for combustion. All of the basic requirements for combustion must be met or there will be no combustion.

An important factor to keep in mind when making adjustments involving gaseous fuels is the limits of flammability, which are stated in percentages of the gas in air of the mixture that would allow combustion to take place. To simplify this, if there is too much gas in the air the mixture will be too rich to burn. If there is too little gas in the air the mixture will be too lean to burn. The upper and lower limits are shown for the more common gaseous fuels in Table 2-3.

Complete combustion can be obtained only when all of the combustible elements are oxidized by all the oxygen with which they will combine. The products of combustion are harmless when all of the fuel

Table 2-3
Fuel gas limits of flammability

GAS	UPPER LIMIT	LOWER LIMIT
Methane	14	5.3
Ethane	12.5	3.2
Natural	14	3
Propane	9.5	2.4
Butane	8.5	1.9
Manufactured	29	4

is completely burned. These products are carbon dioxide (CO_2) and water vapor (H_2O).

The rate of combustion, or burning, depends on three factors.

1. The rate of reaction of the substance with the oxygen.
2. The rate at which the oxygen is supplied.
3. The temperature due to the surrounding conditions.

All the oxygen supplied to the flame is not generally used. This is commonly called *excess oxygen,* or *excess air.* This excess oxygen is expressed as a percentage, usually 50%, of the air required for the complete combustion of a fuel. An example of this is that natural gas requires 10 ft^3 of air for each ft^3 of gas. When 50% excess air is added to this figure, the quantity of air supplied is calculated to be 15 ft^3 of air to each ft^3 of natural gas. There are several factors governing the excess air requirements. They are the uniformity of air distribution and mixing; the direction of gas flow from the burner; and the height and temperature of the combustion area. Excess air constitutes a loss and should be kept to a minimum. However, it cannot usually be less than 25–35% of the air required for complete combustion.

Excess air has both good and bad effects in the combustion area function. It is added as a safety factor in case the 10 ft^3 of required air is reduced for some reason, such as dirty burners, improper primary air adjustments, or a lack in the supply of primary air. The adverse effect is that the nitrogen in the air does not change chemically and tends to reduce the burning temperature and the flue gas temperature, thereby reducing the efficiency of the heating equipment. The air supplied for combustion contains about 79% nitrogen and 21% oxygen.

The products of combustion produced when 1 ft^3 of natural gas is completely burned are: 8 ft^3 of nitrogen, 1 ft^3 of carbon dioxide, and 2 ft^3 of water vapor (Figure 2–7). These products are harmless to human

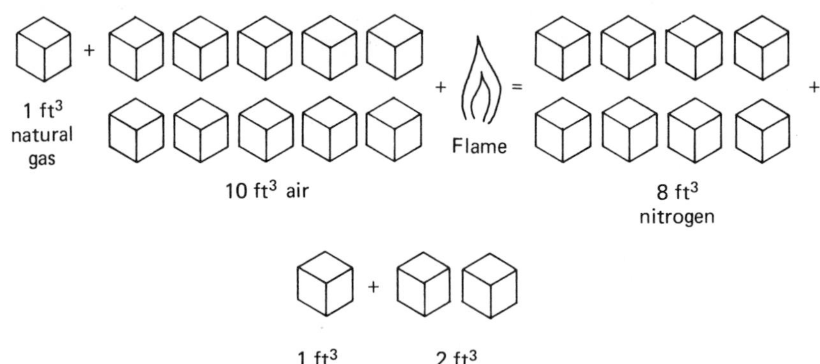

Figure 2-7. Elements of combustion.

beings. In fact, carbon dioxide is the ingredient added to water that makes soft drinks fizz.

The byproducts of incomplete combustion are carbon monoxide—a deadly product; aldehydes—a colorless, inflammable, volatile liquid with a strong pungent odor and an irritant to the eyes, nose, and throat; keytones—used as paint removers; oxygen acids; glycols; and phenols. These byproducts are harmful and must be guarded against by the proper cleaning and adjustment of heating equipment.

Probably the most important step in maintaining good combustion is the proper adjustment of the ratio of primary air to secondary air. This can be accomplished, to any degree of efficiency, only by flue gas analysis, covered in detail in Chapter 6.

When considering the combustion requirements for any given installation, the air supply to the equipment must be calculated. The air supply is governed by: size of the equipment by Btu rating; type of fuel; size of equipment room; building construction tightness; exhaust operation; and the city code.

Heating equipment regulated by experienced personnel will produce clean, economical, efficient combustion. However, when the designing of the equipment and the fundamentals of combustion are ignored, a potential hazard exists.

REVIEW QUESTIONS

1. What are the basic sources of energy for heating?
2. What is the burning temperature of natural gas?
3. What is the most popular grade of fuel oil used for domestic heating?

4. How many grades of fuel oil are there?
5. What specifications govern fuel oil?
6. What are the two main types of gaseous fuels?
7. What is the Btu content of 1 ft^3 of natural gas?
8. Does butane or propane have a higher pressure at 30° F?
9. What is the Btu content of 1 ft^3 of propane?
10. To which gas is an odorant added?
11. Is resistance heating more efficient than a heat pump?
12. Name three types of resistance heating elements.
13. How many Btu are emitted by 1 kW input of electricity?
14. What are the three requirements for combustion?
15. Do all fuels require the same amount of air for combustion?
16. Name the products of complete combustion.
17. When is complete combustion obtained?
18. What is excess oxygen?
19. What percent excess air is usually calculated for operation?
20. How many cubic feet of water vapor are released when 1 ft^3 of natural gas is burned?
21. How many cubic feet of air are required for each cubic foot of natural gas?
22. What are the products of incomplete combustion?

3

Gas Orifices, Burners, and Flames

The objectives of this chapter are:

- To introduce you to the path through which the gas flows on entering a gas heating unit.
- To familiarize you with the purpose and operation of the orifices used in gas heating equipment.
- To acquaint you with the requirements, types, and operation of the gas burners used in a gas heating system.
- To cause you to become aware of the various types of flames encountered in gas heating.
- To further acquaint you with the principles of combustion.
- To acquaint you with the different burner adjustments to obtain the proper flame.

After the gaseous fuel has left the meter loop, or tank piping, and has been distributed through the house piping, it comes to the heating equipment. As the gas progresses through the various controls, such as the gas pressure regulator—not needed on LP gas—and the main gas valve, it enters the equipment gas manifold. The gas pressure in the manifold must be kept as near as possible to 3½ inches water column for natural gas and 11 inches water column for LP gases. The gas then proceeds out of the manifold through the orifice and into the main burner. From the main burner it goes into the combustion chamber of the heating equipment where it is ignited by the pilot burner. The combustion chamber of the heating equipment is where the heat is given up to the air or water that provides heat for our homes.

ORIFICES

An *orifice* (Figure 3–1) may be defined as the opening through which gas is admitted to the main gas burner. It is mounted on the gas manifold by means of pipe threads and projects into the burner (Figure 3–2). It is normally referred to as the orifice spud.

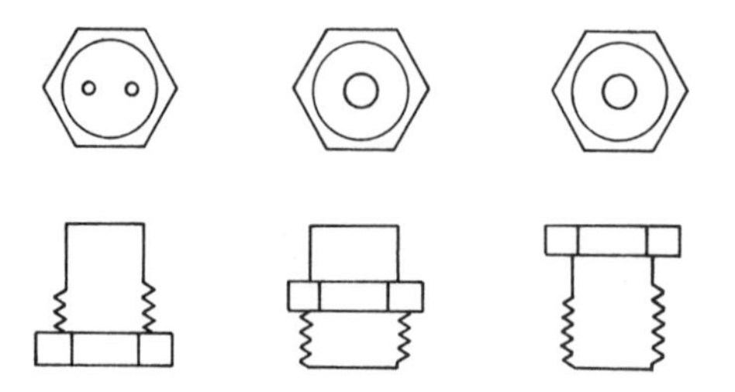

Figure 3–1. Orifice spuds.

For a burner to operate satisfactorily it must be supplied with the proper Btu input. This is the function of the orifice. The size of the orifice, the gas manifold pressure, and the gas density determine the rate of gas flow to the burner. The orifice and burner must be matched in Btu ratings. If the type of gas is changed, say from natural to LP, etc., the orifice size must also be changed. When the orifice is oversized, enough primary air cannot be drawn into the burner. Thus, incomplete combustion results and a yellow, inefficient flame is produced.

An orifice that is sized too small will also have adverse effects on the burner. There will be delayed ignition accompanied by a loud boom when the burner is ignited. Also, the burner will not operate to its full capacity, resulting in poor heating.

The orifice must direct the gas stream exactly down the center of the burner (Figure 3–3). The velocity of gas down the burner tube causes

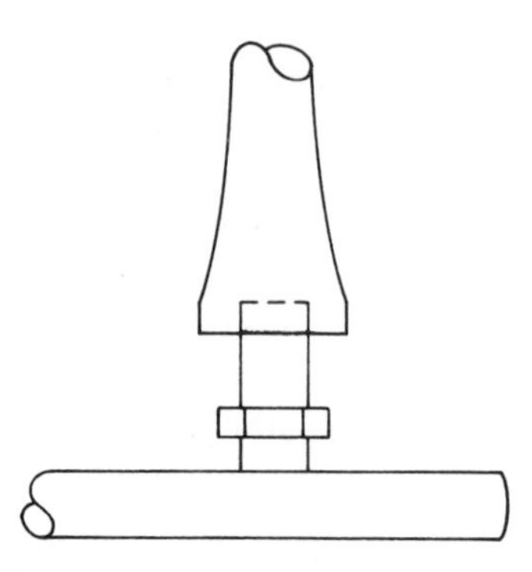

Figure 3–2. Orifice location.

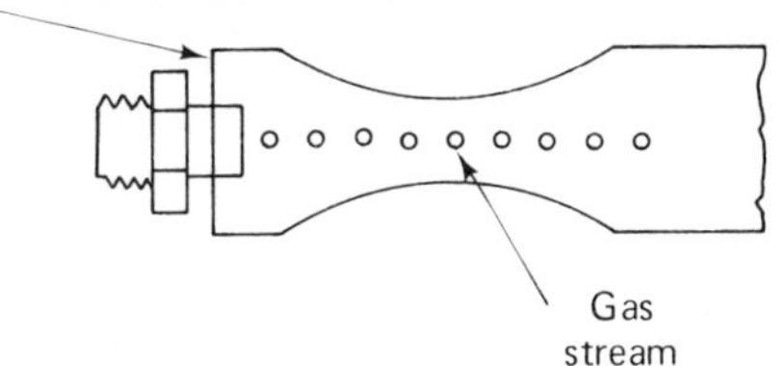

Figure 3–3. Orifice in relation to burner.

the primary air to be drawn in through the burner face and mixed with the gas. If the gas velocity is reduced, insufficient primary air will be the result. Therefore, the orifice must be drilled straight and in the exact center of the spud. This may be accomplished easily by drilling the hole from the rear (Figure 3–4). The *V* shape will insure the desired hole position.

The preceding information should establish that the orifice is a precision piece of equipment and should be treated as such when service is required. To clean an orifice, a soft instrument, such as a broom straw, wire brush bristle, etc., must be used. Care must be taken not to change the shape or size of the hole.

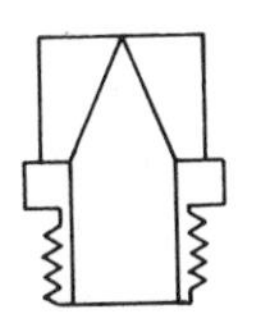

Figure 3–4. Orifice cutaway.

The following table of orifice sizes can be used to insure the proper orifice for the job (see Table 3–1).

MAIN GAS BURNERS

A *gas burner* is defined as a device that provides for the mixing of gas and air in the proper ratio to insure satisfactory combustion.

The first burners were used for lighting. The carbon in the flame was a result of incomplete combustion that produced the light. This flame was not good for heating because of the lower temperature produced.

The blow pipe was the first burner to produce a blue flame and, therefore, any great heat intensity. The blue flame was produced by

Table 3-1
Orifice Capacities for Natural Gas,
1000 Btu/ft^3,
Manifold Pressure 3½ in. Water Column.

Wire Gauge Drill Size	Rate ft^3/hr	Rate Btu/hr
70	1.34	2,340
68	1.65	1,650
66	1.80	1,870
64	2.22	2,250
62	2.45	2,540
60	2.75	2,750
58	3.50	3,050
56	3.69	3,695
54	5.13	5,125
52	6.92	6,925
50	8.35	8,350
48	9.87	9,875
46	11.25	11,250
44	12.62	12,625
42	15.00	15,000
40	16.55	16,550
38	17.70	17,700
36	19.50	19,500
34	21.05	21,050
32	23.70	23,075
30	28.50	28,500
28	34.12	34,125
26	37.25	37,250
24	38.75	39,750
22	42.50	42,500
20	44.75	44,750

blowing primary air into the base of the flame (Figure 3-5). The modern blow pipe accomplishes this same thing by blowing compressed air in the same end that the gas enters. The gas and air is thus mixed in the pipe as in the burners in modern day furnaces.

Modern furnaces make use of four different types of main gas burners. These burners are: (1) *drilled port;* (2) *slotted port; (3) ribbon;* and (4) *inshot* (Figure 3-6). They are made of either cast iron or stamped steel, depending on the type and purpose. Usually the drilled port burner is made of cast iron. The others are made by forming steel into the desired shape. The type of burner used depends on the particular equipment

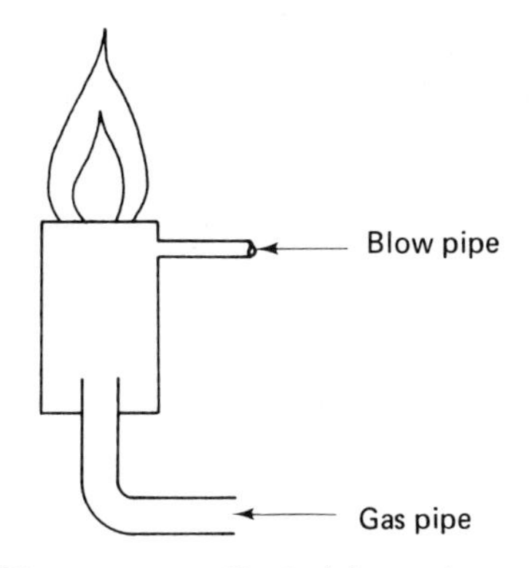

Figure 3-5. Basic blow pipe.

design. Each manufacturer will use the burner that best lends itself to the requirements of the equipment.

Burner Design

The very first part of the burner that is put to use is the burner face (Figure 3-7). The face of the burner is called the air mixer. It is through this opening that the primary air is admitted to the burner.

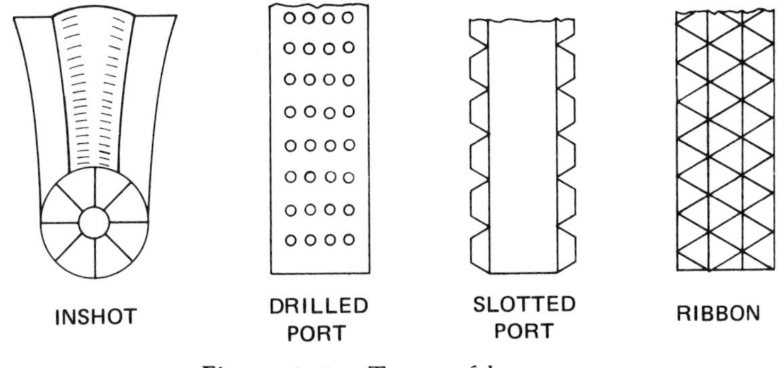

Figure 3-6. Types of burners.

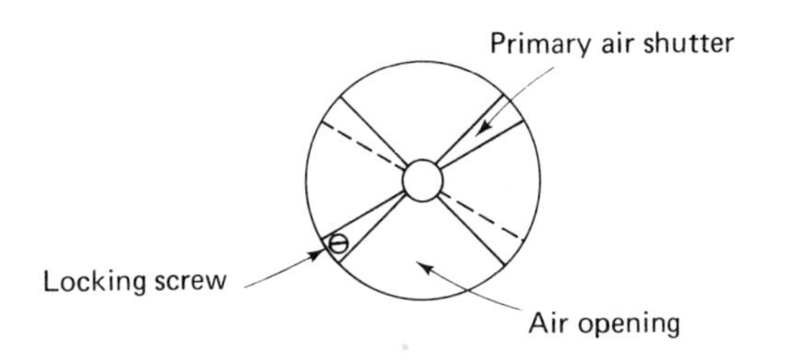

Figure 3-7. Burner face.

The *burner mixing tube* (Figure 3-8) is the next part of the burner with which the gas and air mixture comes into contact. The mixing tube provides the necessary space for the proper mixing of the gas and air and extends the full length of the burner. The venturi type of mixing tube is much more successful than a straight piece of pipe. The venturi reduces the turbulence of the mixture and allows a more even distribution to the burner ports.

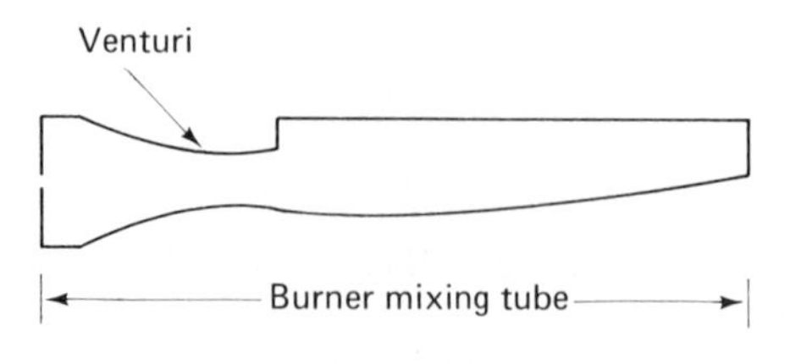

Figure 3-8. Side-view of burner.

The *burner head* is the last part of the burner involved in the mixing and distribution of the gas and air mixture. The burner head is the top part of the burner (Figure 3-9). It is here that the ports are placed that determine the type of burner (drilled port, slotted port, ribbon, or inshot). The proper design of the burner head is necessary if complete combustion of the gas is to be obtained. The burner ports are openings that release the gas and air mixture into the combustion chamber.

There should not be more than two rows of ports on a burner head without providing sufficient room for the secondary air to circulate over the burner head. This circulation of air cools the burner and helps to prevent burner burn-out. For this reason it is important to keep the rust and scale formations removed from the burners.

The control of primary air is accomplished on the face of the main burner. There are several methods used for this purpose. The adjustable shutter, however, is the most common method used (Figure 3-10). The adjustment is made by enlarging or reducing the size of the opening

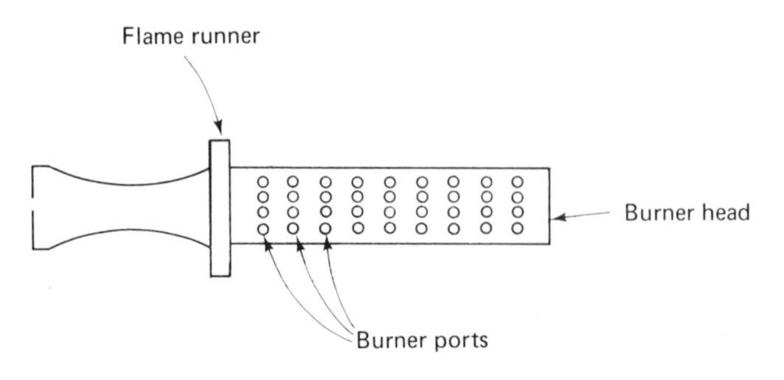

Figure 3-9. Burner head.

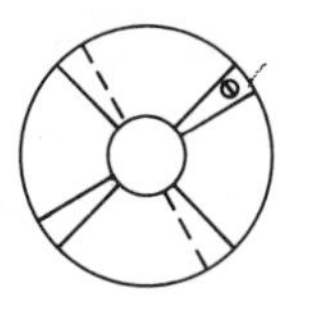

(a) Adjustable shutter

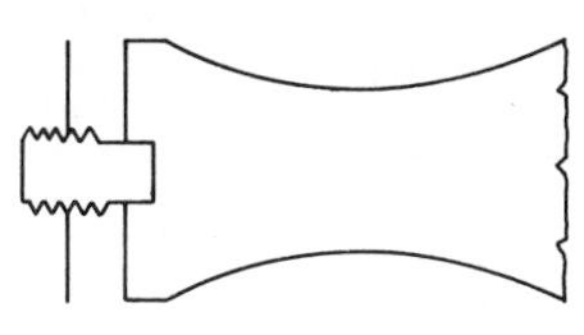

(b) Threaded shutter

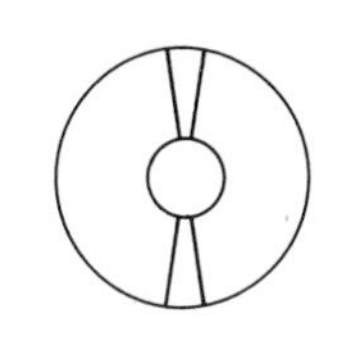

(c) Nonadjustable shutter

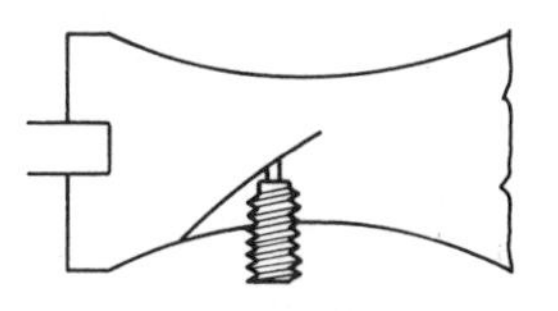

(d) Deflector shutter

Figure 3-10. Primary air adjustment methods.

through which the air is admitted. The threaded shutter is less common and is used only on cast iron burners (Figure 3-10). The deflector type, or adjustable baffle (again see Figure 3-10) is gaining acceptance in the industry. This burner has larger air openings than necessary. This is an advantage because it will not clog up as fast as the other types. The primary air adjustment is made by pushing the deflector into the gas stream. This added restriction reduces the gas velocity and reduces the intake of primary air. Some burners have fixed openings and no adjustment can be made. These openings are sized according to the burner Btu rating and the proper flame is automatically obtained when the correct orifice spud is used.

FORCED DRAFT BURNERS

The burners discussed previously have been of the atmospheric type. Another type that is gaining popularity, especially with the advent of the roof mount type units, is the *forced draft burner*. The equipment using this type of burner makes use of a small blower to provide the combustion air (Figure 3-11). The combustion area is sealed from atmospheric conditions. The air is forced into the combustion area in which there is a positive pressure. The amount of air supplied and the inside pressure are controlled by an adjustment on the flue outlet. By forcing a fixed amount of air into the combustion chamber, preset combustion conditions are maintained.

Forced draft is ideal for outdoor heating equipment where unusual conditions prevail and where vent pipe heights are restricted. The main

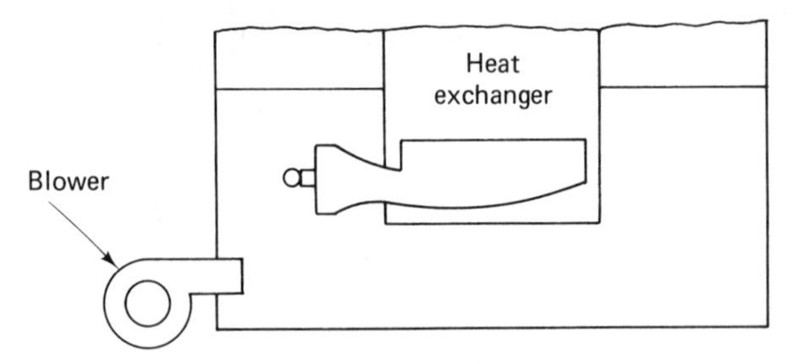

Figure 3-11. Forced draft burner.

gas valve cannot open on these units until a definite pressure is obtained inside the combustion area. When making adjustments involving this pressure, the manufacturer's specifications must be consulted. This pressure will vary from manufacturer to manufacturer and sometimes on different models of the same manufacturer.

POWER GAS BURNERS

Another type of burner that is being used is known as the *power gas burner* (Figure 3-12). These burners are small, highly efficient, and suited for use with natural gas. Power burners have a very wide range of capacities. The maximum capacity is 500,000 Btu and the minimum is 120,000 Btu. They are applicable to any heating application where oil or an equivalent capacity is presently being used or as an integral component of a new boiler-burner system. They can also be directly installed and fired through existing doors of cast iron or steel firebox boilers without requiring further brickwork or modification of the boiler base.

Figure 3-12. Power gas burner. (*Courtesy Ray Burner Co.*)

Power burners furnish 100% of the combustion air through the burner for firing rates up to their rated capacities. These burners have locked air shutters, but with a reduced input start by means of a slow opening gas valve. Power burners are easy to install and to adjust due to the use of an orifice to meter the gas and a dial to adjust the air shutter to match the gas orifice used. An automatic electric pilot ignition is used that incorporates a 6,000 volt (V) transformer to provide the necessary spark to ignite the pilot gas.

In operation, these burners are more noisy than the atmospheric-type burner. Power burners are applicable to small boilers and to residential and industrial furnaces. They are also used as antipollution after-burners. These burners are automatic and are equipped with the latest types of safety and operating controls.

PILOT BURNERS

A *pilot burner* is defined as a small burner used to ignite the gas at the main burner ports and a source of heat for the pilot safety device. Approximately 50% of all heating equipment failures can be attributed to this burner.

The automatic pilot is probably the most important safety device used on modern gas heating equipment. During normal operation, the pilot flame provides the heat necessary to actuate the pilot safety device. Also, the pilot is located in the combustion chamber in proper relation to the main burners so that the flame will ignite the gas admitted when the thermostat demands heating. Should something happen to cause the pilot to become unsafe, the pilot safety will prevent any unburned gas from escaping into the combustion chamber.

There are basically three types of pilot burners: (1) the *millivolt,* (2) *the bimetal,* and (3) the *liquid filled.*

The *millivolt* type can be divided into two types: the *thermocouple* and the *thermopile.* The thermocouple is an electricity producing device made of dissimilar metals. One end of the metal is heated and is called the *hot junction,* while the other end remains cool and is called the *cold junction* (Figure 3–13). The output of a thermocouple is approximately 30 millivolts (mV) when heated to a temperature of about 3,200° F. The thermopile is a series of thermocouples wired so that the ouput voltage will be approximately 750 mV (Figure 3–14). The greater the number of thermocouples in series, the higher will be the output voltage.

The *bimetal pilot* is named so because the sensing element is made of two dissimilar metals welded together so that they become one piece (Figure 3–15). Because the two metals have different expansion rates, they bend when heated. As the bimetal element is heated by the pilot flame, a movable contact is pushed toward a stationary contact. When

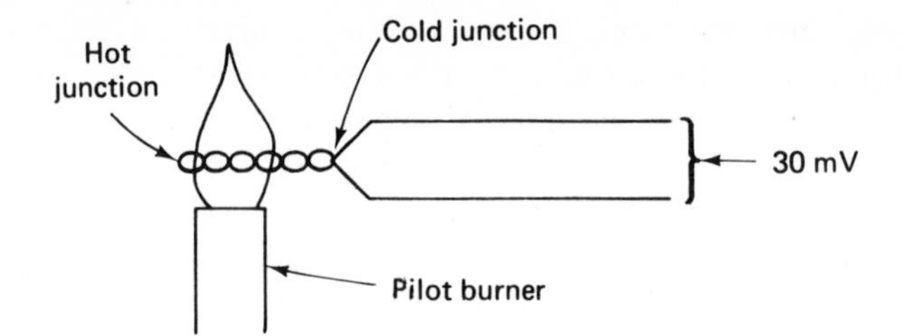

Figure 3-13. Thermocouple principle.

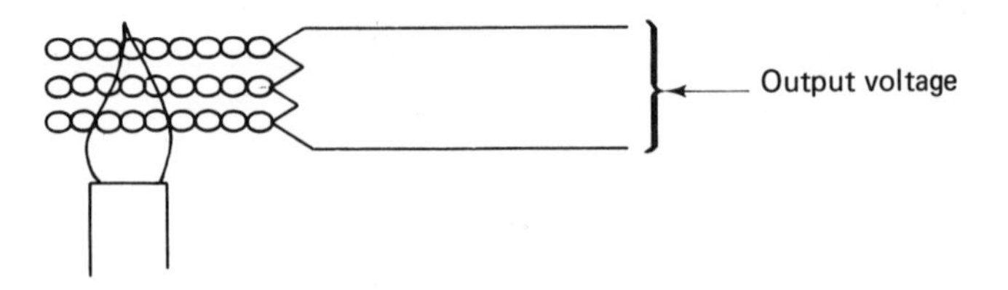

Figure 3-14. Thermopile principle.

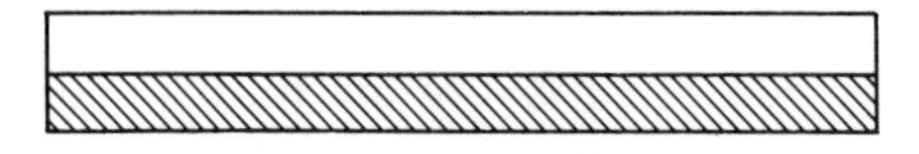

Figure 3-15. Bimetal element.

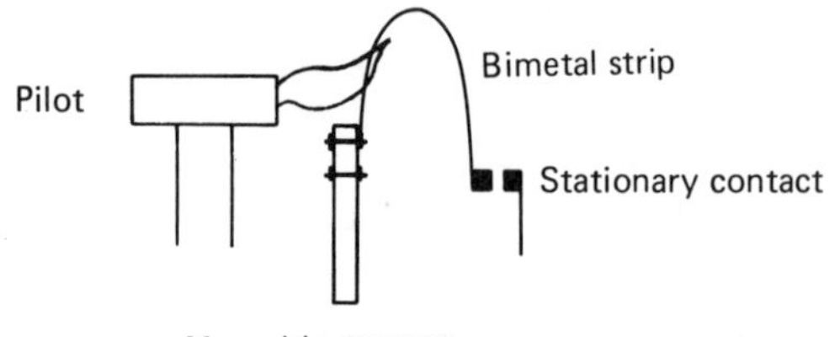

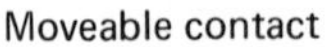

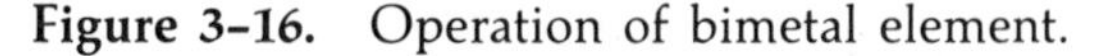

Figure 3-16. Operation of bimetal element.

these contacts touch, an electrical circuit is completed through them (Figure 3-16). This control is usually employed to operate in the control, or low voltage, circuit.

The *liquid filled pilot* functions from the pressure exerted when a liquid is heated and vapor is formed. This liquid and vapor is contained in a closed system so that more accurate control can be obtained. The bulb is located in the pilot flame (Figure 3-17) and when the designated temperature and pressure is reached, a circuit is completed and the heating unit can function.

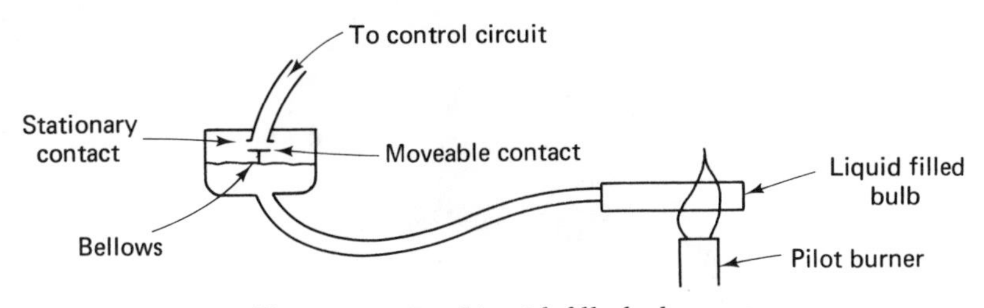

Figure 3-17. Liquid filled element.

ELECTRIC IGNITION

Several types of electric ignitors are in use on modern heating equipment. The main types are the glow coil and the direct spark ignitor. Some state and local laws require the use of automatic ignitors and have done away with the standard standing pilot on new equipment.

Glow Coil Ignitors

These ignitors are used to light or relight pilot burners in the case of flame failure (Figure 3-18). These units are assembled in a draft protected holder which provides means for either horizontal or vertical mounting to a pilot burner. Glow coil type ignitors are connected to an electrical power source of 2.5 volts. The coil is heated because of electrical resistance. After the pilot is lighted, the electrical power is interrupted to the glow coil. Continuous electrical power to the glow coil will cause it to burn out and become inoperative. Their use is limited to the lighting of pilot burners only and should not be used for main burner ignition.

Figure 3-18. Glow coil ignitor. (*Courtesy ITT General Controls*)

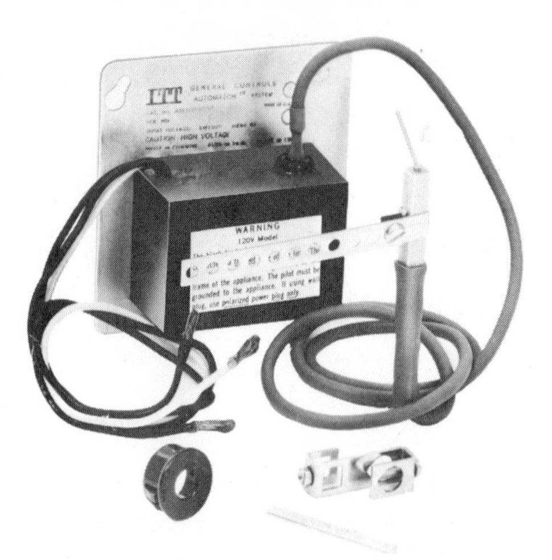

Figure 3-19. Direct spark ignitor. (*Courtesy ITT General Controls*)

Direct Spark Ignitors

These systems (Figure 3-19) may be used as direct replacement pilot lighters on equipment that has been certified by an approved testing agency with the pilot lighter as part of the original equipment. Where automatic pilot relighters are not certified as an original component of the unit, their application must be limited to roof top heating units, space heaters, open bay heaters or on installations where the unburned gases are quickly vented.

These units have a spark frequency of approximately 100 sparks per minute. They may be used on either 24 volts or 115 volts. They are satisfactory for use in a temperature range of −40° F to +160° F.

FLAME TYPES

There are basically two types of flames—the *yellow* and the *blue* flame. There are different variations of these two flames that are produced by changes in the primary air supplied to gaseous fuels. Primary air is the air that is mixed with the gas before ignition. The counterpart of primary air is secondary air which is mixed with the flame after ignition (Figure 3–20).

The following is a list of flames and their indications that are encountered in heating systems using gas as a fuel. The first will be the yellow flame and we will progress to the blue flame.

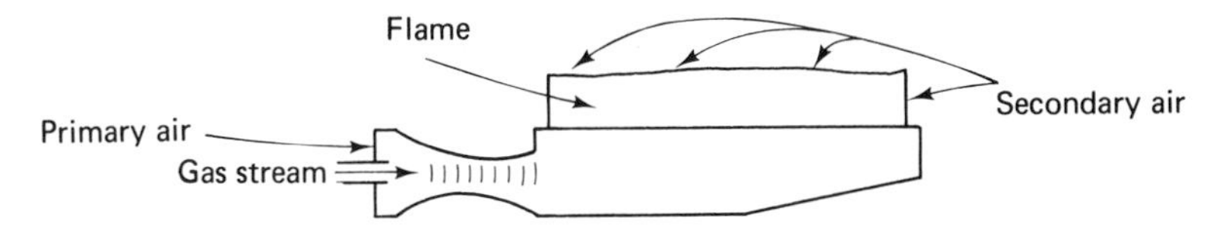

Figure 3-20. Primary air and secondary air supporting combustion.

Yellow Flame

This flame has a small blue-colored area at the bottom of the flame (Figure 3-21). The outer portion, or outer envelope, is completely yellow and is usually smoking. This smoke is unburned carbon from the fuel. The yellow, or luminous, portion of the flame is caused by the slow burning of the carbon that is being burnt. Soot is the outcome of the yellow flame, which produces a lower temperature than does the blue flame. A yellow flame is an indication of insufficient primary air. Incomplete combustion is also indicated by a yellow flame—a hazardous condition that cannot be allowed.

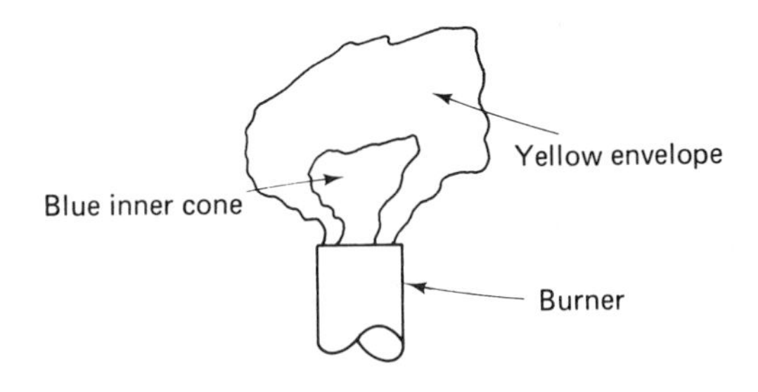

Figure 3-21. Yellow flame.

Yellow-Tipped Flame

This condition occurs when not quite enough primary air is admitted to the burner. The yellow tips will be on the upper portion of the outer mantle (Figure 3-22). This flame is also undesirable because the unburned carbon will be deposited in the furnace flues and restrictors, thus eventually restricting the passage of the products of combustion.

Orange Color

Some orange or red color in a flame, usually in the form of streaks, should not cause any concern. These streaks are caused by dust particles in the air and cannot be completely eliminated (Figure 3-23). However,

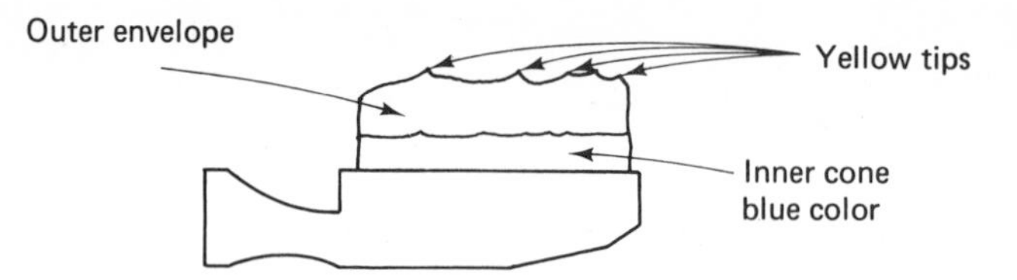

Figure 3–22. Yellow-tipped flame.

when there is a great amount of red or orange in the flame, the furnace location should be changed or some means provided for filtering the combustion air to the equipment.

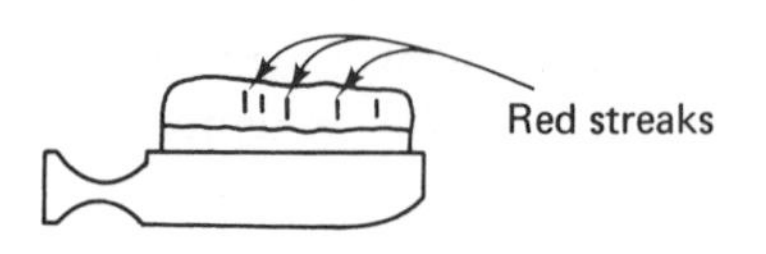

Figure 3–23. Orange streaks in flame.

Soft, Lazy Flame

This condition appears only when just enough primary air is admitted to the burner to cause the yellow tips to disappear. The inner cone and the outer envelope will not be as clearly defined as they are in the correct blue flame (Figure 3–24). This type of flame is best for high-low fire operation. It is not suitable, however, wherever there may be a shortage of secondary air. This flame will burn blue in the open air; however, when it touches a cooler surface, soot will be deposited on that surface.

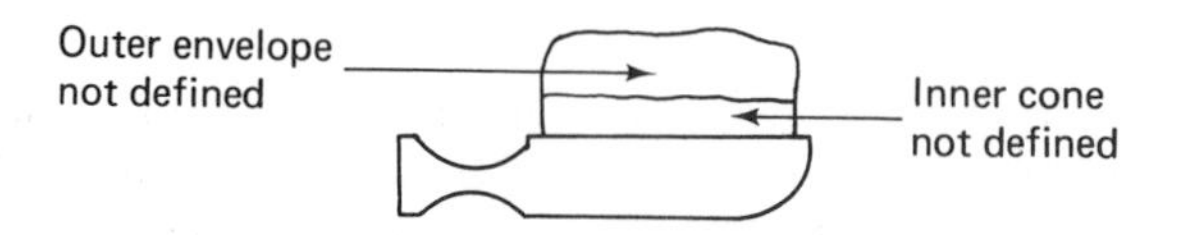

Figure 3–24. Soft, lazy flame.

Sharp, Blue Flame

When the proper ratio of gas to air is maintained, there will be a sharp, blue flame (Figure 3–25). Both the outer envelope and the inner cone will be pointed and the sides will be straight. The flame will be resting on the burner ports and there will not be any noticeable blowing

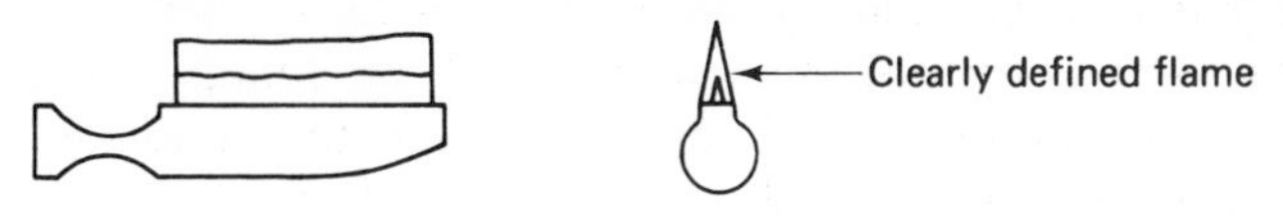

Figure 3-25. Sharp flame.

noise. The flame will ignite smoothly on demand from the thermostat. Also, it will burn with a nonluminous flame. This is the most desirable flame for heating purposes on standard heating units.

Lifting Flame

When too much primary air is admitted to the burner, the flame will actually lift off the burner ports. This flame is undesirable for many reasons. When the flame is raised from the ports, there is a possibility that intermediate products of combustion will escape to the atmosphere (Figure 3-26). The flame will be small and will be accompanied by a blowing noise, much like that made by a blow torch. There will be rough ignition and, if enough secondary air is available, ignition may be impossible. Intermediate products of combustion are the same as the products of incomplete combustion which must be avoided.

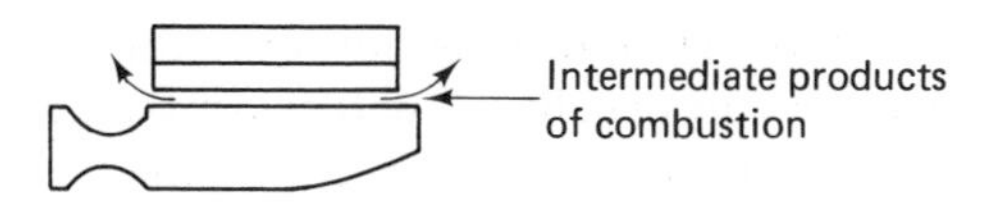

Figure 3-26. Lifting flame.

Floating Flame

A floating flame is an indication that there is a lack of secondary air to the flame. In severe cases, the flame will leave the burner and have the appearance of a floating cloud (Figure 3-27). Again, the intermediate products of combustion are apt to escape from between the burner and the flame. The flame is actually floating from place to place wherever sufficient secondary air for combustion may be found.

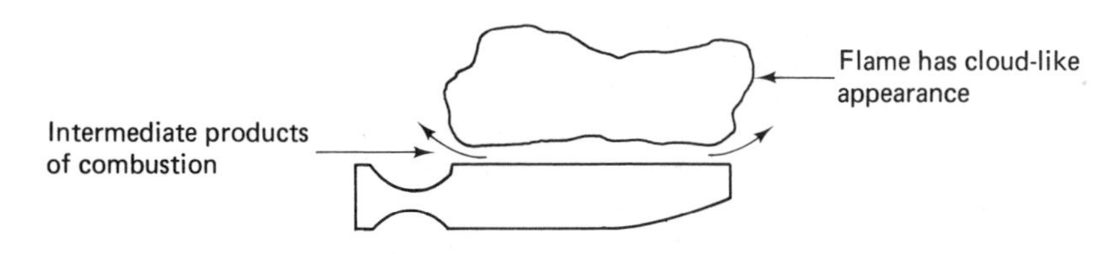

Figure 3-27. Floating flame.

This flame is sometimes hard to detect because, as the door to the equipment room is opened, sufficient air will be admitted to allow the flame to rest on the burner properly. However, the obnoxious odor of aldehydes may still be detected. Another possible cause of this flame is an improperly operating vent system. If the products of combustion cannot escape, fresh air cannot reach the flame. This is a hazardous situation that must be eliminated regardless of the cause.

COMBUSTION TROUBLESHOOTING

Heating service technicians are called on to make fast, accurate analyses and corrections of heating equipment in order to provide the customer with safe, economical, and adequate heat. The major situations dealing with the flame that he will be called upon to analyze are (1) delayed ignition, (2) roll-out ignition, (3) flashback, (4) resonance, (5) yellow flame, (6) floating main burner flame lifting off the burners, and (7) main burner flame too large. The following paragraphs will discuss these problems and make some recommendations for their solution.

Delayed Ignition

Delayed ignition is caused by improper or poor flame travel to the main burner or by poor flame distribution over the burner itself. When this situation occurs, it can usually be detected by a noisy ignition—that is, a light explosion, or puff, on ignition of the main burner.

The possible causes and correction of delayed ignition are:

1. *Distorted burner or distorted carry-over wing shots.* Both the carry-over wing slots and the main burner slots should be uniformly shaped and of the proper size (Figure 3-28). If not, the distortion could cause the adjacent burner to have faulty ignition accompanied by an accumulation of gas, which causes a noisy and dangerous ignition.

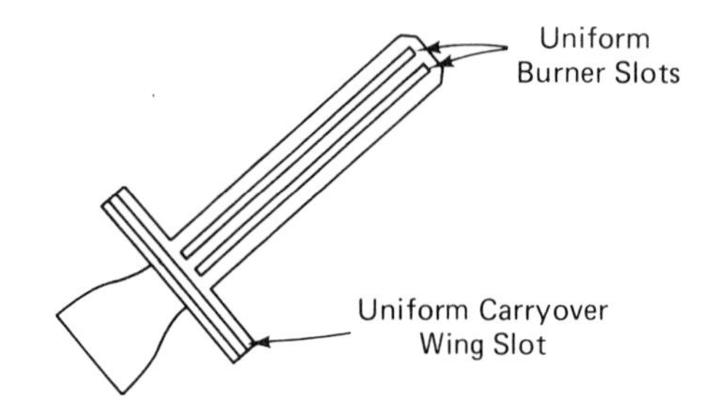

Figure 3-28. Carry-over wing and burner slots.

2. *Misaligned carry-over wings.* The purpose of the carry-over wing is to direct the gas for ignition to the main burner. The carry-over wing on each burner must be aligned with the adjacent burner so that the flame path is no more than 1/16 inch above or 1/8 inch below the adjacent burner. A delayed ignition could result if these limits are not met (Figure 3-29). This problem can usually be solved by loosening the clamp aligning the burner and retightening the clamp.

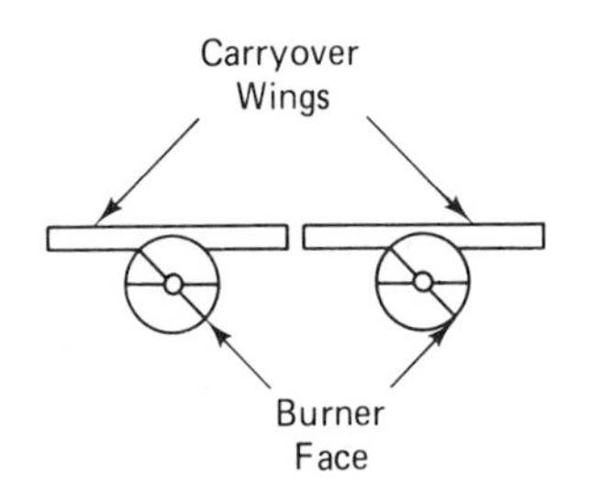

Figure 3-29. Carry-over wing adjustment.

3. *Painted-over or rusted carry-over slots.* Occasionally, paint will get into the carry-over wings during their manufacture. Also, the burners and heat exchanger will rust and this rust will accumulate in the ports and slots, preventing proper flame travel during the ignition period. This poor flame travel results in delayed ignition. The solution to this problem is to clean the ports and slots of paint and rust. If rust is the problem, the owner should be instructed to turn off the pilot during summer operation to reduce the possibility of rust recurring if condensation forms on the cooler surfaces of the furnace.

4. *Low manifold gas pressure.* If the supply gas pressure to the furnace should be too low, the flow of ignition gas may be slow from one burner to the next, possibly resulting in delayed ignition of some of the burners. This possibility exists on any installation, especially during cold weather when the demand for gas is high. All furnaces are equipped with a pressure tap on the main gas valve or the manifold to measure the gas pressure. Use either a water manometer or a gas manifold pressure gauge (Figure 3-30). The manifold gas pressure for natural gas is 3½ to 4 inches of water column. Propane equipment requires an 11-inch water column. These pressures should be taken with all other gas appliances on the same gas main in operation.

5. *Main gas valve equipped with step opening regulator.* When delayed ignition occurs on a furnace equipped with a step opening gas valve, the remedies are:

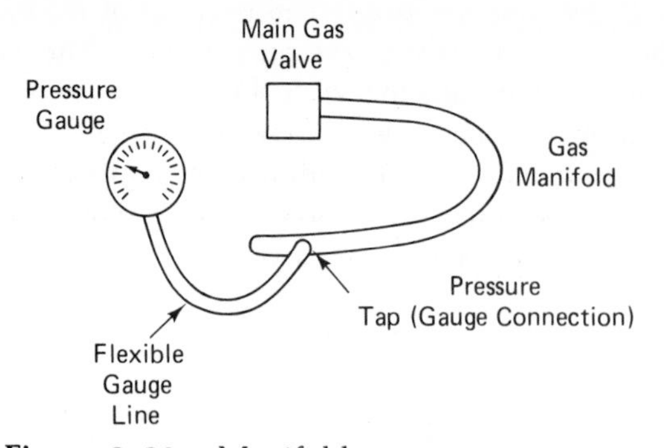

Figure 3–30. Manifold pressure connection.

(a) carefully check and correct any of the previously mentioned possibilities.

(b) next, measure the first step opening of the gas pressure with a manometer or manifold pressure gauge. If the first step pressure opening has less than 2 inches water column, there are two options which may be taken. (1) The first step pressure may be adjusted to at least 2 inches water column. (2) Replace the step opening regulator with a standard regulator.

Roll-Out Ignition

Roll-out is usually indicated by a puff or whish sound on burner ignition. This situation is usually caused by two conditons: (a) soot or some other obstruction in the heat exchanger causing a restriction; (b) a fast or quick opening gas valve which allows an accumulation of unburned gas in the combustion zone to ignite late, causing delayed ignition. The solution is to remove the soot or obstruction from the heat exchanger. In the case of a fast opening gas valve, a surge arrester may be installed in the bleed line from the regulator (Figure 3–31). The surge arrestor causes the rate of bleed-off to be reduced, resulting in a slower opening gas valve and thus providing a smoother ignition and a minimum roll-out on ignition.

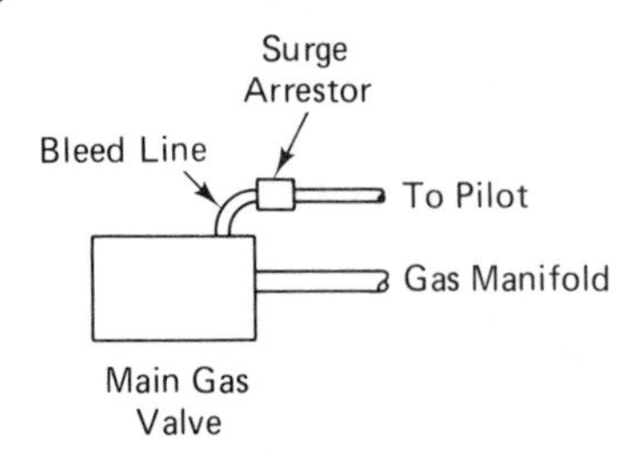

Figure 3–31. Surge arrestor location.

Flashback

Flashback occurs when the flow of the gas-air mixture entering the burner is slower than the velocity of the flame travel. This results in a popping sound and sometimes a flame burning at the orifice. It is usually caused by a low manifold gas pressure, a change in the gas-air mixture, or a sudden down draft on the flame pattern. The following are possible causes of flashback:

1. Extremely hard flame
2. Distorted burner, or carry-over wing slots
3. Low manifold gas pressure
4. Defective burner orifice
5. Misalignment of the burner and orifice
6. Erratic operation of the main gas valve
7. Unstable gas supply pressure
8. Rust or soot in the burner
9. Wrong type of LP gas, or gas-air supply mixture.

Extremely Hard Flame: In this situation, close off on the primary air shutter until a small yellow tip appears on the flame. A yellow tip of up to three inches can be tolerated by most furnaces when they are hot.

Distorted Burner or Carry-Over Wing Slots: When this situation occurs, check the burner ports and carry-over wing slots for damage. Correct any damage by repairing or replacing the burner.

Low Manifold Gas Pressure: To correct this condition, check the manifold pressure with a manometer or manifold pressure gauge after the burners have been fired for at least five minutes. Correct any pressures that do not meet the recommendations for the type of gas used.

Defective Burner Orifice: To determine whether or not the burner orifice is bad, turn off the main gas valve, remove the burners, then remove each individual orifice from the manifold. Check the outside of the orifice spuds for dents or misshaping of the orifice hole. Check the inside of the orifices for foreign matter such as dirt, grease, pipe dope, etc. Remove any foreign material, being careful not to damage the orifice. Replace the orifices in the manifold and check for gas leaks with a soap and water solution or other liquid type leak detector. Never use a match for leak testing gas piping.

Misaligned Burner Orifices: If flashback is persistent on any one burner, first check the orifices for damage and foreign matter. If the problem still exists, completely remove the manifold assembly from the furnace, and apply water pressure to the assembly. This procedure will indicate any orifice misalignment by a crooked spray of water from a misaligned orifice. (See Figure 3-32.) Make any necessary repairs to correct the problem. In some cases, repair may require the replacement of the entire assembly.

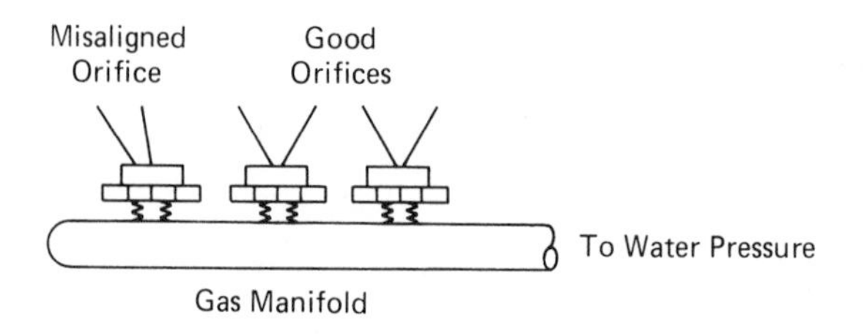

Figure 3-32. Orifice misalignment.

Unstable Gas Supply Pressure: This condition is more likely to occur on LP gas furnaces. This condition can be identified by use of a manifold pressure gauge or a manometer connected to the gas line.

This situation is usually caused by a chattering regulator or the main gas line exposed to extreme cold temperatures. It is sometimes necessary to install two-stage gas pressure regulators to eliminate this problem.

Rust or Soot in the Burner: When this situation occurs, the main burners must be removed and all foreign matter removed from inside them. The cause of rusting should be determined and corrected. Sometimes rusting is caused by condensation if the pilot burner is left on during summer operation. If so, the user should be instructed to turn off the pilot during the cooling season.

Sooting of a furnace may be caused by low gas pressure and/or an improper setting of the primary air adjustment. Either of these conditions is generally accompanied by a large yellow flame. Precautions must be taken to prevent sooting of a furnace heat exchanger.

Wrong Type LP Gas, or Gas Mixture: The furnace name plate must be checked to be certain that the furnace is designed for the type of gas being supplied to it. Mixtures of butane and propane gases used in furnaces not approved for liquefied petroleum gas could be the cause of poor burner operation. They could result in flashback, sooting and burner damage because of overheating. A quick check for this situation is to check the relief valve setting. Tanks equipped with relief valve settings of 100 to 150 psi will accommodate gas mixtures. Relief valve settings of 200 to 250 psi are used on tanks equipped for pure propane.

Resonance

Resonance can be identified by a loud rumbling noise or a pure tone buzz or hum. Resonance is most common when butane is used as the fuel. This situation may be caused by excessive primary air being supplied to the main burner or by a defective main burner spud. If excessive primary air is the cause, adjust the primary air shutter until a slight yellow tip appears on the flame. Lock the shutter in this position. If defective spuds are suspected, remove the spuds and check for nicks or dents on the outer edge of the orifice. Also, check for dirt or other foreign material inside the orifice. Clean or replace as necessary.

Yellow Flame

A large yellow, or luminous, flame is a good indication that almost all of the oxygen required for combustion is being taken from the secondary air. Complete combustion will occur as long as enough oxygen is supplied and nothing interferes with the flame. If the flame should touch a cooler surface, however, soot and toxic products will be released. Therefore, large yellow flames must be avoided.

There are five possible causes of large yellow flames:

1. Primary air shutter closed off too much; to correct, adjust the primary air shutter

2. Partially clogged main burner ports or orifices; to correct, remove the spud and inspect for damage

3. Misaligned burner spuds; to correct, check alignment and correct as necessary

4. Soot or foreign material inside of the heat exchanger; to correct, clean the heat exchanger to allow proper combustion.

5. Poor venting, down draft, or improper combustion air supply; to correct, eliminate any poor vent piping. It sometimes may be necessary to install an outside air duct to provide the required combustion air.

Floating Main Burner Flame

This situation is not very common, but may occur under the following conditions:

1. The heat exchanger is blocked with soot; to correct, clean the soot from the heat exchanger and correct cause of poor combustion

2. Air blowing into the heat exchanger; to correct, check for a leaking or cracked heat exchanger that may allow circulating air into the heat exchanger and correct

3. Negative interior pressure in the furnace room; to correct, re-check the combustion air requirements. Also, check for exhaust fans in the furnace room. Correct as necessary.

Main Burner Flame Too Large

The possible causes and the corrective action to be taken are:

1. Excessive gas manifold pressure; to correct, check pressure with a manifold pressure gauge or manometer and reduce to proper pressure.

2. Defective gas pressure regulator; to correct; replace the gas valve or pressure regulator

3. Orifice size too large; to correct, replace the orifice with proper size. Do not solder orifices closed and redrill because the solder may melt out and cause more problems.

REVIEW QUESTIONS

1. What is the manifold pressure for natural gas?
2. Define an orifice.
3. What factors determine the orifice size?
4. What are the functions of a pilot burner?
5. What is the purpose of the main burner?
6. Give one indication of a too small main burner orifice.
7. Name four types of main burners.
8. Name the four types of primary air adjustments.
9. Define a main burner.
10. What is an indication of too much primary air?
11. Name three types of pilot burners.
12. What is the output voltage of a single thermocouple?

13. Why must the orifice size be changed when the type of gas is changed?
14. What is another name for the burner face?
15. Why should the rust and scale be removed from burner heads?
16. What is the name of a burner that uses a blower for supplying combustion air?
17. Define a pilot burner.
18. What are the main types of pilot ignitors used on modern heating equipment?
19. What is the voltage supplied to a glow coil ignitor?
20. What is the approximate spark frequency of direct spark ignitors?
21. What is an indication of too little secondary air?
22. What is an indication of too little primary air?
23. What does an orange or red color in a flame indicate?
24. What is hazardous about a lifting flame?
25. What four conditions can cause delayed ignition?
26. How is roll-out ignition usually indicated?
27. When does flashback occur?
28. How is resonance identified?
29. What is a good indication that almost all of the oxygen required for combustion is taken from the secondary air?
30. What are the five possible causes of a yellow flame?
31. Is a floating main burner flame a very common problem?
32. What are the possible causes of too large a main burner flame?

4

Fuel Oil Burners

The objectives of this chapter are:

- To introduce you to the different types of fuel oil burners.
- To acquaint you with fuel oil nozzles.
- To provide you with the operating principles of the most popular types of fuel oil burners used in comfort heating systems.
- To familiarize you with the ignition principles used on fuel oil burners.
- To acquaint you with the necessity of proper analysis and treatment of fuel oil.
- To introduce you to the piping systems involved in fuel oil heating systems.
- To familiarize you with the operating principles of combination gas-oil burners.

Fuel oils, like natural and LP gases, are excellent heating fuels. Among the many differences, the lighting and burning of fuel oil requires special equipment. Fuel oil in its liquid form will not burn; it must be either atomized or vaporized. From this we can define a fuel oil burner as a mechanical device that prepares oil for combustion. The actual burning of the fuel takes place in the firebox, with the atomizing method being the most popular. The atomizing type of burner prepares the oil for burning by breaking it up into a fog-like vapor. This is accomplished in three ways: (1) by forcing the oil under pressure through a nozzle—air or steam may be forced through with the oil or the oil may be forced alone; (2) by allowing the oil to flow out the end of small tubes that are rapidly rotated on a distributor; or (3) by forcing the oil off the edge of a rapidly rotating cup that may be mounted either vertically or horizontally.

FUEL OIL BURNERS

A typical *high-pressure* oil burner is comprised of an electric motor, blower, and fuel unit. The motor is mounted on the motor shaft and the fuel unit is directly connected to the end of the motor shaft. The fuel unit consists of a strainer, a pump, and a pressure regulating valve.

In operation, when the thermostat demands heat for the structure, the fuel oil is drawn from the storage tank. The oil is drawn through a strainer where the solid particles are removed, passed through the fuel pump and forced to the pressure regulating valve. The oil hesitates here until the pressure has reached 100 psig. When this pressure is reached, the pressure regulating valve opens and allows the fuel oil to proceed to the burner nozzle where it is atomized. After the fuel leaves the nozzle, it is mixed with the proper amount of air and is ignited by a high voltage transformer—approximately 10,000 V and 25 milliamperes (mA) (Figure 4–1).

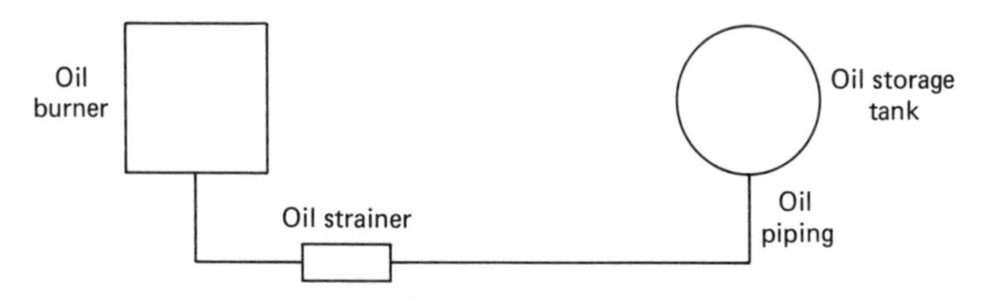

Figure 4–1. Basic fuel oil system.

When sufficient heat has been supplied to the structure, the thermostat will signal the burner motor to stop. As the motor reduces its speed, the oil pressure from the fuel pump also is reduced. As the oil pressure is reduced below 90 psig, the pressure regulating valve stops the flow of oil to the nozzle and the flame is extinguished because of the lack of fuel. The oil burner will remain at rest until the structure requires additional heat, at which time the sequence is started again.

The *low-pressure* fuel oil burner is similar in appearance to the high-pressure burner (Figure 4–2). It also operates much the same as the

Figure 4-2. A low-pressure oil burner (*Courtesy The Carlin Co.*)

high-pressure burner except for the method used to atomize the oil. In the low-pressure burner, the oil is atomized by compressed air, similar to a paint spray gun. The primary air and the fuel oil are forced through the nozzle at the same time with 1-15 pounds of pressure. The secondary air is supplied in the same manner as the air used by the high-pressure burner (Figure 4-3).

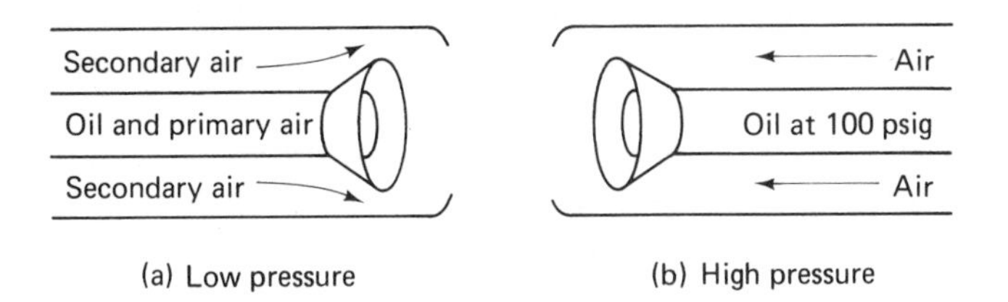

Figure 4-3. Oil burner feed principle.

The *rotary-type* oil burner incorporates a completely different principle from either the high- or low-pressure burners. The rotary burner employs a cup to atomize the oil. The oil is forced to the edge of the cup in small droplets because of the high speed of the cup. As the droplets leave the edge of the cup, they are dispersed into a current of rapidly moving air provided by a centrifugal blower (Figure 4-4). These droplets are further atomized by this air and the fog-like fuel is projected parallel to the axis of the cup. The atomized oil is then ignited in much the same manner as any other type of oil burner. In some burners the cup may be replaced with small tubes extending to the face of the rotating member. The oil is forced to the ends of the tubes where centrifugal force distributes the oil into the air stream.

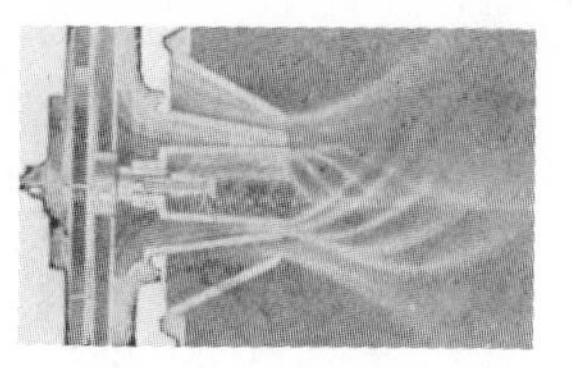

Figure 4-4. Rotary oil burner principle. (*Courtesy Ray Burner Co.*)

Nozzles

To better achieve an understanding of how a nozzle fits into the overall performance of an oil burner, we shall first review the steps of efficient combustion.

Like all combustible matter, the oil must first be vaporized. That is, converted to a vapor or a gas. This requirement must be fulfilled before combustion can come about. This vaporization is usually accomplished by the application of heat.

The oil vapor must then be thoroughly mixed with air so that the necessary oxygen will be present for combustion. It is then necessary to raise the temperature of this mixture above the ignition point.

A continuous supply of air and fuel must be provided for continuous combustion. The products of combustion are then removed from the combustion chamber.

These same steps are necessary for all types of oil burners—the low-pressure gun type, the high-pressure gun type, as well as the rotary burners.

One of the functions of a nozzle is to atomize the fuel, or break it up into tiny droplets that can be vaporized in a short period of time. The atomizing nozzle performs three basic and vital functions for an oil burner.

1. *Atomizing.* It speeds up the vaporization process by breaking up the oil into tiny droplets. There are about 55 billion droplets per gallon of oil at a pressure of 100 psig. The exposed surface of 1 gallon of oil is expanded to approximately 690,000 in.^{2} of burning surface. The individual droplet sizes range from .002–.010 in. The smaller droplets are necessary for the fast, quiet ignition and to establish a flame front close to the burner head. The larger droplets take longer to burn and help fill the combustion chamber with flame.

2. *Metering.* A nozzle is so designed and dimensioned that it will deliver a fixed amount of atomized fuel to the combustion chamber within a plus or minus range of 5% of its rated capacity. This means that

nozzles must be available in many flow rates to satisfy a wide range of industry needs. Under 5 gallons per hour (gph), for example, 21 different flow rates and 6 different spray angles are considered as standard.

3. *Patterning.* A nozzle is also expected to deliver atomized oil to the combustion chamber in a uniform spray pattern and a spray angle best suited to the requirements of a specific installation.

Now that the operation of a nozzle is known, a cut-away showing the functional parts of a typical nozzle is shown in Figure 4-5. The flow rate, spray angle, and pattern are directly related to the design of the tangential slots, swirl chamber, and orifice.

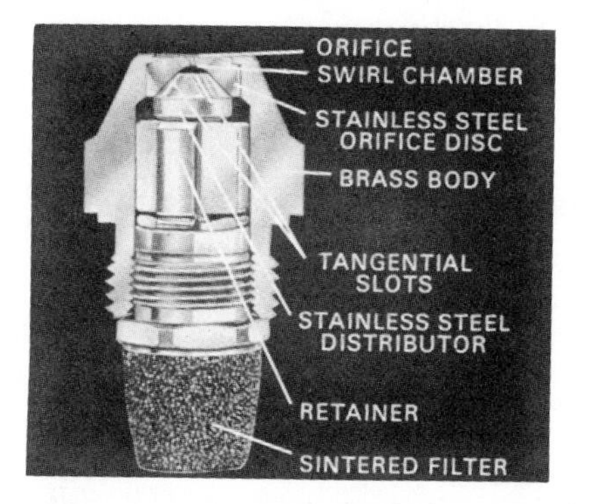

Figure 4-5. Typical high pressure nozzle. (*Courtesy Delavan, Inc.*)

In operation, a source of energy is needed to break up the oil into small droplets. Therefore, pressure is supplied to the nozzle from the oil pump, usually at 100 psig for high-pressure nozzles or 1-15 psig for low-pressure nozzles (Figure 4-6). However, pressure energy alone is not

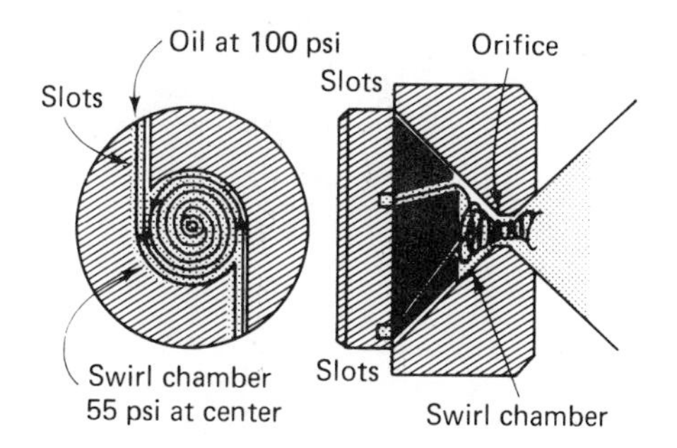

Figure 4-6. How a nozzle works. (*Courtesy Delavan, Inc.*)

enough to do the job. It must be converted to velocity energy. This is accomplished by directing the pressurized fuel through a set of slots. These slots are cut in the distributor at an angle, or tangentially, to produce a high velocity rotation within the swirl chamber. At this point, about half of the pressure energy is converted to velocity energy.

As the oil swirls, centrifugal force is exerted against the sides of the chamber, driving the oil against the orifice walls, leaving a void or core of air in the center. The oil then moves forward out of the orifice in the form of a hollow tube. The tube then becomes a cone-shaped film of oil as it emerges from the orifice, ultimately stretching to a point where it ruptures and throws off droplets of liquid.

Normally, 100 psig is satisfactory for the fixed pressure supplied to a high-pressure nozzle, and all manufacturers of this type nozzle calibrate their nozzle for that pressure.

It is interesting to observe the sprays for a nozzle at various pressures (Figure 4–7). At low pressure, the cone-shaped film is long and the droplets breaking off it are large and irregular [Figure 4–7(a)]. As the pressure is increased, the spray angle becomes better defined. Once a stable pattern is formed, any increase in pressure does not affect the basic spray angle, measured directly in front of the orifice.

At higher pressures, however, note that beyond the area of the basic spray angle, the amount of droplets does make a slight change in direction, inward. This change at this point is because the droplets have a lower velocity due to air resistance, and the air drawn into the spray, by the spray itself, moves the droplets inward. This is the same phenomenon that causes a shower curtain to be drawn into the spray.

Pressure has another predictable effect on nozzle performance. An

Spray at 10 psi pressure. (a)

Spray at 100 psi pressure. (b)

Spray at 300 psi pressure. (c)

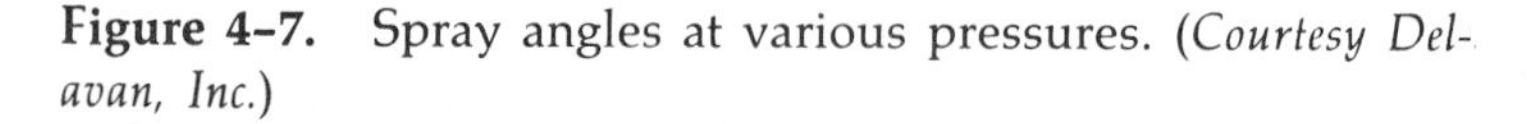

Figure 4–7. Spray angles at various pressures. (*Courtesy Delavan, Inc.*)

increase in pressure causes a corresponding increase in the flow rate of a nozzle, assuming all other factors remain equal. This relationship between pressure and flow rate is best shown in Table 4-1.

An increase in pressure also reduces the droplet size in the spray. For example, an increase from 100-300 psig reduces the average droplet diameter about 28%.

If the pressure is too low, the burner may be underfired. Efficiency may also drop sharply because droplet size is larger and the spray pattern may be changed. If the pressure is not carefully checked, the marking on the nozzle becomes meaningless.

Regardless of the number of spray patterns offered by manufacturers, nozzles can be grouped into two basic classifications for the purpose of this discussion (Figure 4-8).

In the hollow cone, as the name implies, the greatest concentration of droplets is at the outer edge of the spray, with little or no droplet distribution in the center. In general, the hollow-cone spray can be recommended for use in the smaller burners, particularly those firing 1 gph and under, regardless of air pattern. Hollow-cone nozzles have a more stable spray angle and pattern under adverse conditions than solid-cone nozzles of the same flow rate. This is an important advantage in fractional gallonage nozzles where high velocity fuel may cause a reduction in spray angle and an increase in droplet size.

In the solid cone, by definition, the distribution of droplets is fairly uniform throughout the pattern (Figure 4-9). This type of spray pattern is recommended where the air pattern of the burner is heavy in the center, or where long fires are required. They are also recommended for smoother ignition in most burners firing above 3 or 4 gph. An interesting characteristic of solid-cone patterns is that the cone tends to become more and more hollow as the flow rate increases, particularly above 8 gph.

The all-purpose nozzle is neither a true hollow-cone nor a true solid-cone nozzle. At the lower rates, it tends to be more hollow than solid, and as the flow rate increases, the pattern becomes more like a solid cone (Figure 4-10). It can be used in place of the solid or the hollow cone nozzles between .4 and .8 gph, and is suitable for most burners, regardless of the air pattern.

Nozzle Filter

The nozzle filter, or strainer, is designed to prevent dirt or other foreign matter from getting into the nozzle and clogging its passages. Nozzle manufacturers specify the type and size, based on protecting the smallest passage in the nozzle, the slots. In all nozzle sizes, the strainer

Table 4-1
Effects of Pressure on Nozzle Flow Rate

Nozzle Rating at 100 psi	Nozzle Flow Rates in Gallons per Hour (Approx.)					
	80 psi	120 psi	140 psi	160 psi	200 psi	300 psi
.50	0.45	0.55	0.59	0.63	0.70	0.86
.65	0.58	0.71	0.77	0.82	0.92	1.12
.75	0.67	0.82	0.89	0.95	1.05	1.30
.85	0.76	0.93	1.00	1.08	1.20	1.47
.90	0.81	0.99	1.07	1.14	1.27	1.56
1.00	0.89	1.10	1.18	1.27	1.41	1.73
1.10	0.99	1.21	1.30	1.39	1.55	1.90
1.20	1.07	1.31	1.41	1.51	1.70	2.08
1.25	1.12	1.37	1.48	1.58	1.76	2.16
1.35	1.21	1.48	1.60	1.71	1.91	2.34
1.50	1.34	1.64	1.78	1.90	2.12	2.60
1.65	1.48	1.81	1.95	2.09	2.33	2.86
1.75	1.57	1.92	2.07	2.22	2.48	3.03
2.00	1.79	2.19	2.37	2.53	2.82	3.48
2.25	2.01	2.47	2.66	2.85	3.18	3.90
2.50	2.24	2.74	2.96	3.16	3.54	4.33
2.75	2.44	3.00	3.24	3.48	3.90	4.75
3.00	2.69	3.29	3.55	3.80	4.25	5.20
3.25	2.90	3.56	3.83	4.10	4.60	5.63
3.50	3.10	3.82	4.13	4.42	4.95	6.06
4.00	3.55	4.37	4.70	5.05	5.65	6.92
4.50	4.00	4.92	5.30	5.70	6.35	7.80
5.00	4.45	5.46	5.90	6.30	7.05	8.65
5.50	4.90	6.00	6.50	6.95	7.75	9.52
6.00	5.35	6.56	7.10	7.60	8.50	10.4
6.50	5.80	7.10	7.65	8.20	9.20	11.2
7.00	6.22	7.65	8.25	8.85	9.90	12.1
7.50	6.65	8.20	8.85	9.50	10.6	13.0
8.00	7.10	8.75	9.43	10.1	11.3	13.8
8.50	7.55	9.30	10.0	10.7	12.0	14.7
9.00	8.00	9.85	10.6	11.4	12.7	15.6
9.50	8.45	10.4	11.2	12.0	13.4	16.4
10.00	8.90	10.9	11.8	12.6	14.1	17.3
11.00	9.80	12.0	13.0	13.9	15.5	19.0
12.00	10.7	13.1	14.1	15.1	17.0	20.8
13.00	11.6	14.2	15.3	16.4	18.4	22.5
14.00	12.4	15.3	16.5	17.7	19.8	24.2
15.00	13.3	16.4	17.7	19.0	21.2	26.0
16.00	14.2	17.5	18.9	20.2	22.6	27.7
17.00	15.1	18.6	20.0	21.5	24.0	29.4
18.00	16.0	19.7	21.2	22.8	25.4	31.2
19.00	16.9	20.8	22.4	24.0	26.8	33.0
20.00	17.8	21.9	23.6	25.3	28.3	34.6
22.00	19.6	24.0	26.0	27.8	31.0	38.0
24.00	21.4	26.2	28.3	30.3	34.0	41.5

Table 4-1 (cont.)
Effects of Pressure on Nozzle Flow Rate

26.00	23.2	28.4	30.6	32.8	36.8	45.0
28.00	25.0	30.6	33.0	35.4	39.6	48.5
30.00	26.7	32.8	35.4	38.0	42.4	52.0
32.00	28.4	35.0	37.8	40.5	45.2	55.5
35.00	31.2	38.2	41.3	44.0	49.5	60.5
40.00	35.6	43.8	47.0	50.5	56.5	69.0
45.00	40.0	49.0	53.0	57.0	63.5	78.0
50.00	44.5	54.5	59.0	63.0	70.5	86.5

(Courtesy Delevan, Inc.)

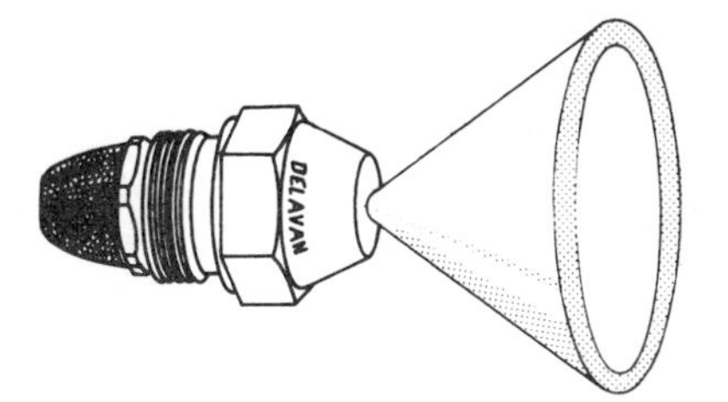

(a) Hollow cone spray pattern

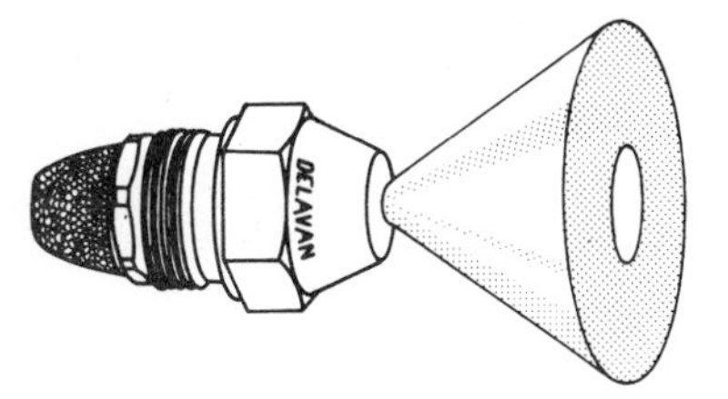

(b) All-purpose spray pattern

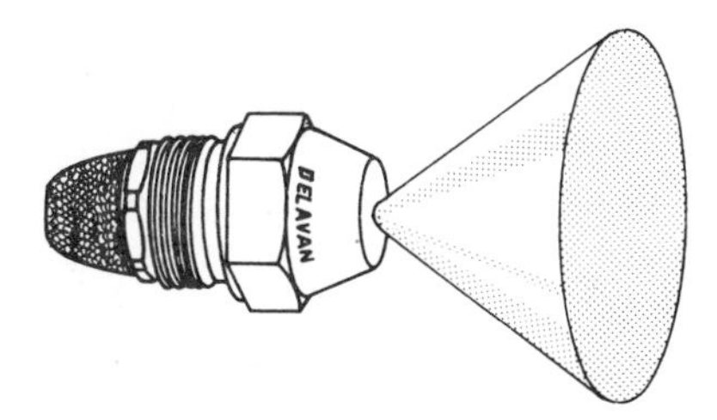

(c) Solid cone spray pattern

Figure 4-8. Spray patterns. *(Courtesy Delavan, Inc.)*

or filter openings are not larger than one-half the size of the smallest passage. For example, a 200-mesh screen has openings of .0029 inch and is used on .50 gph nozzles in which the smallest passage is .006 inch (Figure 4-9).

Porous bronze filters may have the same size openings as strainer screens, but the extra depth provides better filtering. They are also made in several densities.

The nozzle is the very last part of the burner with which the fuel oil comes into contact. It is located on the top of the nozzle tube which is installed inside the combustion chamber (Figure 4-10).

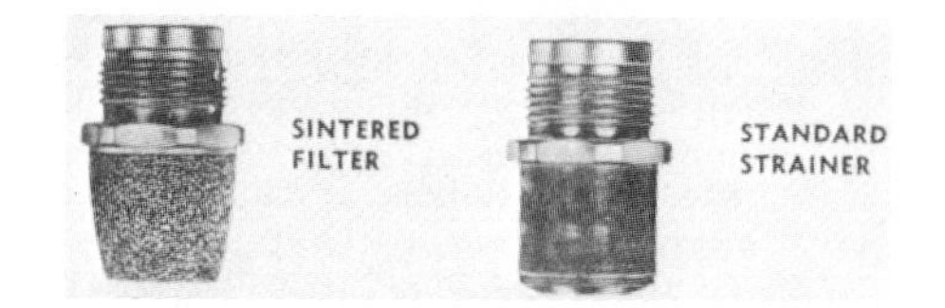

Figure 4–9. Filters and strainers. (*Courtesy Delavan, Inc.*)

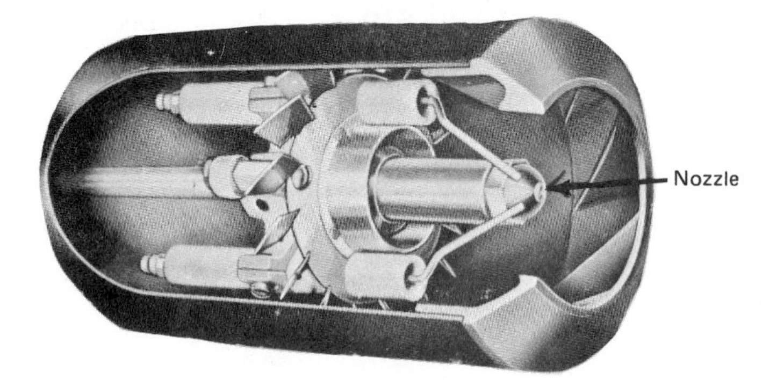

Figure 4–10. Nozzle location. (*Courtesy Monarch Manufacturing Works, Inc.*)

BURNER OPERATION

The foregoing discussion has dealt mainly with high-pressure burners and nozzles. However, for a more complete discussion of the operation and servicing of fuel oil burners, the Ray rotary burner will be used as an example. First, we will discuss the principles of operation of oil burners, and then a more complete component description will follow.

The *rotary oil burner* is a horizontal atomizing unit consisting of only one moving part, a hollow steel shaft, on which are mounted the atomizing cup, the fan, and the motor rotor. These parts comprise the shaft assembly, which rotates on two ball bearings. The fuel oil is introduced by means of a stationary fuel tube extending through the center of the hollow main shaft of the burner into the atomizing cup (Figure 4–11).

The atomizing cup, in the smaller sizes, is made of brass. In the larger sizes, the atomizing cups are made of a special metal alloy selected to meet the requirements of furnace temperatures. All Ray atomizing cups are of the long taper design that permit the oil to form a film on the inside of the cups before it reaches the rim and results in an even discharge. In the Ray rotary oil burner, the oil is introduced into the tail end of the fuel tube, passes through the tube, and flows out of the fuel

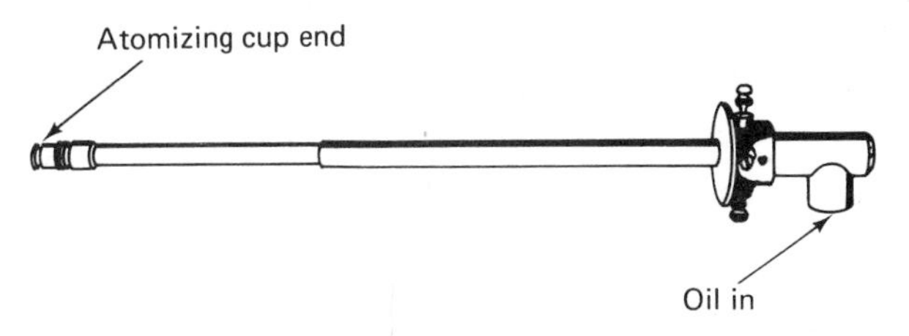

Figure 4–11. Fuel tube. (*Courtesy Ray Burner Co.*)

tip into the back portion of the atomizing cup. The fuel tube and tip are stationary and the oil passages are of such size that high pressure is not required to obtain the rated oil capacity. The atomizing cup is carefully machined on the inside to a gradual smooth taper and rotates at high speed in the counter-clockwise direction as viewed from the tail end. The oil, which flows into the rear of the atomizing cup, is forced by centrifugal action to form a thin uniform film on the inside of the atomizing cup and gradually moves forward, absorbing heat from the fire, and flows off the sharp edge of the cup in a thin film directly into a high velocity air stream. The primary air is discharged from the angular vane nozzle in a rotational direction opposite to that of the oil. The impact of the high velocity air on the film of oil instantly breaks the latter into a mist of fine particles that are readily evaporated. When intimately mixed with the proper amount of air, the vapor is quickly ignited and burns with a dense turbulent flame. Such a flame gives a maximum of heat without smoke or soot and with a minimum of refractory deterioration.

The fan, which is mounted on the hollow shaft, supplies a relatively small amount of the total air necessary for combustion, its principal purpose being to atomize the oil and to control the shape of the flame. The remainder of the air necessary for complete combustion enters the combustion chamber as secondary air.

Nozzles

The nozzle on the Ray horizontal rotary burner is an air nozzle that serves a dual purpose: (1) to provide the primary air at the proper velocity, and (2) to shape the contour of the flame.

The nozzles are furnished in four sizes. They are extra large, large, small, and extra small. The size refers to the bore of the nozzle controlling the maximum air quantity and affecting its velocity. The actual velocity, however, is controlled by the primary air butterfly setting.

Since the delivery rate of the fan is reduced with a smaller nozzle, it should be understood that the firing capacity of the burner is automatically reduced when the smaller sizes are employed.

The larger size nozzles, in general, are used in cases where the burner is required to operate at its maximum capacity, whereas the smaller sizes are used when reduced burner capacities are required. However, there are some instances when the fixed rule does not apply, such as abnormal draft conditions and extreme grades of oil.

The nozzles for Ray rotary burners are also furnished in a number of angulations. The term angulations refers to the position of the vanes within the nozzles that serve to produce a flame of some definite width. The angulation of these vanes range from 50–90° in steps of 5° and 10°.

The 50° nozzles produce a short, wide flame, while the 90° nozzles produce a long, narrow flame. It should be noted also that the shape of the flame may also be influenced by the shape of the combustion chamber and by the direction of the flow of secondary air through the combustion chamber (Figure 4–12).

The standard nozzles furnished with the various sizes of burners are indicated in Table 4–2. These were selected as standard because it was learned that these sizes serve best in the greater majority of cases, and where recommended firebox sizes can be used.

Where recommended firebox dimensions cannot be followed, the selection of the nozzles should be altered accordingly. However, in cases where the firebox is short and wide, it is necessary to provide additional height from the center line of the burner to the firebox floor to accommodate the wide nozzles properly.

There are times when the combustion of the oil will not be stable with the large size nozzles. This can be corrected by installing a nozzle of smaller size.

When any such changes are made, it is well to select a nozzle that is one size wider. For example, when a large 65° nozzle is to be replaced with a smaller nozzle, a 60° nozzle should be selected, if the same flame length is to be maintained. This results from the change in air velocity. Conversely, when changing from a small to a large nozzle, the large nozzle should be one size narrower. For example, use a large 70° nozzle when changing from a small 65° nozzle.

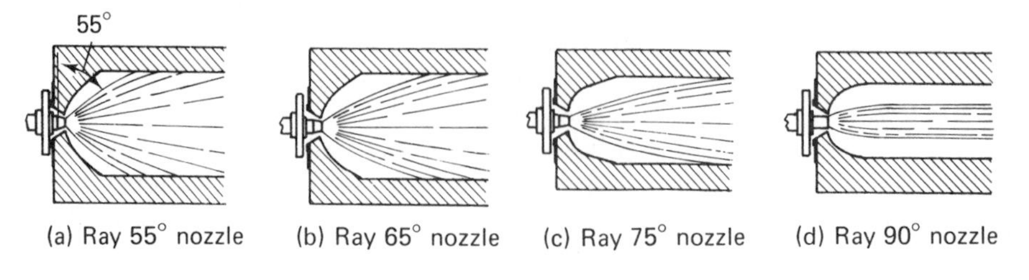

Figure 4–12. Comparison of flame contour produced by Ray angular vane nozzles. (*Courtesy Ray Burner Co.*)

Table 4-2
General Nozzle Specifications

BURNER SIZE	NOZZLE SIZE	OIL 550	GAS 547	GAS 550	COMBINATION 547	COMBINATION 550	F.D. SPECS. 131-134 141-144	F.D. SPECS. 101-104	VANE ANGLES*	NOZZLE DIA. (Inches)
RCR	Standard	9363-5 •	X	X	X	X	X	X	65°	1-13/64
000	Extra Small	4228-XS-5 •	X	X	X	X	X	X	65° – 70°	1-3/16
	Small	4228-S	X	X	X	X	X	X	"	1-1/4
0	Extra Small	4228-XS	4228-XS	X	4228-XS	X	X	X	55° – 90°	1-3/16
	Small	4228-S	4228-S	X	4228-S	X	X	X	"	1-1/4
	Large	4228-L-5 •	4228-L	X	4228-L-15 •	X	X	X	"	1-3/8
	Extra Large	4228-XL	4228-XL-15 •	X	4228-XL	X	X	X	"	1-1/2
1	Extra Small	4232-XS	4232-XS	X	4232-XS	X	4232-XS	4232-XS	55° – 90°	1-7/16
	Small	4232-S	4232-S	9506-2-25 •	4232-S	9506-2-25 •	4232-S	4232-S	"	1-9/16
	Large	4232-L-5 •	4232-L	X	4232-L-15 •	X	4232-L-5 •	4232-L-6 •	"	1-11/16
	Extra Large	4232-XL	4232-XL-15 •	X	4232-XL	X	4232-XL	4232-XL	"	1-13/16
2	Extra Small	4237-XS	4237-XS	9506-3-25	4237-XS	9506-3-25	4237-XS	4237-XS	55° – 90°	1-1/2
	Small	4237-S	4237-S	9506-1-25 •	4237-S	9506-1-25 •	4237-S	4237-S	"	1-5/8
	Large	4237-L-5 •	4237-L	X	4237-L-15 •	X	4237-L-5 •	4237-L-6 •	"	1-3/4
	Extra Large	4237-XL	4237-XL-15 •	X	4237-XL	X	4237-XL	4237-XL	"	1-7/8
3	Extra Small	4241-XS	4241-XS	9427-2-25	4241-XS	9427-2-25	4241-XS	4241-XS	55° – 90°	1-3/4
	Small	4241-S	4241-S	9427-1-25 •	4241-S	9427-1-25 •	4241-S	4241-S	"	1-7/8
	Large	4241-L-5 •	4241-L	9427-3-25	4241-L-15 •	9427-3-25	4241-L-4 •	4241-L-6 •	"	2
	Extra Large	4241-XL	4241-XL-15 •	X	4241-XL	X	4241-XL	4241-XL	"	2-1/8
3-30		4241-S-4 •	X	9427-1-24 •	X	9427-1-24 •	4241-S-4 •	4241-S-4 •	"	1-7/8
5	Extra Small	4245-XS	4245-XS	9537-2-25	4245-XS	9537-2-25	4245-XS	4245-XS	55° – 90°	2-3/16
	Small	4245-S	4245-S	9537-1-25 •	4245-S	9537-1-25 •	4245-S-5 •	4245-S-5 •	"	2-5/16
	Large	4245-L-5 •	4245-L	X	4245-L-15 •	X	4245-L	4245-L	"	2-7/16
	Extra Large	4245-XL	4245-XL-15 •	X	4245-XL	X	4245-XL	4245-XL	"	2-9/16
5-45		4245-XS-4 •	X	9537-2-24 •	X	9537-2-24 •	4245-XS-4 •	4245-XS-4 •	"	2-3/16
6	Extra Small	4249-XS	4249-XS	9534-2-25	4249-XS	9534-2-25	4249-XS	4249-XS	50° – 90°	2-3/4
	Small	4249-S	4249-S	9534-1-25 •	4249-S	9534-1-25 •	4249-S-5 •	4249-S-5 •	"	2-7/8
	Large	4249-L-5 •	4249-L	X	4249-L-15 •	X	4249-L	4249-L	"	3
	Extra Large	4249-XL	4249-XL-15 •	X	4249-XL	X	4249-XL	4249-XL	"	3-1/8
6-60		4249-XS-4 •	X	9534-2-24 •	X	9534-2-24 •	4249-XS-4 •	4249-XS-4 •	"	2-3/4
7	Extra Small	4253-XS	4253-XS	9564-2-25	4253-XS	9564-2-25	4253-XS	4253-XS	50° – 90°	3-3/8
	Small	4253-S	4253-S	9564-1-25 •	4253-S	9564-1-25 •	4253-S-5 •	4253-S-5 •	"	3-1/2
	Large	4253-L-6 •	4253-L	X	4253-L-16 •	X	4253-L	4253-L	"	3-5/8
	Extra Large	4253-XL	4253-XL-16 •	X	4253-XL	X	4253-XL	4253-XL	"	3-3/4
7-85		4253-XS-4 •	X	9564-2-24 •	X	9564-2-24 •	4253-XS-4 •	4253-XS-4 •	"	3-3/8
8	Extra Small	4257-XS	4257-XS	9565-2-25	4257-XS	9565-2-25	4257-XS	4257-XS	50° – 90°	3-7/8
	Small	4257-S	4257-S	9565-1-25 •	4257-S	9565-1-25 •	4257-S-5 •	4257-S-5 •	"	4
	Large	4257-L-6 •	4257-L	X	4257-L-16 •	X	4257-L	4257-L	"	4-1/8
	Extra Large	4257-XL	4257-XL-16 •	X	4257-XL	X	4257-XL	4257-XL	"	4-1/4
8-130		4257-XS-4 •	X	9565-2-24 •	X	9565-2-24 •	X	X	"	3-7/8
9	Extra Small	4261-XS	4261-XS	X	4261-XS	X	4261-XS	4261-XS	50° – 90°	4-1/4
	Small	4261-S	4261-S	X	4261-S	X	4261-S-5 •	4261-S-5 •	"	4-3/8
	Large	4261-L-6 •	4261-L	X	4261-L-16 •	X	4261-L	4261-L	"	4-1/2
	Extra Large	4261-XL	4261-XL-16 •	X	4261-XL	X	4261-XL	4261-XL	"	4-5/8
10	Extra Small	4265-XS	4265-XS	X	4265-XS	X	4265-XS	4265-XS	50° – 90°	5-1/8
	Small	4265-S	4265-S	X	4265-S	X	4265-S	4265-S	"	5-1/4
	Large	4265-L-6 •	4265-L	X	4265-L-16 •	X	4265-L-5 •	4265-L-5 •	"	5-3/8
	Extra Large	4265-XL	4265-XL-16 •	X	4265-XL	X	4265-XL	4265-XL	"	5-1/2
12	Extra Small	4265-S	4265-S	X	4265-S	X	X	X	50° – 90°	5-1/4
	Small	4265-L	4265-L	X	4265-L	X	X	X	"	5-3/8
	Large	4265-XL-6 •	4265-XL-16 •	X	4265-XL-16 •	X	X	X	"	5-1/2
	Extra Large	4265-XXL	4265-XXL	X	4265-XXL	X	X	X	"	5-5/8

VANE ANGLE CODE								
90°	80°	75°	70°	65°	60°	55°	50°	
-0	-2	-3	-4	-5	-6	-7	-8	Model 550 Oil Burners
-10	-12	-13	-14	-15	-16	-17	-18	Model 547 Gas and Combination Burners
-20	-22	-23	-24	-25	-26	-27	-28	Model 550 Gas and Combination Burners

• Standard nozzle on burner unless specified otherwise.

* Available in 5° increments.

X = Not Available

All nozzles interchangeable with older model burners.

(Courtesy Ray Burner Co.)

The amount of primary air used through the nozzle should be just enough to produce good atomization, and also sufficient enough to prevent the oil from breaking through the air stream.

It is good practice to select a nozzle size just large enough to furnish the amount of air required for good atomization at full fire rate and without pulsation, with the butterfly valve nearly full open.

When the oil is breaking through the air stream, the flame will be smoky near the front of the firebox, which will usually result in carbon forming on the refractories at this point.

An excess of primary air usually has a tendency to cause the flame to burn away from the nozzle. This frequently results in flame pulsations.

The best possible firing results are obtained when the primary air delivery through the nozzle has exactly the proper volume and velocity.

In cleaning the nozzle, it is suggested that all carbon or oil residue be removed with a knife or a scraper. When this is done, the nozzle should be thoroughly cleaned with kerosene or solvent, making certain that the passages between the vanes are not obstructed.

The vanes should be checked to be certain that they are not bent, loose, or broken. Bent vanes should be straightened and in cases where the vanes are broken, the nozzle should be replaced.

Atomizing Cup

The purpose of the atomizing cup, in conjunction with the air blast from the nozzle, is to atomize the fuel oil.

The inside of the atomizing cup is tapered slightly. This taper, coupled with the speed at which the cup rotates, causes the oil to move forward to the edge of the cup in a thin film.

To promote good atomization, it is important that the inside of the cup be smooth and also that the cup be kept clean.

It is equally important that the edge of the cup be trimmed and belled properly, and also that there be no ragged edges. Any of these imperfections will cause the oil film to be irregular and will, therefore, promote poor atomization.

It is also important that the outside of the cup be kept clean so that the air delivery from the nozzle will not be obstructed or deflected from its normal course.

The atomizing cup can be cleaned or polished, both internally and externally without removing it. Ordinarily the cup can be cleaned by wiping it with a cloth or with the use of kerosene or solvent and a cloth.

In certain cases, when burning oil of low viscosity, such as number 2 oil, the thin edge of the cup may be belled out slightly using a heavy

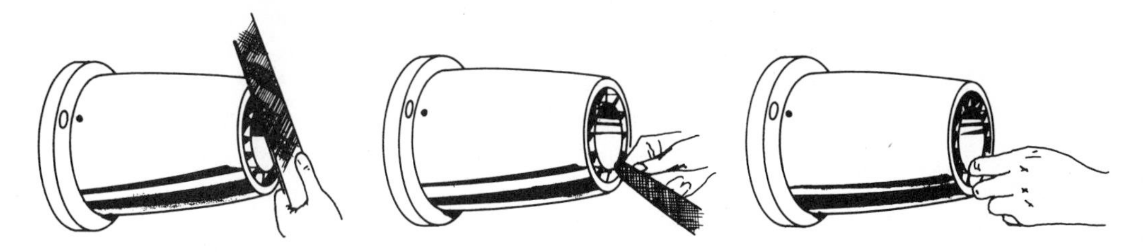

Figure 4–13. Dressing atomizing cups. (*Courtesy Ray Burner Co.*)

round tool such as a piece of shafting. This will help to stabilize a flame that otherwise may be difficult to control.

When replacing atomizing cups (Figure 4–13), the following is the proper method of trimming and dressing. First, consult the proper data sheet for exact measurements. If the cup is too long, dress it down with a file held to the spinning cup. Then taper the inside of the lip of the cup as shown and polish it with emery cloth. It is important, for perfect atomization, that this lip be absolutely smooth and correctly rounded on the inside, but sharp on the outer edge.

Fan Housing Cover

The fan housing cover serves to enclose the fan housing and to provide a mounting for the nozzle. It is equipped with a baffle and vanes that direct the air from the discharge side of the fan to the nozzle. It is important that the space between the baffle and the cover be kept clean so that the flow of air from the fan will not be retarded or reduced.

To simplify locating the cover when it is to be replaced, before it is removed, mark it with a center punch or a crayon to serve as a guide. Unless the cover is properly positioned, it will not be possible to position the nozzle properly.

Before replacing the cover be certain that it is thoroughly clean and that all of the rivets holding the baffle are tight. To insure proper alignment, the machine surfaces of the housing and the cover must be clean.

Fan

The fan delivers the primary air to the burner nozzle. Since the amount of air delivered by the fan materially affects the firing capacity of the burner, it is important that the fan, the fan housing, the air inlet housing, and the primary air butterfly be kept clean.

The fan is properly balanced at the factory to insure freedom from vibration. Any fouling of the fan may result in an unbalanced condition which will be reflected in the operation of the burner. The parts should be inspected at least once a season. Inspection of the fan housing can be made by removing the nozzle and the fan housing cover, or by removing the atomizing cup. In the event these parts are fouled and the fan or fan housing requires cleaning, the fan should be removed.

Air Inlet Housing and Primary Air Butterfly

The air inlet housing serves as a duct to supply the air to the inlet of the fan. The primary air butterfly serves as a throttle to regulate the amount of air drawn into the fan and consequently also regulates the amount of air furnished to the nozzle (Figure 4-14).

It is important that these parts be kept clean. The air inlet housing ordinarily requires cleaning only once every season, except in cases where the air is fouled with dust or lint. It is good practice to wipe the butterfly once each month to insure positive air regulation and delivery, and thereby maintain more uniform firing conditions. This should be done only when the burner is idle, otherwise the cloth may be drawn into the fan, clogging the intake.

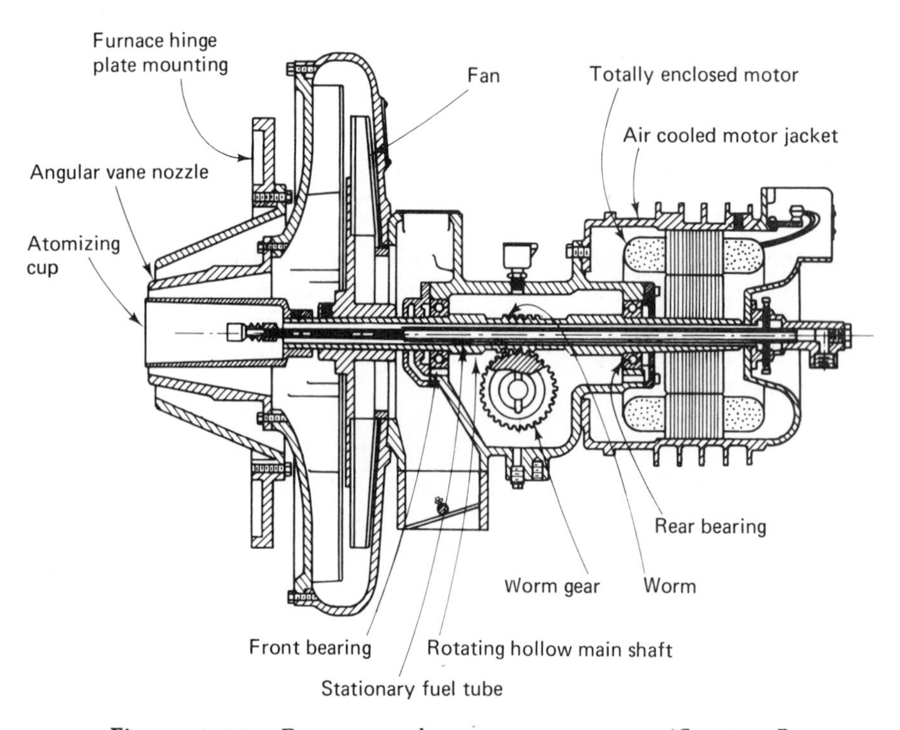

Figure 4-14. Ray rotary burner components. (*Courtesy Ray Burner Co.*)

Worm Gear and Worm Gear Shaft

The worm gear and the worm gear shaft drive the fuel oil pump and are driven by the worm gear on the hollow main shaft. The worm gear also distributes the lubricating oil, lubricating the bearings on the shaft (Figure 4-15).

The worm gear shaft is provided with a driving pin that meshes with the slots on the hub of the worm gear from which it is driven.

These parts require no adjustment, and so long as the burner is properly lubricated, the life of the worm gear will be unlimited. A lack of proper lubrication, however, will cause the worm gear to overheat and will materially shorten its life.

If the worm gear is damaged or the driving pin in the worm gear shaft is broken, it is recommended that the pump on the burner, or the two-stage pumps in the reservoir, be checked to be certain that they operate freely before either of these parts is replaced and the burner put into operation again. If this is not done, the new parts may be broken or damaged.

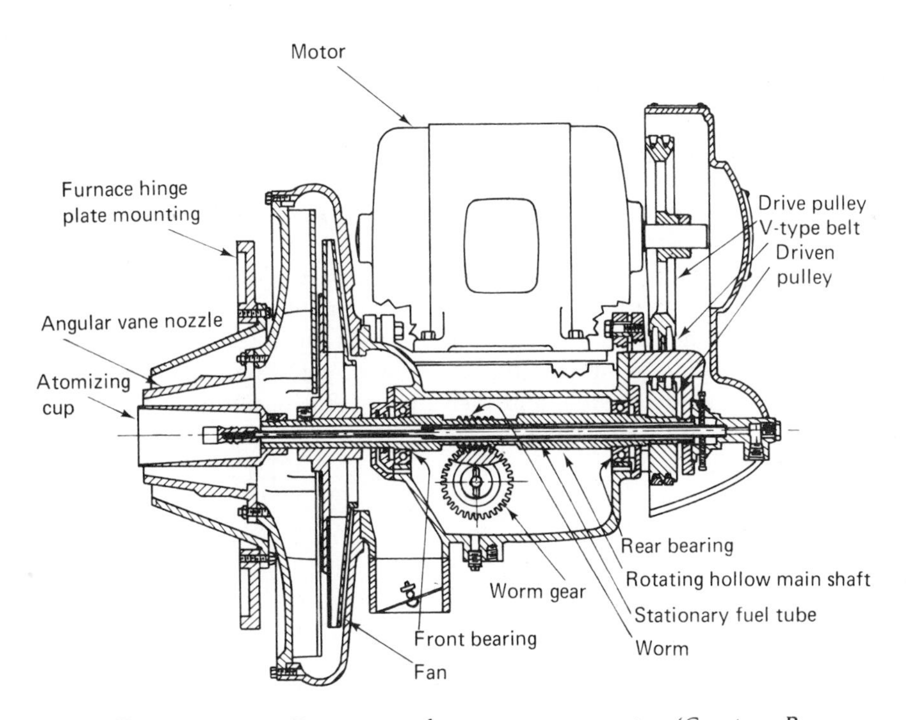

Figure 4-15. Ray rotary burner components. (*Courtesy Ray Burner Co.*)

Gear Housing

The gear housing encloses the bearings of the worm gear and the worm gear shaft. It also serves as a reservoir for the lubricating oil. The worm gear cover is provided with an oil gauge cup, used to check the amount of lube oil in the housing.

When the burner is idle, the gauge cup should be filled to within ⅛ inch of the top of the cup. Motor oil of number 30 S.A.E. should be used for lubrication and should be added as often as is necessary. In cases where the surrounding temperature is low, it may be advantageous to use number 20 S.A.E. oil. The oil in the gear housing should be drained once each season, and the housing refilled with new oil.

Fuel Feed Assembly

The fuel feed assembly consists of a tailpiece, fuel tube, and tip assembly (Figure 4-16). Some assemblies include a vibration dampening assembly that screws on to the end of the tip. This assembly permits the transportation of the fuel oil from the burner piping to the atomizing cup through the hollow main shaft, without the use of stuffing boxes or special seals.

It is important that the fuel feed assembly be properly centered in the hollow main shaft and also that both the tube and the inside of the shaft be clean, so that fuel tube rattling will not occur. Rattling of the fuel tube will cause the burner to be very noisy in its operation, and may, after prolonged rattling, cause the fuel tube to break. For centering the fuel tube in the hollow shaft, four adjusting screws with locknuts are provided in the tailpiece.

To check the fuel tube alignment, look at it from the front or atomizing cup end of the burner. The fuel tube should be centered in the hollow shaft as accurately as possible. A fairly accurate check of its location can be made by moving the fuel tube tip with the ends of the fingers, first from side to side, and then up and down, noting the motion required to contact the shaft at each point from the free standing position.

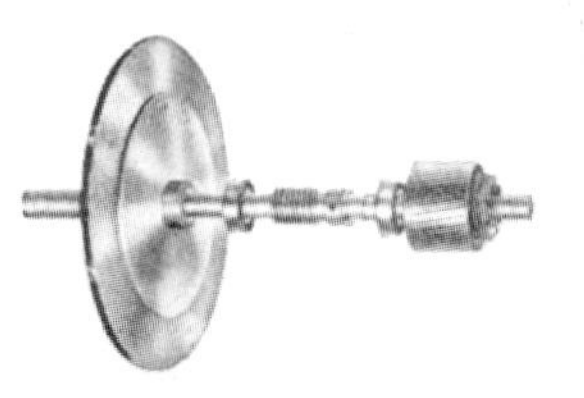

Figure 4-16. Fuel feed assembly. (*Courtesy Ray Burner Co*)

Prior to making any adjustments, it is suggested that the tailpiece mounting screws be checked to be sure that the tailpiece is properly secured, as these will affect the alignment of the tube.

If the fuel tube should rattle when it is found to be centered, the fuel feed assembly should be removed to determine:

1. If the hollow shaft is clean.
2. If the fuel tube is clean externally.
3. If the fuel tube is bent.

With all of these in order and with the fuel tube centered, no rattling should occur.

The fuel tube should be cleaned thoroughly and then checked to be sure that it is smooth and absolutely straight. It is suggested that the straightedge be placed on all four sides of the fuel tube in turn.

The fuel tube is made of steel and can be straightened through bending, but extreme care should be exercised.

The hollow shaft should then be checked for cleanliness. If necessary, it should then be cleaned by drawing a piece of cloth saturated with kerosene or solvent through it, in much the same manner as cleaning a gun barrel. At no time should a fuel feed assembly be installed without first of all checking the hollow shaft for cleanliness.

Fuel Tip

The fuel tip on the end of the fuel tube assembly distributes the fuel oil on the inner surface of the cup and prevents the oil from running back along the tube into the hollow shaft. (Refer to Figure 4-14.)

The orifice or discharge hole in the fuel tip should be at the top in every case to eliminate the possibility of the fuel draining from the fuel tube when the burner is idle.

Except for replacements or for relocation of the orifice, there is no occasion for removing the fuel tip, because it can be withdrawn through the hollow shaft with the feed assembly.

The fuel tip baffle of the nozzle type should be installed with the nozzle at the bottom and the tip orifice at the top, as stated previously.

Motor Jacket Cover

The motor jacket cover encloses the motor housing and provides a mounting for the tailpiece (Figure 4-17). Incorporated in the motor jacket cover is a terminal box for the connections of the motor, oil valve, etc. Since the motor jacket cover provides a support for these various items, it is important that it be properly positioned and properly secured.

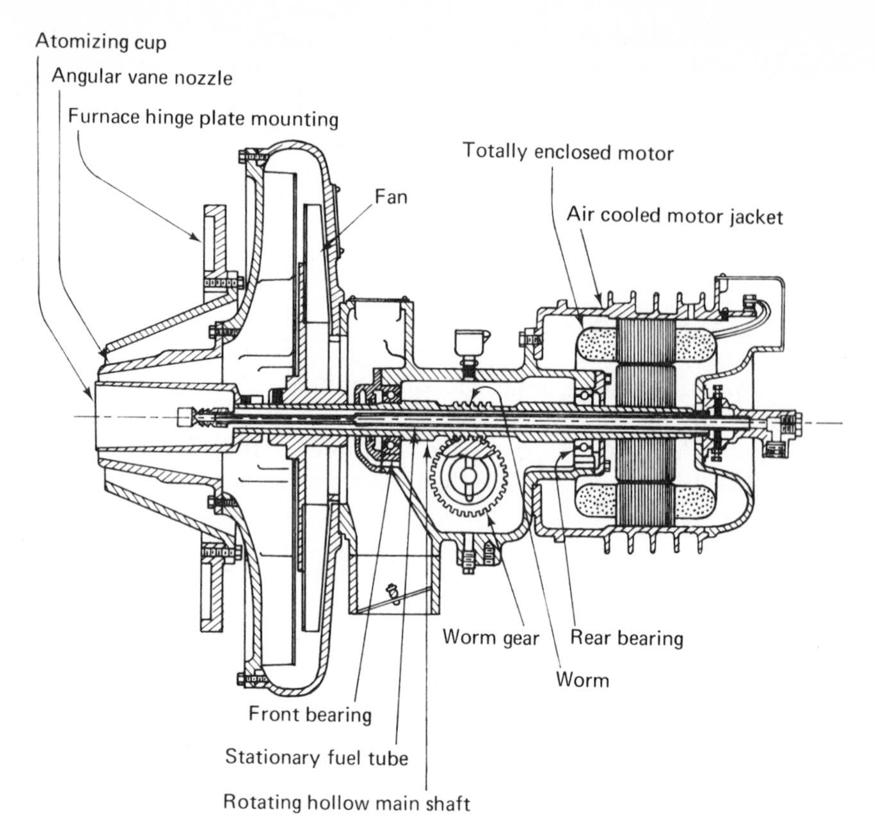

Figure 4-17. Ray burner rotary burner components. (*Courtesy Ray Burner Co.*)

The motor jacket cover is secured to the motor jacket with four screws. Prior to removing the motor jacket cover, the fuel feed assembly should be removed.

Motor Jacket

The motor jacket serves as a housing for the motor parts. It is secured to the gear housing with four screws, and is provided with setscrews or a clamping screw to secure the stator in position. A drain hole is provided to drain off any lubricating oil which might leak from the bearing cover. In replacing the motor jacket, it is important that the hole be placed at the bottom. To insure proper stator alignment, the bore of the motor jacket must be clean and smooth. This will simplify installation of the stator.

Hollow Shaft and Bearings

The hollow shaft is the main burner shaft and is equipped with two ball bearings (Figure 4-18). The shaft has a worm gear to drive the worm gear in the pump. The shaft also supports the rotor, the fan and the atomizing cup. This assembly revolves as a single unit.

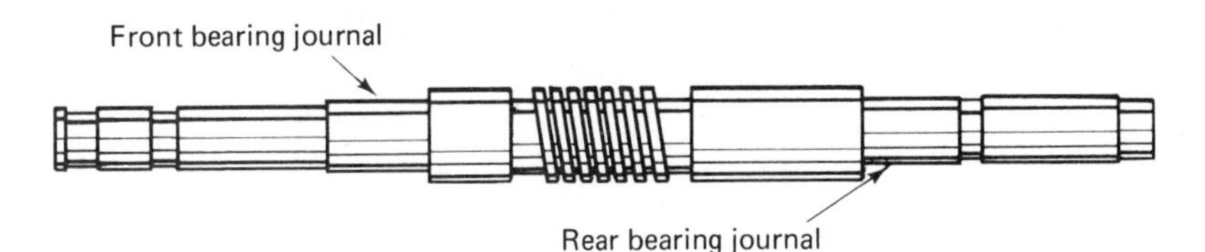

Figure 4-18. Main burner shaft. (*Courtesy Ray Burner Co.*)

The ball bearings are located at each end of the gear housing, enclosed by a bearing cover with gaskets. They are fitted to the shaft with only a light tap fit and not a tight pressed fit. The fitting of the bearings in the housing is a light push fit and should permit assembly by pressing in with the two thumbs.

When the shaft and bearings are fitted on the assembly into the gear housing, there should be just a slight amount of end play, usually about .010 inch. The forward bearing assembly includes a thrust spring that forces the shaft assembly to the rear.

In replacing the bearing cover, the drain groove should always be located at the bottom, and the bearing covers should also clear the shaft to eliminate binding or wear. The front bearing cover is provided with a breather port that serves as a vacuum break to eliminate oil being drawn from the front bearing by the vacuum created by the fan. It is important that this port remain open, and that it be cleaned when any repairs are made.

Ray Viscosity Valve

The Ray viscosity valve, in conjunction with the two-stage pump and reservoir and modulating system, automatically adjusts the fuel oil rate to match the load requirements. With a single initial adjustment, this valve automatically compensates for any fluctuations in viscosity, and meters the oil at a prescribed and modulated rate of flow (Figure 4-19).

The primary pump *P* delivers oil from the storage tank to the reservoir in excess of the pumping rate of the secondary pump *S*. The non-variable flow of oil from the secondary pump *S* enters the inlet port of

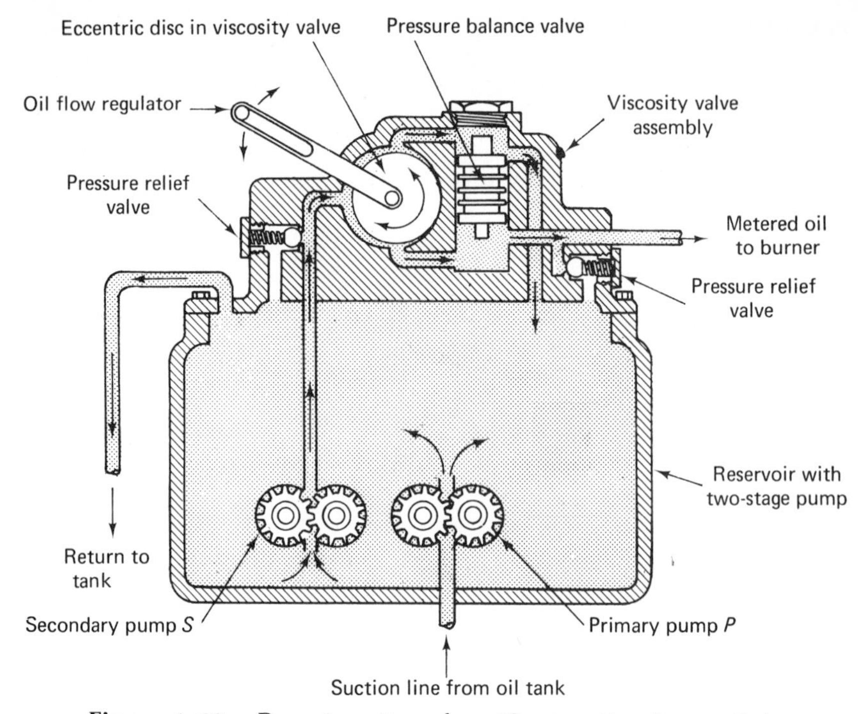

Figure 4-19. Ray viscosity valve. (*Courtesy Ray Burner Co.*)

the viscosity valve. The oil passes through the valve in two paths, the flow through each of which is controlled by positioning the eccentric disc in the viscosity valve assembly. Since the flow is viscous, the rate through each side remains constant regardless of oil viscosity, the pressure automatically varying to compensate for changes in flow resistance.

The oil from each path enters opposite ends of the pressure balancing valve in which the free-floating piston proportions the area of the outlet ports from the cylinder, maintaining equalized pressures at both ends. This assures a viscous flow rate to the burner unaffected by changes in pressure in the return line to the tank or in the line to the burner nozzle. The metered oil from one path is delivered to the burner and the remainder, from the other path, back to the reservoir.

Two pressure relief valves, one at the pump discharge and one at the outlet to the burner nozzle, operate only to relieve excessive pressure when operating against a closed valve or in case of incorrect adjustment. For an adjusted position of the eccentric disc, therefore, the flow rate to the burner remains constant, independent of oil temperature changes or line pressure variations.

When adjusting the viscosity valve, it should be noted that the end of the metering shaft has a slot for screwdriver setting. There may also be a position-indicating notch in the end of the shaft or a pin indicator through the shaft. As viewed from the shaft extension side, as above, the notch or pin must operate within the arc represented; that is approximately one o'clock to five o'clock (Figure 4-20).

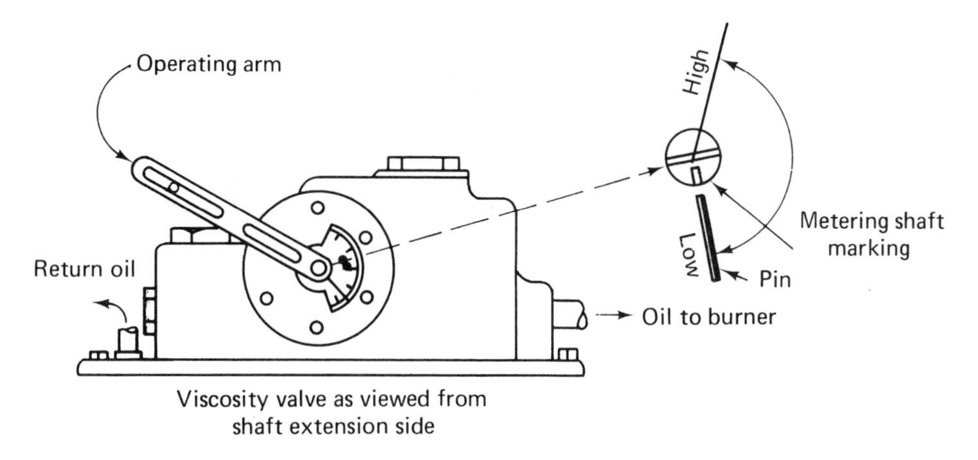

Figure 4-20. Adjusting Ray viscosity valve. (*Courtesy Ray Burner Co.*)

Single-Stage Pump

The single pump used on manual or semiautomatic burners is the rotary gear type. It serves a dual purpose: (1) to draw the oil from the tank; (2) to deliver the oil at some predetermined pressure to the oil regulating valve (Figure 4-21).

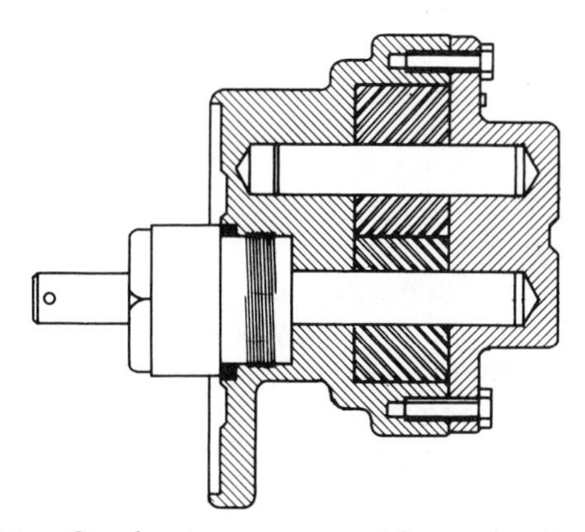

Figure 4-21. Single-stage pump with mechanical seal. (*Courtesy Ray Burner Co.*)

The pump is designed to operate at low speed (200–300 rpm) to eliminate wear and to insure long life. When properly primed and in good condition, it should generate a vacuum of approximately 27 inches of mercury at sea level.

Ordinarily, the pump requires no service except for renewing the mechanical seal. If the pump does not generate sufficient vacuum after a long period of service, this condition can usually be corrected by removing one or two thin gaskets between the pump body and cover.

Care should be exercised when gaskets are removed to insure that the adjoining gaskets are not broken or torn. When it is necessary to remove any gaskets or to make any repairs to the pump requiring the removal of the pump cover, it is recommended after assembly that the burner shaft be rotated manually to insure that the pump operates freely before applying power to the burner motor.

Should the pump become badly worn after years of service, it should be replaced. Replacing the gears and shafts in an old body is not recommended because, if the gears and shafts are badly worn, it is certain that some wear has occurred in the pump body or the pump cover. Consequently, the new gears and shafts will not correct the fault.

If the pump is removed from the burner or a new pump is to be installed, it is very important that the pump be properly aligned with the worm gear shaft. Pump aligning tools are furnished by the Ray Oil Burner Company for this purpose, and these should be used in making such repairs or replacements.

There may arise some cases where the heavier grades of Bunker fuel oils are used or when changing from light to heavy fuel oils, in which the pump is fitted too close. An indication of this form of trouble will be the breaking of the driving pin in the worm gear, the breaking of the shear pins in the pump coupling, or damage to the fiber worm gear. To correct it an extra gasket may be added under the pump cover. Usually the addition of one gasket of .001–.0015-inch thickness will correct the trouble.

Two-Stage Pump

The two-stage pump is of the same design as the single pump, except that the two pumps are contained in a single housing with pump covers at each end, and are driven by a single shaft (Figure 4–22). The primary pump has greater capacity than the secondary pump. It serves only as a suction pump and is used only to draw the fuel oil from the storage tank and to discharge it into the reservoir. The secondary pump serves as a pressure pump, drawing its oil from the reservoir and discharging it under pressure into the valve assembly.

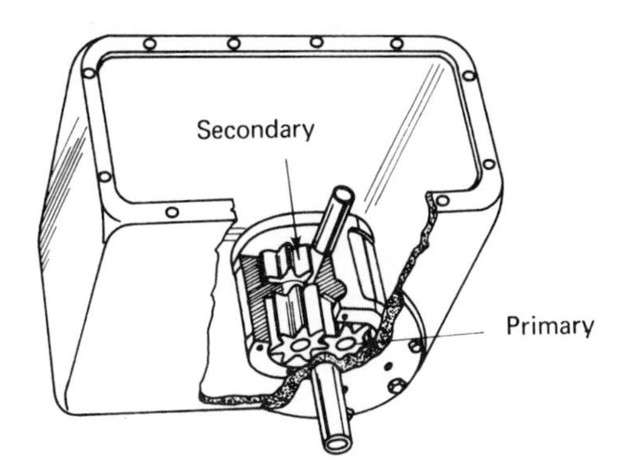

Figure 4–22. Two-stage pump. (*Courtesy Ray Burner Co.*)

The only test required for the secondary pump is a pressure test. If the secondary pump will generate and maintain the desired operating pressure, it is in proper working condition. The general instructions and suggestions covering the single pump will also apply to the two-stage pump.

Reservoir

The reservoir serves as a housing for the two-stage pump to which the viscosity valve assembly or the reservoir cover and relief valve assembly are connected. The oil from the primary pump is discharged into the reservoir (Figure 4–23).

The secondary pump draws its oil from the reservoir at a point about one inch from the bottom of the reservoir. The amount of oil in excess of that amount consumed by the burner that is delivered by the secondary pump is discharged back into the reservoir. When the reservoir is filled and overflowing, this oil is discharged through the hinge and the return line to the storage tank. Thus, no oil will be discharged to the tank unless the reservoir is filled and overflowing.

There is a tendency for some small particles of dirt to filter through the strainer basket, and since these particles are heavier than the oil, some of this oil settles to the bottom of the reservoir. Should it be allowed to accumulate for long periods of time, it can be drawn into the secondary pump. In order to eliminate this form of trouble, the reservoir should be drained and flushed out occasionally. The reservoir is provided with a clean-out hole and cover through which the cleaning of the reservoir is simplified or with plugged holes in the bottom through which dirt may be flushed out. It is suggested that the reservoir be cleaned or flushed out once each season to insure trouble-free burner operation.

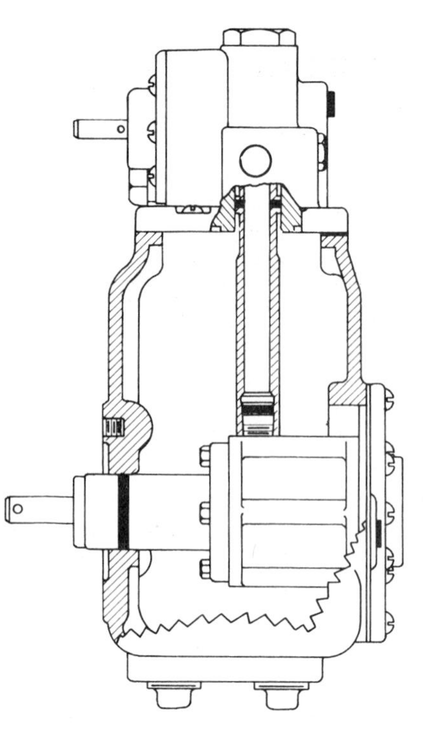

Figure 4-23. Two-stage pump and reservoir. (*Courtesy Ray Burner Co.*)

Pump Seals

A mechanical seal is used; this is an adjustable screw-type packing gland that uses a redesigned stuffing box in place of the spring loaded type. Because of the direction in which the shaft rotates, the threads on the gland nut are left-handed (Figure 4-24).

To replace the mechanical seal, the following steps should be followed (Figure 4-25). All threads on the hub are left-handed. Hold the sleeve with pipe pliers on the recess to avoid marring the ground surfaces. Unscrew the seal retainer (left-hand thread). The stationary seat and O ring will come out with the retainer. Push the stationary seal out and insert new seat into O ring. Do not scratch the seat face and keep it clean. Unscrew sleeve (left-hand) with pipe pliers. Grasp brass body of rotating seal and slide off end of pump shaft. Slide drive washer off the shaft. Inspect pressed-on truarc ring. It should be at a depth of ½ inch from the near edge of the O ring groove. Normally, the drive washer and truarc ring do not need replacing.

If the truarc ring is broken, install a new one by wedging a screwdriver into the truarc slot just barely enough to install the ring on to the

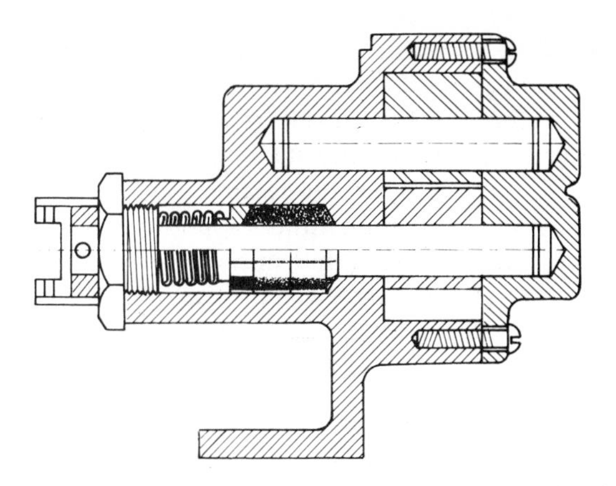

Figure 4-24. Single-stage pump with packing gland. (*Courtesy Ray Burner Co.*)

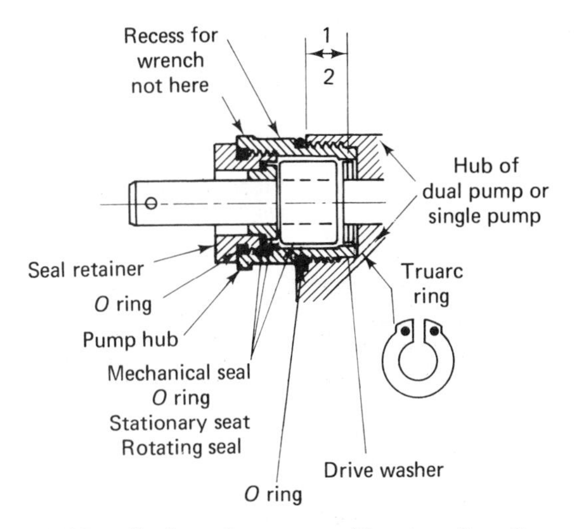

Figure 4-25. Seal replacement. (*Courtesy Ray Burner Co.*)

shaft end. Overstressing will weaken it and allow it to slip on the shaft. If this ring slips or is out of position, the correct pressure will not be maintained between the two polished sealing surfaces and leakage may result.

Remove screwdriver and press the truarc ring down to the ½ inch dimensions as shown, using a piece of ½ inch iron pipe with a good square end. Install the driver washer with the 45° bent finger inserted into the slot in the truarc ring.

Oil the shaft lightly and slide the new rotating seal on to the shaft indexing two slots on two fingers of the drive washer. The polished carbon face is positioned outward to run against the stationary polished steel face. Keep the carbon clean and unscratched.

Reinstall the sleeve or hub. Install retainer with new stationary seat positioning the O ring between the stationary seat and the retainer and the intermediate O ring between the retainer and the sleeve. The outer O ring seals the complete assembly into the pump hub or body. Remember that seal faces must be clean and unscratched for proper operation.

Three O rings of different sizes are involved. They must all be carefully examined and, if defective in any way, must be replaced.

Solenoid Oil Valve

The solenoid oil valves are used to open or shut off the oil supply to the burner. When the solenoid coil is energized, the oil valve will open; when de-energized, the valve will be closed (Figure 4-26).

Solenoid oil valves are designed for some specific operating pressures, and these pressures should not be exceeded if positive valve action is to be expected.

Oil valve failure can usually be traced to one of the following causes:

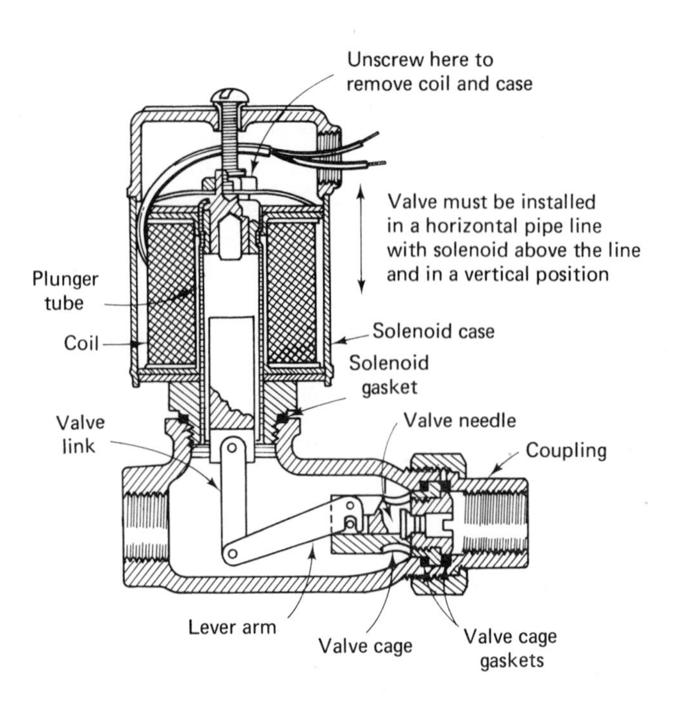

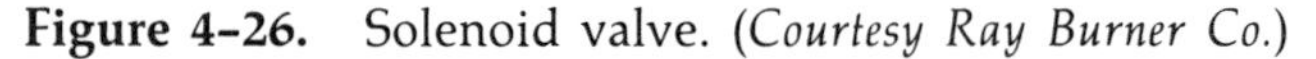
Figure 4-26. Solenoid valve. (*Courtesy Ray Burner Co.*)

1. Improper power characteristics.

2. Valve parts fouled with dirt or foreign matter.

3. Valve parts worn.

4. Valve mechanism damaged by dropping, bumping, or rough handling.

In making repairs to solenoid oil valves, follow the manufacturer's instructions.

Electric Oil Heater

The purpose of the electric oil heater is solely to maintain an initial supply of oil at a suitable temperature for starting purposes, and not for heating and maintaining the oil that is normally consumed by the burner at some required temperature (Figure 4-27).

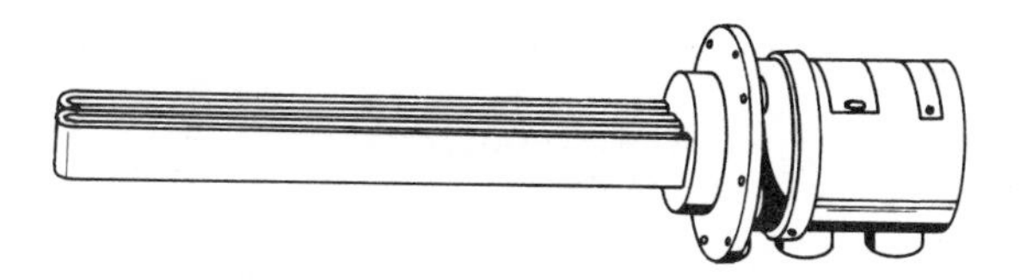

Figure 4-27. Electric oil heater. (*Courtesy Ray Burner Co.*)

The oil heater should be adjusted to maintain an oil temperature as low as practical. Excessive temperatures may cause gasification or carbonizing, and may also cause unsatisfactory starting conditions.

To insure the best possible results from the heater, the housing and the heater element should be cleaned at least once a season. In cases where the fuel oil contains excessive amounts of sludge, sediment, or water, these parts should be cleaned more frequently.

The heater is usually equipped with an auxiliary switch head that provides a control circuit with adjustable contacts, closed in the hot position. This circuit is usually wired into the burner circuit to allow starting only after a predetermined temperature is reached.

Nozzle Protector

The nozzle protector provides a port for the burner nozzle, and also serves to protect the nozzle and provide some secondary air for combustion, which also cools the nozzle.

When making the installation or later when making repairs, it is important that the nozzle protector be installed concentric with the nozzle so that the secondary air will be admitted uniformly. Unless this is done, the flame may become distorted.

It is also important that the plastic refractory be installed flush with the nozzle protector, and completely surround it so that damage to this part may be avoided (see Table 4–3).

Ignition Bracket Assembly

The ignition bracket assembly consists of a bracket, ignition transformer, solenoid gas valve, ignitor, fire safety switch, twist lock receptacle, terminal block, and burner latch. In mounting the ignition bracket assembly, care should be exercised to insure that the igniters are centered in the igniter protectors (Figure 4–28).

The wiring to the ignition bracket should be made with flexible conduit to permit removing the ignition bracket assembly as a unit for making inspection or repairs to the igniters, etc.

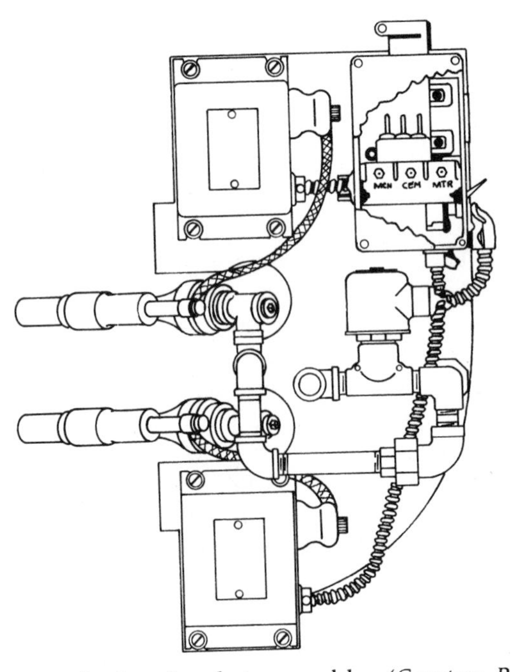

Figure 4–28. Igniter bracket assembly. (*Courtesy Ray Burner Co.*)

Fire Safety Switch

As the name implies, the fire safety switch is a device that operates when overheating or external fire causes the safety string to burn. In some cases, a wire and fusible link is used to replace the string. With either of these, a fusible link rated at 160° F should be used. The fire safety switch is a spring-loaded, normally open contact switch. It must be held closed with the safety string.

Igniter Assembly

Igniter assemblies of two types were originally furnished with Ray burners. Both were known as the gas-electric type, employing an electric spark at 5,000 V to ignite the gas. With these igniters, the ignition remains on constantly during the ignition period.

The *raw gas igniters* were specified and used where manufactured, mixed or natural gas was used for ignition purposes (Figure 4–29). The *bottled gas igniters* are now standard and specified for use with bottled gases, propane, butane and similar gases, and also for natural gas (Figure 4–30). Where bottled gases are used, it is recommended that a vapor switch of the reverse action, manual reset type be used on the gas line to insure that the burner will not attempt to start when the bottled gas supply falls below a safe operating pressure.

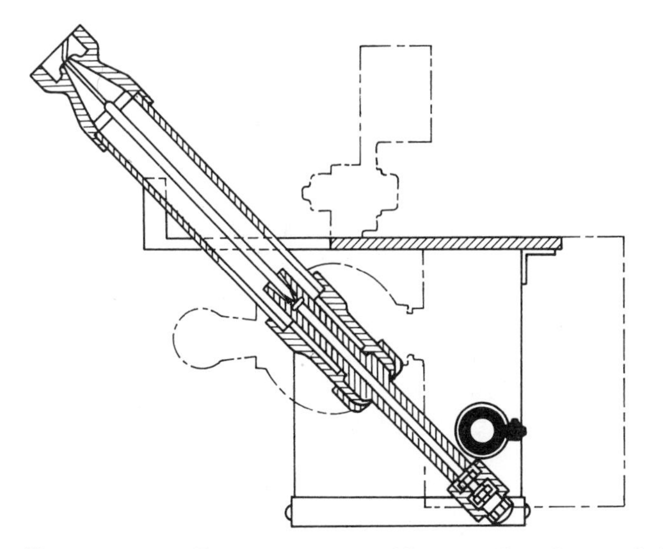

Figure 4–29. Raw gas igniter. (*Courtesy Ray Burner Co.*)

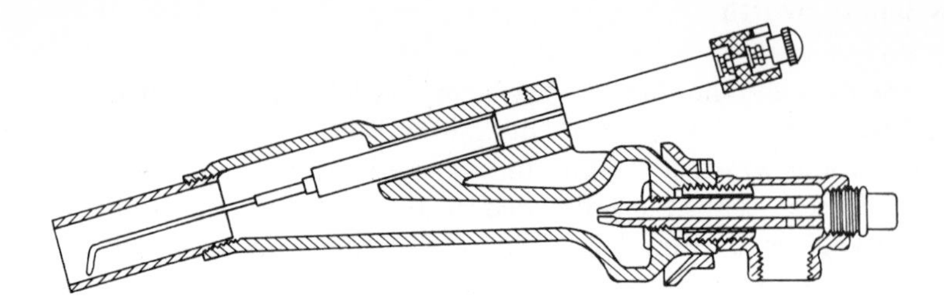

Figure 4-30. Bottled gas igniter. (*Courtesy Ray Burner Co.*)

Vapor switches are also used as an added safety device on installations employing any other type of gas for ignition, and are recommended wherever complete protection is desired.

Ignition Transformers

The ignition transformer, commonly known as a *step-up transformer*, is used to step up the voltage to 5,000 V in order to jump the gap at the end of the ignition electrode. Ignition transformers are quite trouble-free and require only that the primary voltage correspond to that of the transformer, that the electrical connections be clean and tight, and that the secondary terminals be kept clean.

If the secondary terminal post becomes fouled with soot, dust, or foreign matter, the high-tension spark can easily short circuit to the transformer case, resulting in no spark at the ignition electrode. A transformer in good condition should cause the spark to jump a gap of not less than ¼ inch.

Oil Strainer

The oil strainer serves to remove any dirt or foreign matter that might be injurious to the pump on a pump type burner, and also to remove the smaller particles of dirt, etc., that might cause the fouling of the viscosity valve or metering valve.

So that these are properly protected, the mesh of the strainer must be of proper size. With the lighter grades of oil, and also with the heavier grades of oil, where the firing rates are small, the strainer mesh should be 60 or even 80 mesh. With the larger firing rates up to 60 gph the basket should not exceed 30 mesh. Strainers coarser than 20 mesh should not be used on automatic burners. If cleaning is too frequently required, a larger strainer or two strainers in parallel should be used.

It is important that the strainer be kept clean so that the flow of oil to the pump or to the burner will be normal. There is, of course, no fixed period for cleaning the strainer basket. With some fuel oils cleaning may not be required for several weeks, while with other oils it may be necessary to clean the strainer baskets daily.

Where frequent cleaning is necessary, it may be desirable to clean the tank and the lines or in some cases to treat the fuel oil with a sludge solvent. When cleaning the strainer, check the gasket each time to be certain that it is intact.

Where the strainer is installed in a gravity feed or pressure feed, line leaks at the gasket are very easily detected. However, when the strainer is in the suction line, a leaky gasket is not easy to detect and can be responsible for faulty oil pressure or lack of pump capacity.

Relief Valve

The relief valve is a spring loaded valve consisting of a valve seat, valve, spring, and adjusting stem or screw, that is used for regulating the oil pressure from the pump to the metering valve (Figure 4-31). The oil

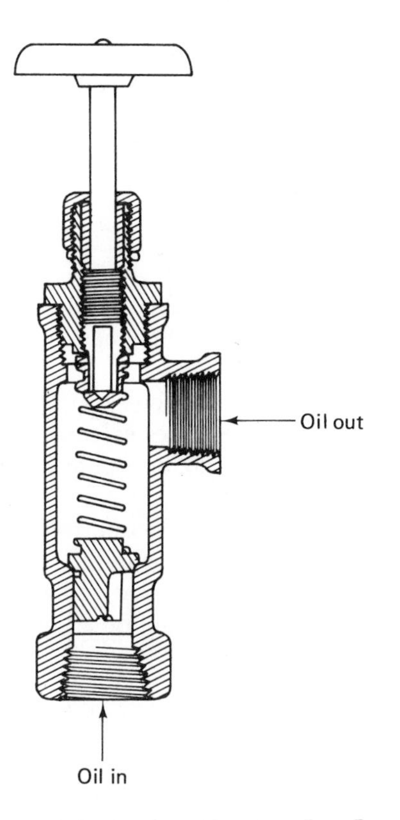

Figure 4-31. Relief valve (*Courtesy Ray Burner Co.*)

enters from the underside of the valve, forcing the valve from its seat. The oil pressure is regulated by the amount of pressure exerted on the valve by the spring. Consequently, the greater the spring compression, the higher the oil pressure.

The adjusting spring or screw serves to increase or decrease the spring tension. Turning the adjusting stem or screw clockwise will increase the oil pressure; counterclockwise will decrease it. Once the oil pressure has been established at the proper value, there should be no changes required.

It should definitely be understood that the oil pressure is a function of spring tension on the valve plus the resistance of oil that is discharged through the valve. With the same spring tension on the valve, the oil pressure will be somewhat greater at the higher discharge rates. Therefore, with the burner operating with no flame, the oil pressure will be greater than when the flame is established and at its normal size.

Except for wear on the seat of the valve, or in cases of a weak or broken spring, the relief valve requires no attention. In the event the oil pressure is not stable or proper, the trouble can usually be traced to faulty oil lines, fouled strainer basket, or to a faulty oil pump. Therefore, if trouble is experienced, these various items should be checked before any changes or repairs are made to the relief valve.

FUEL OIL TREATMENT

The treatment of fuel oil is a complicated operation. There are many types of cure-alls on the market for this purpose. Unfortunately, cure-alls for this purpose are no better than cure-alls for other types of equipment.

Fuel oil treatment should be done by persons with the necessary facilities. To properly treat oil requires extensive laboratory tests. At any rate, some of the problems that are encountered with fuel oil cannot be immediately overcome.

To indicate the complexity of fuel oil treatment, let us explore the composition of only one ingredient, sludge. All the material found on the tank bottom is generally considered sludge. If a sample of sludge were taken and analyzed it would contain all or any combination of the following items:

1. It may be a mixture of water and heavy fuel oil.

2. There may be insoluble solid compounds formed due to the chemical reaction of the fuel oil with the surrounding air.

3. There may be rust, dirt, and scale from many sources.

4. There may be organic precipitation caused by the blending of distillate oil with residual oil.

5. There may be some settling of suspended heavy chemical compounds due to the high cracking of residuals. This causes some coke and free carbon along with a small amount of heavy insoluble compounds.

When we consider the foregoing elements of sludge, only one source of trouble, we can readily conceive that the treatment of fuel oil is complicated. In view of this problem, the following suggestions are recommended:

1. Apply common sense in considering cure-alls.

2. Have a chemical analysis test made on the fuel oil in question.

3. Obtain the help of reputable fuel oil dealers.

4. Know exactly what the problem is. Be sure that fuel treatment will solve the problem.

PIPING SYSTEMS

Two types of piping systems are used in conveying the fuel oil to the burner. They are the *single-pipe* system and the *two-pipe system* (Figure 4–32). The single-pipe system is recommended whenever the bottom of the fuel tank is above the burner or at the same level as the burner. This includes outdoor fuel tanks that are at such levels. The two-pipe system is recommended when the fuel tank is below the level of the burner, and the fuel unit must pull (lift) the fuel up to the burner. For two-pipe installations the bypass plug must be installed.

Table 4–3 shows, for the standard single-stage fuel unit, the allowable lift and lengths of ½ and ⅝ inch OD tubing for both suction and return lines in the two-pipe systems.

When using the optional two-stage fuel unit, a greater amount of lift is attainable, as shown in Table 4–4.

Be sure that all oil line connections are absolutely airtight. Check all connections and joints. Flared fittings are recommended. Do not use compression fittings.

For typical component and connection data see Figure 4–33.

For typical combustion head and electrode adjustments, see Figure 4–34.

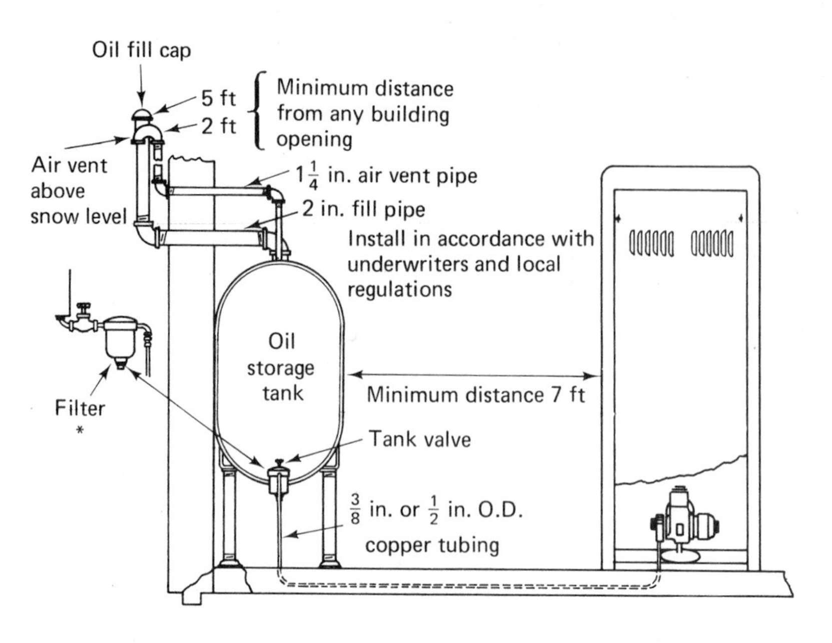

(a)

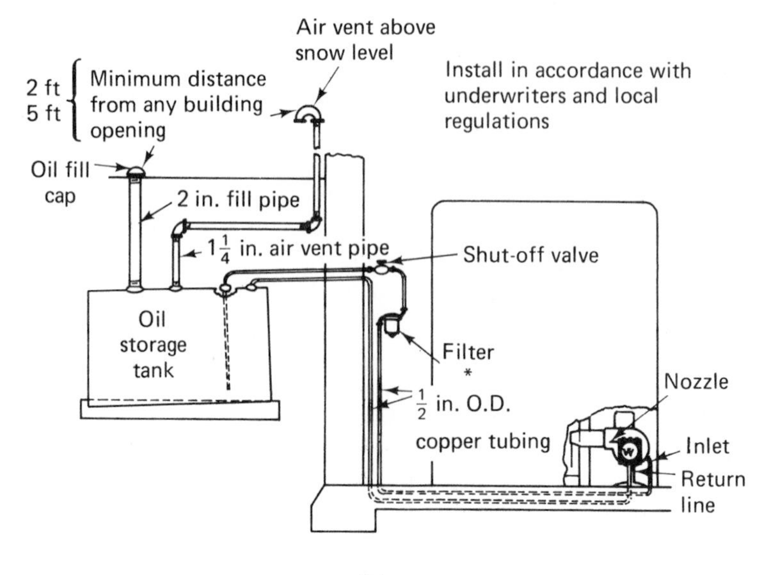

(b)

Figure 4–32. Oil piping system. (*Courtesy The Carlin Co.*)

Table 4–3
Single-Stage Units
Two-Pipe Systems
(Sundstrand JA2BB-100)

LIFT (FEET)	LENGTH OF TUBING (FEET) ½ IN. OD	⅝ IN. OD
0	100	100
2	100	100
4	84	100
6	66	100
8	48	100
10	30	83

(Courtesy The Carlin Co.)

Table 4–4
Two-Stage Units
Two-Pipe Systems
(Sundstrand H2PB-100)

LIFT (FEET)	LENGTH OF TUBING (FEET) ½ IN. OD	⅝ IN. OD
0	100	100
2	88	100
4	78	100
6	69	100
8	59	100
10	49	100
12	39	100
14	29	82
15	24	68

(Courtesy The Carlin Co.)

GAS-OIL BURNERS

Power pressure burners are fully automatic for natural gas or light fuel oil. Fully automatic combination gas-oil pressure burners are manufactured to cover a wide range of heating requirements. They have particular value when continuous uninterrupted heat is desired (Figure 4–35). Within their capacity range, they are ideal for homes, apartments, churches, schools, hospitals, stores, shops, power boilers, and for many other heating applications. This type of burner is indispensable where gas utilities demand standby oil heating facilities.

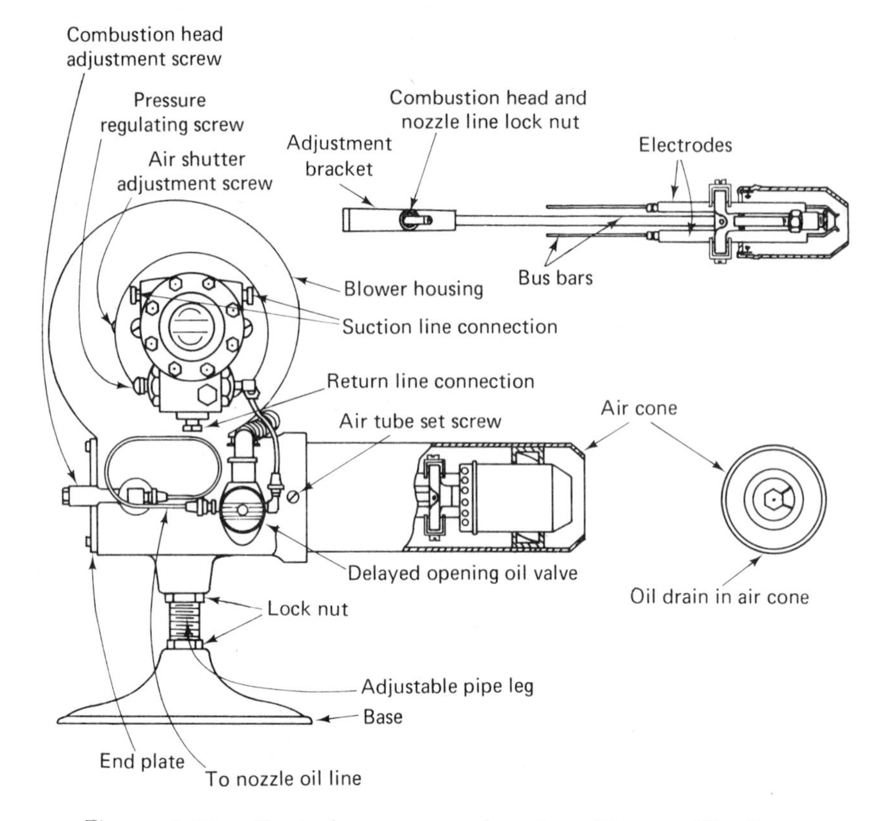

Figure 4–33. Typical component location. (*Courtesy The Carlin Co.*)

Combination burners furnish 100% of the combustion air through the burner for firing rates up to their rated capacity. The oil is atomized by pressures generated by a gear-type high-speed fuel pump assembled as a unit with an oil strainer and pressure regulating valve, which is an integral part of the burner. Injection of the oil by pressure through the nozzles produces a very fine oil spray. The air is delivered by a multi-vane fan through a steel nonreverberating blower tube equipped with a combustion head and is intimately mixed with the oil mist for efficient combustion. The gas is introduced through a double wall blower tube and injected into the air just before it enters the combustion chamber. Because of the use of larger diameter fans and the air pressure differential through the combustion head, draft fluctuations have little effect on the operation and efficiency of this type burner. Extremely high CO_2 settings can be realized with this equipment if desired.

The igniter for oil firing is an electric spark from a 10,000 V transformer mounted on the burner. These burners use a separate gas pilot system for ignition when gas is being used as the main fuel. This pilot

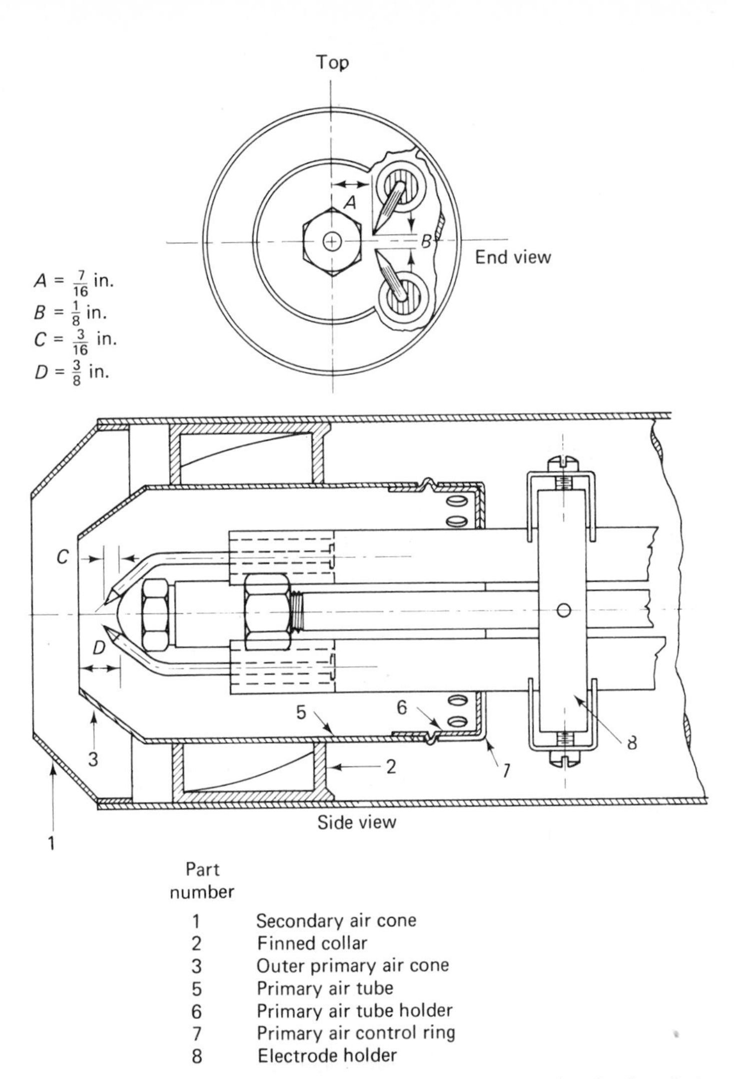

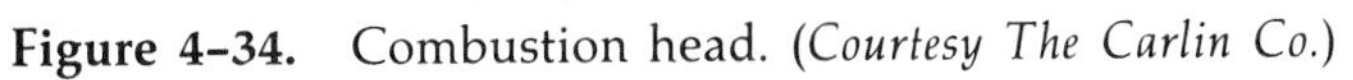

Figure 4-34. Combustion head. (*Courtesy The Carlin Co.*)

Figure 4-35. Typical gas-oil burner. (*Courtesy Ray Burner Co.*)

system consists of the pilot assembly, pilot solenoid valve, and a 6,000 V ignition transformer all mounted and wired on the burner. This unique pilot is supplied with air under pressure from the burner blower so its operation does not depend on firebox draft for its air supply. This also allows the pilot to operate even in forced draft conditions where positive firebox pressure would normally preclude use of an atmospheric-type pilot.

An air supply safety switch is mounted on these burners to insure operation of the fan before the gas valve can operate. It serves to assure the air supply for safe combustion. This switch immediately shuts off the main gas valve in the event of air supply failure.

A gas-oil selector switch is mounted on the burner and the selection of fuel to be burned is made by merely moving the selector switch to either GAS or OIL. No other operation is necessary to change from one fuel to another.

REVIEW QUESTIONS

1. Why does fuel oil require special lighting equipment?
2. Name two methods for preparing fuel oil for burning.
3. Name three types of fuel oil burners.
4. At what point in the fuel oil system are the solid particles removed?
5. At what pressure does the regulating valve open?
6. What is the approximate secondary voltage of an ignition transformer?
7. At what pressure does the regulating valve shut off?
8. At what pressures does the low-pressure oil burner operate?
9. How is the primary air injected into the fuel oil in the high-pressure fuel oil burner?
10. How is the fuel oil distributed with a rotary oil burner?
11. Name three functions of the atomizing nozzle.
12. What two types of energy are used in atomizing fuel oil?
13. Once a stable spray pattern is established, does it help to increase the pressure?
14. Does an increase in pressure increase or decrease the droplet size?
15. Name two types of cone shapes available with nozzles?

16. What is the standard pressure for high-pressure nozzles?
17. How do manufacturers recommend nozzle filters?
18. How are porous bronze filters better than mesh filters?
19. What is the only moving part in the Ray rotary burner?
20. Name the components of a fuel unit.
21. How is the oil atomized in a low-pressure burner?
22. How is the oil atomized in the rotary oil burner?
23. Why are small droplets necessary in oil burners?
24. Which cone has a more stable operation under adverse conditions?
25. What nozzle is best suited for use between .4 and 8 gph?
26. Describe the best flame obtained in burning fuel oil.
27. What is the principal purpose of the fan on the Ray rotary burner?
28. How is the actual air velocity controlled in the Ray rotary burner?
29. What does the term *angulation* refer to?
30. What is the result of oil breaking through the air stream on the Ray rotary burner?
31. What is apt to cause flame pulsations in the Ray rotary burner?
32. What is of prime importance when cleaning the nozzle on the Ray rotary burner?
33. What is the purpose of the Ray atomizing cup?
34. What fluids are used when cleaning the atomizing cup?
35. What conditions are necessary concerning the lip of the atomizing cup?
36. What purpose does the worm gear and shaft perform?
37. What is the purpose of the secondary pump in a two-stage pump?
38. What caution should be observed when installing a rotary seal?
39. What temperature is the oil heater adjusted to maintain?
40. What care should be taken in the placement of igniters?
41. What is the purpose of the fire safety switch?

5

Heat Exchangers

The objectives of this chapter are:

- To show you the purpose of a heat exchanger.
- To acquaint you with the flow of the heating medium through a heat exchanger.
- To introduce the different types of heat exchangers to you.
- To familiarize you with the gas flow through the heat exchanger.

PRIMARY HEAT EXCHANGERS

A *heat exchanger* is defined as a device used to transfer heat from one medium to another.

The heat exchanger (Figure 5-1) is the heart of the heating plant. It is designed to transfer heat from the combustion gases to the heating medium flowing through the passages. It also serves as the combustion area. Openings in the lower section permit installation of the main burners and allow secondary air to reach the flame. As the fuel is burned, the flue gases rise through the vent passages. Restrictions are incorporated in the heat exchanger to control the flow of combustion air and reduce the amount of excess air to the 35-50% limits. The restriction is produced in several ways: (1) the top section is reduced in cross-sectional width; (2) by baffles in the exhaust openings at the top of the heat exchanger (Figure 5-2); and (3) a combination of both. These restrictions also permit maximum heat transfer by reducing the vent gas velocity that allows the maximum heat to be extracted from it. The gases then leave the heat exchanger and enter the venting system.

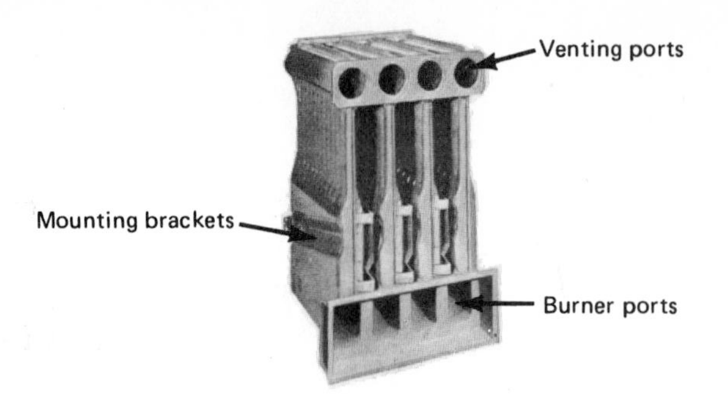

Figure 5-1. A forced air gas heat exchanger. (*Courtesy Dearborn Stove Co.*)

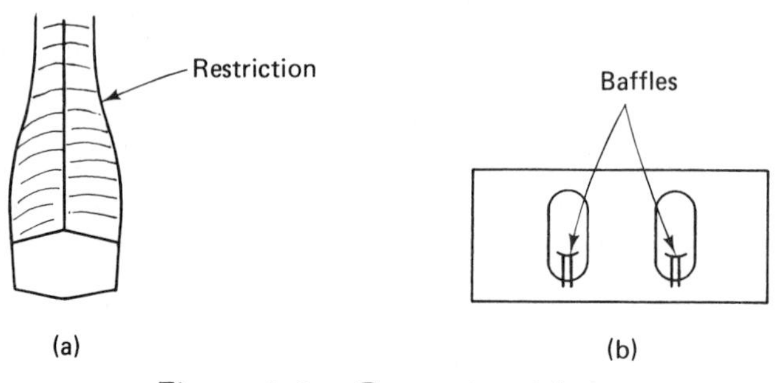

Figure 5-2. Gas vent restrictions.

There are two general classifications of heat exchangers: (1) the barrel and (2) the sectionalized (Figure 5-3). They may be constructed of either cast iron or steel. Due to the weight, expense, and slow response to temperature change of cast iron, steel has almost dominated the forced air furnace heating industry in recent years.

The major problem with steel is the noise that sometimes accompanies the expansion and contraction associated with the heating and cooling of a heat exchanger. This noise, however, can be reduced by placing dimples and ribs in the metal during the stamping operation. Ceramic coating is also used to help eliminate noise while providing corrosion protection. This ceramic coating does not hinder the transfer of heat from the heat exchanger.

The application of the heat exchanger determines its physical shape. The heat exchanger in a horizontal forced air furnace (Figure 5-4) would need to be shaped differently from one used in an upflow furnace (Figure 5-5). In most instances, there is very little difference from the upflow to the downflow heat exchanger.

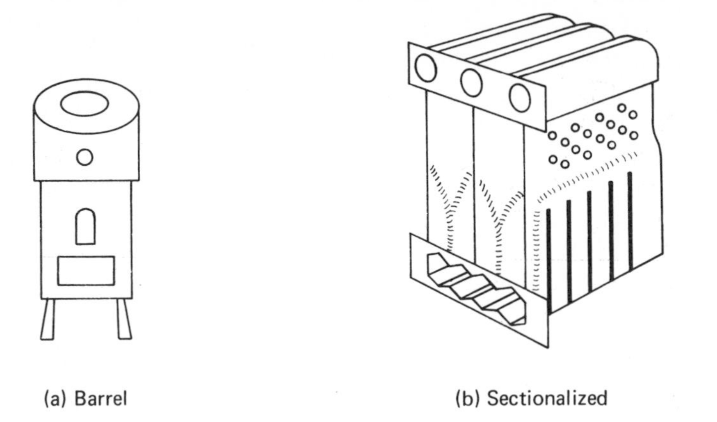

Figure 5-3. Types of heat exchangers.

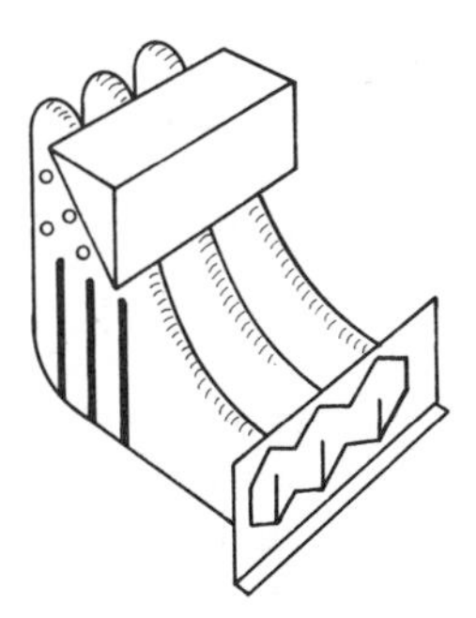

Figure 5-4. Horizontal heat exchanger.

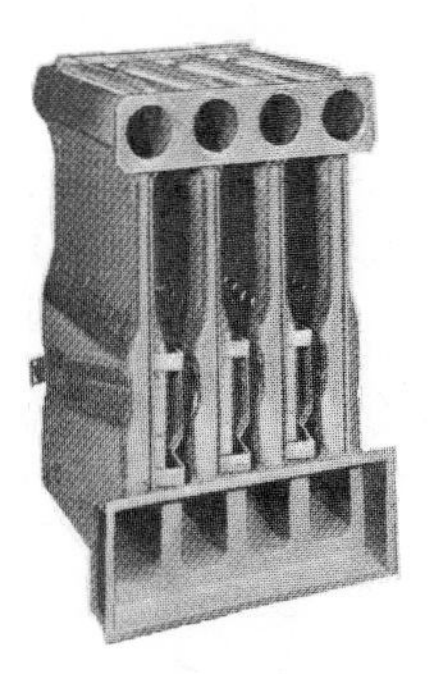

Figure 5-5. Upflow heat exchanger. (*Courtesy Dearborn Stove Co.*)

There are two flow paths through a furnace heat exchanger. Both the heated air and combustion gases pass through the unit without any mixing of the two. The heated air flows around the outside of the heat exchanger (Figure 5-6), while the flue gases pass through the inside.

The heat exchanger is normally an extremely trouble-free apparatus, especially when the burners are installed and adjusted properly. From time to time, however, due to the continuous expansion and contraction, a hole will develop in the metal. A hole in the heat exchanger presents a dangerous situation. When the flame is burning before the blower starts, carbon monoxide could be admitted to the air stream. After the blower starts, the flame will be agitated due to the greater pressure in the duct system. The amount of agitation will, naturally, depend on the size of the hole. A large hole could possibly cause the flame to be blown from the front of the furnace and catch fire to anything in the immediate area, along with depositing the products of combustion in the home. The only difference between a large hole and a small hole is that the small hole is more difficult to detect. In any case a leaking heat exchanger must be replaced.

When main burners that are dirty or out of adjustment have been used for a period of time, the flue passages in the heat exchanger may become clogged with soot. This soot is hazardous because it retards the combustion process and does not allow the removal of all the products of combustion. Soot is a deposit of unburned fuel; therefore, it is combustible. Because soot is combustible, caution should be exercised in the use of highly combustible sprays or fogs for its removal. Sometimes when these products are used the soot will float out of the combustion area and be a hazard to the surroundings. At times the smoldering soot can be seen floating from the vent above the roof. To properly desoot a heat exchanger, the main burners must be removed and cleaned. While the burners are removed, the flue passages must be cleaned of all soot

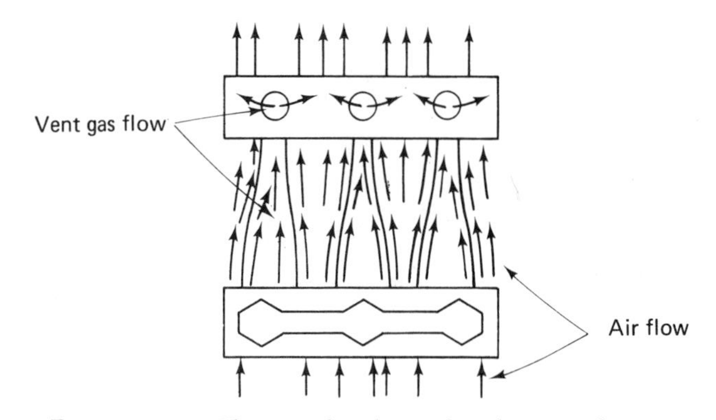

Figure 5-6. Flow paths through a heat exchanger.

deposits. This is a costly, time consuming, and dirty job that could possibly be avoided by keeping the main burners properly adjusted. An example of the proper alignment of burner to heat exchanger may be seen in Figure 5-7.

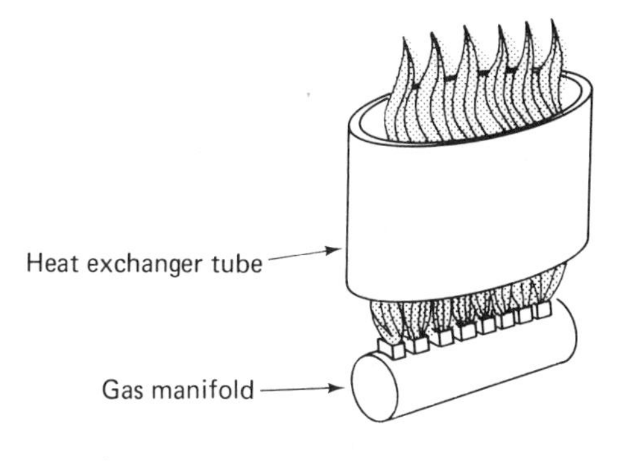

Figure 5-7. Alignment of burner and heat exchanger. (*Courtesy Modine Manufacturing Co.*)

SECONDARY HEAT EXCHANGERS

Up to this point, only heat exchangers used on gas or oil fired units have been discussed. They are called *primary heat exchangers.* This type of heat exchanger by no means comprises all the elements used for heat transfer.

Hydronic heating also enjoys a fair share of the heating industry. The boilers used for hydronic heating also employ the same type of primary heat exchanger as forced air units. However, water is forced through its passages in place of air (Figure 5-8).

When a boiler is used, a single heat exchanger will not transfer the desired heat to the structure. There must be another (secondary) heat transfer element placed in the room. When hydronic heating (hot water or steam) is used, the heating medium is passed from the boiler to the room heating element through a system of piping and returned to the boiler where it is reheated and recirculated. In some installations the secondary heating element is attached to a duct system and used as a central heating unit.

Natural draft hydronic heating elements, sometimes referred to as radiators, are manufactured for mounting in a variety of locations. They are an assembly of increments, each composed of vertical heating fins and a dividing plate that is reinforced with a steel tube beading added for strength and protection.

The heating element consists of a heavy wall copper tube formed

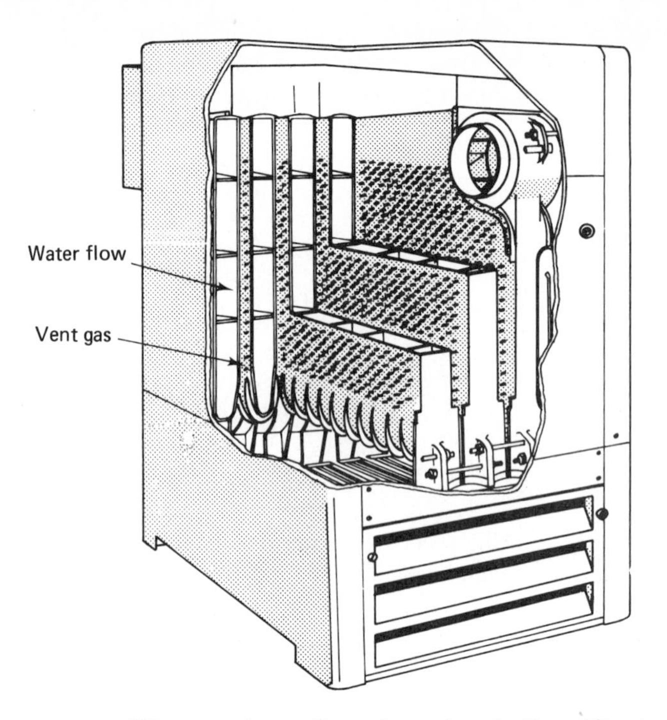

Figure 5-8. Water and gas flow through a boiler. (*Courtesy Weil-McLain Co., Inc., Hydronic Division*)

into a continuous serpentine coil. In turn, the coil passes through each heating fin and is expanded under 3,750 psig of hydraulic pressure to lock all elements into a one-piece unit. Such bonding is ideal for high pressure-high temperature systems since there is no solder or soft metal to break away.

Another type of secondary heat exchanger is the *embedded coil.* This type element has a pipe embedded in the floor, wall, or ceiling (Figure 5-9). This system of piping is connected to the boiler in the conventional manner. The coil has no fins and employs the surrounding structure to radiate the heat. It is ideal in play rooms or dens where a warm floor is desired. These elements, naturally, should be installed during construction of the room.

The efficient operation of any hydronic heating element requires that all air be kept from collecting in the piping. Most installations have air vents installed on top of the coil as well as in the highest point of the entire system. Should these elements become air locked, the heating will be reduced and sometimes completely stopped. In order to maintain maximum output from these units, the air must be continuously removed from the piping, coils, and boilers.

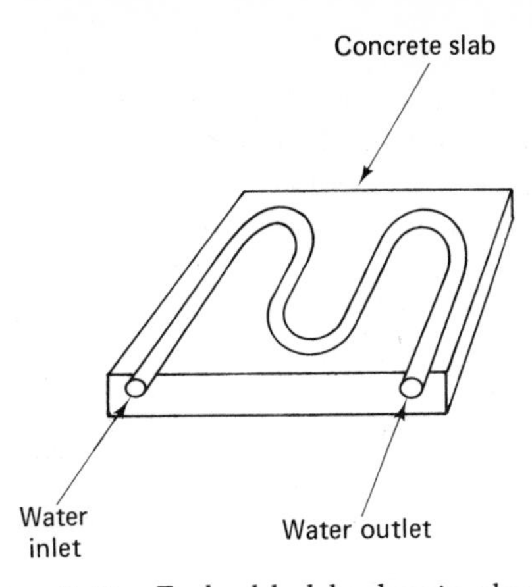

Figure 5-9. Embedded hydronic element.

Each manufacturer designs and fabricates his heat exchanger to fit a definite set of conditions. They will publish the conditions for a specific output rating of their heat exchanger (see Table 5-1). Sometimes the heat exchangers of the same manufacturer will have different conditions and performance ratings. It can be seen that when considering a heat exchanger, several factors must be considered. The difficulty in selecting a heat exchanger is being sure which will deliver the most overall satisfaction.

REVIEW QUESTIONS

1. How many paths are incorporated in a heat exchanger?
2. The bottom section of a gas or oil fired heat exchanger also serves as what area?
3. Why are restrictions designed into a heat exchanger?
4. What are the excess air limits?
5. Name two classifications of gas or oil heat exchangers.
6. Name one major problem encountered with steel heat exchangers.
7. How is this problem reduced?
8. What is the purpose of ceramic coating on a heat exchanger?
9. How many flow paths are there through a furnace?
10. Do the flow paths through a furnace ever mix?
11. Why is a broken heat exchanger dangerous?

Table 5-1
Typical Radiator Selection Chart,
Wall Mounted Position,
Radiator Heat Outputs in Square Feet Equivalent
Direct Radiation

Rad. No.	Length L See Note	Ship. Weight in Pounds	200° Water	220° Water	1 Lb Steam 240° Water	15 Lb Steam 280° Water	35 Lb Steam	50 Lb Steam	100 Lb Steam	150 Lb Steam
Height 14½ inches										
10	25	54	10	12	14	18	23	25	31	35
15	37.5	78	16	20	24	31	38	42	52	59
20	50	102	23	28	33	43	53	59	73	83
25	62.5	127	30	36	43	56	68	76	94	107
30	75	151	37	45	52	68	84	94	116	132
35	87.5	175	43	53	62	81	98	111	137	156
40	100	199	50	61	71	93	114	128	158	180
45	112.5	224	56	69	81	106	128	145	179	204
50	125	248	63	77	90	118	144	162	200	228
Height 23½ inches										
10	25	76	16	20	23	30	37	41	51	58
15	37.5	109	26	32	37	49	60	67	84	95
20	50	142	37	45	52	68	84	94	116	132
25	62.5	176	47	57	67	87	107	120	149	169
30	75	209	57	70	82	107	131	147	181	206
35	87.5	242	67	82	96	126	154	173	214	243
40	100	275	78	94	111	145	178	199	246	280
45	112.5	309	88	107	126	164	201	226	279	317
50	125	342	98	119	141	183	225	252	311	354

(Courtesy Shaw-Perkins Manufacturing Co.)

12. What is the desired method of removing soot from a heat exchanger?
13. Name the main burner conditions which will increase the life of a heat exchanger.
14. Soot is a deposit of what?
15. How many heat exchangers are required for hydronic heating?

16. How many heat exchangers are required for gas or oil heating?
17. Name the heat exchangers used for hydronic heating.
18. What are natural draft heat exchangers termed?
19. Are hydronic heat exchangers used in a variety of locations and positions?
20. What is used in conjunction with pipes in secondary hydronic heat exchangers?
21. What must be done to insure proper heating of hydronic heating systems?
22. What is most difficult in selecting a hydronic heat exchanger?
23. What must be consulted in selecting a hydronic heat exchanger?
24. What is the purpose of any heat exchanger?
25. What objectives does each manufacturer design his heat exchanger to meet?

6

Venting

The objectives of this chapter are:

- To acquaint you with the operating principles of the venting system.
- To point out the safety precautions for a venting system.
- To introduce you to the problems encountered in designing a venting system.
- To acquaint you with barometric draft control design and installation procedures.
- To instruct you in gas burner combustion testing.

American Gas Association (AGA) design certified gas equipment must be capable of venting products of combustion through the draft diverter opening without connection to a vent or chimney.

It, therefore, becomes apparent that the only purpose of a vent for an AGA design certified heating unit is to convey the flue gas products from the draft diverter of the unit to the outside of the building. The vent may be defined as a system of piping used to remove all the products of combustion to the outside air.

VENTING

After the fuel is burned in the combustion area, it is passed through the flue passages of the heat exchanger. On leaving the heat exchanger these products of combustion enter the draft diverter, or the draft control, whichever is used. At this point, the flue gases are mixed with an amount of air equal to that required for combustion air. For example, when 1 ft^3 of natural gas is burned, 15 ft^3 of combustion air are required. At the draft diverter, another 15 ft^3 of air per ft^3 of natural gas are required for dilution air.

DRAFT DIVERTER

The purpose of the draft diverter (Figure 6-1) is to neutralize excessive drafts and downdrafts through the heating unit and to produce an emergency outlet for relieving the flue gases in the event that the chimney or breeching should become obstructed. Chimney draft is not necessary for the proper operation of the unit, which has, as a part of its design, a constant natural draft through the heat exchanger for the proper performance of the burners and pilots. Excessive updraft or downdraft could otherwise adversely affect combustion, resulting in the generation of carbon monoxide (CO), reducing combustion efficiency, causing hazardous ignition, or even extinguishing the safety pilot, thus causing a nuisance shut-down.

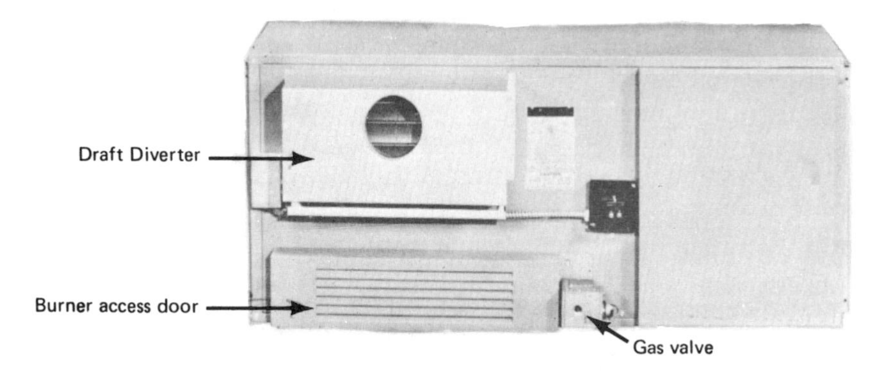

Figure 6-1. Draft diverter installed on a horizontal furnace. (*Courtesy Dearborn Stove Co.*)

The correct installation of the heating unit and its draft diverter is vital for safe operation. All heating units are tested and listed by AGA with their draft diverter in position. Check the following list to be sure that the draft diverter is correctly installed.

1. Do install the draft diverter free of obstructions below and to the side.

2. Do install exactly as specified by the manufacturer's instructions.

3. Do obtain correct draft diverter from the manufacturer if it is missing. If assistance is needed, check with the fuel supplier.

4. Do not alter the pipe lengths between the draft diverter and the unit flue outlet.

5. Do not change the position of the draft diverter on the heating unit.

6. Do not change the design of the draft diverter by adjusting or cutting the baffles, skirt, etc.

7. Do not locate the draft diverter in a different room from the unit.

8. Do not substitute or exchange draft diverters.

VENT PIPING

The venting system not only conveys the gases; it also conveys the air taken in through the relief opening of the draft diverter. This dilution air imposes an added volume of gases that must be handled by the vent system. The vent, therefore, must be capable of conveying the flue gases and dilution air to the outside atmosphere with a minimum of draft resistance. If the vent system has insufficient flue area or height, some flue gases will spill from the relief opening of the draft diverter. Flue gas spillage is a hazard that must be avoided by the proper design and construction of the venting system. It is obvious that the vent system having the least resistance would be a vertical vent taken directly from the outlet of the draft diverter. This low cost hook-up is highly recommended, where applicable, and can be easily used on multiple outlet units as well as on single outlet units.

The installation of a double wall UL listed gas unit is not normally a complicated matter. Most vent installations are relatively simple, requiring the observation of only a few basic rules. Some complex multiunit or multistory vent systems, however, will require the engineered capacities and configurations which are clearly set forth in tables. These tables give recommended vent size for various configurations and unit draft diverter outlet diameters according to the rated Btu input of the unit and the height of the vent. These Gas Vent Institute (GVI) vent capacity tables are based on the use of UL listed double-wall gas vent pipes and fittings from the unit to the vent cap (Figure 6-2). The concept of

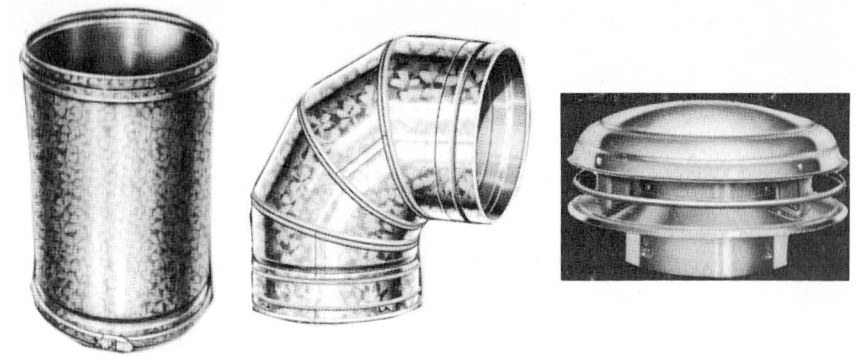

Figure 6–2. Common vent pipe and fittings. (*Courtesy Hart and Cooley Manufacturing Co.*)

using UL listed materials for the entire venting system is consistent with the "General Information" statement issued by the Underwriters' Laboratories on August 5, 1960 (Figure 6–3).

When installing or servicing heating equipment, make sure that all heating units and vents have an adequate supply of air for combustion and venting. (For suggested methods see Figure 6–4.) These suggestions are based on the information contained in *NFPA Bulletin* Number 54, Section 3 and 4, June, 1959. When installing two grills, one should be installed high and the other installed low. The lower grill should be at or below the air access to the unit. The higher grill should be above the relief opening of the draft diverter. For short units, the measurement between the two grills should never be less than 3½ feet. When these grills are connected to a duct, the duct should have a cross-sectional area not less than the free area of the grills. The total air supply to fuel burning equipment is normally three times the amount required for combustion.

Air, as it flows into and through a heating unit and is vented up the stack, serves three vital functions. First, on entering the heating unit as combustion air, it furnishes the oxygen required for the flame. Second, while circulating through the heating unit, air serves as the medium through which the heat is transferred from the flame to the heat exchanger. Third, air is introduced directly into the stack as dilution air. Dilution air serves a double purpose. It carries the spent gases up the vent, and it also serves as a variable through which over-fire draft is governed.

Draft is the force that is applied to the movement of a column of air into and through a heating unit. The force involved is static pressure—the ton-per-square-foot pressure—exerted by the weight of the earth's atmosphere.

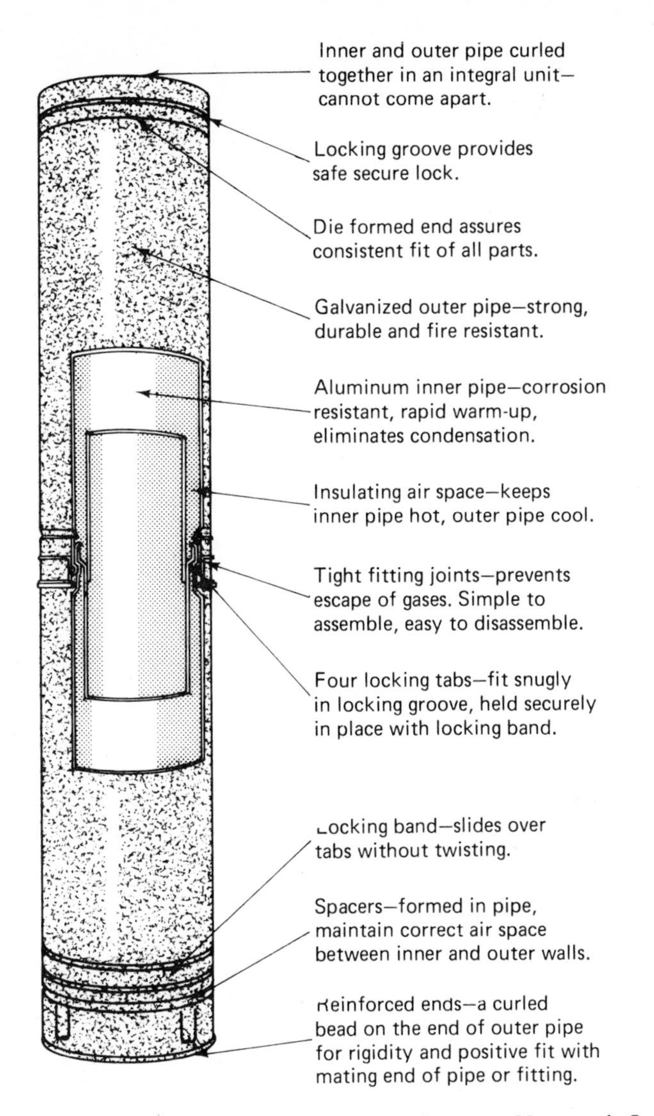

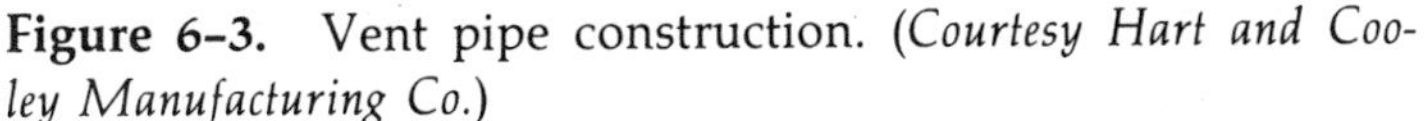

Figure 6–3. Vent pipe construction. *(Courtesy Hart and Cooley Manufacturing Co.)*

The movement of air through a heating unit requires a static pressure differential. To produce this draft, the pressure of the air surrounding the heating unit and vent system must be greater than the pressure exerted by the gases within the unit and vent system.

This pressure differential, and therefore, the degree of draft force exerted at a given barometric reading, depends on two factors.

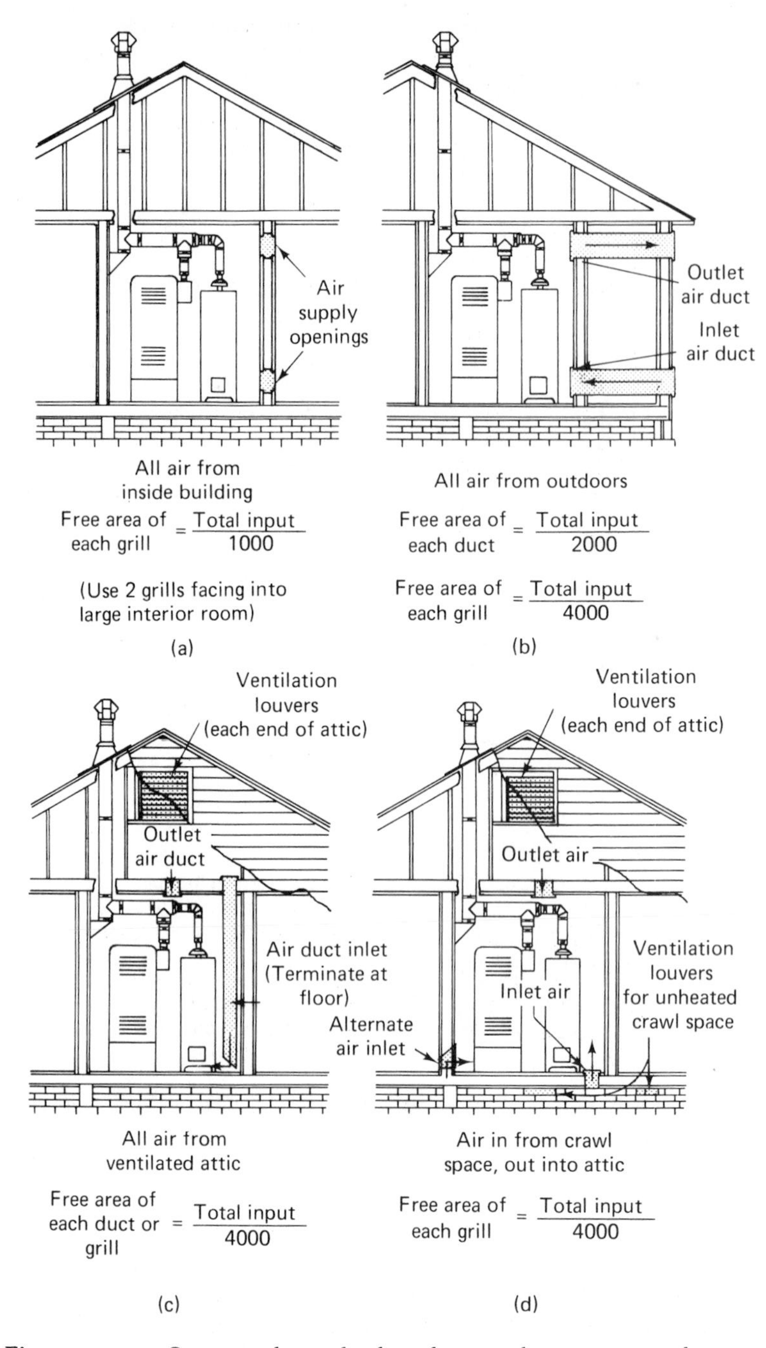

Figure 6–4. Suggested methods of providing air supply. (*Courtesy William Wallace Division, Wallace-Murray Corporation*)

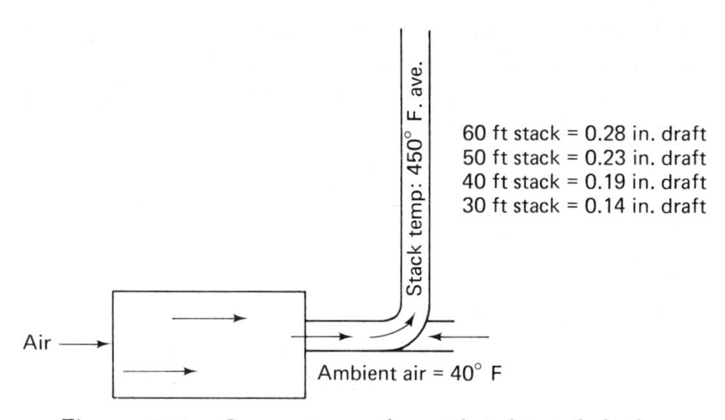

Figure 6-5. Comparison of vent height and draft.

1. The amount of temperature differential both inside and outside the heating unit and vent system.

2. The height to which this temperature differential is stacked, or the vent height (Figure 6-5).

A natural draft cannot be produced without a vent system. Draft cannot be used as a source of oxygen unless the vent is connected to the unit. A vent, no matter how high, can provide no over-fire draft if it is separated from rather than connected to the heating unit.

The draft force produced by a vent system of any given height varies constantly. This is because the conditions which create the draft are subject to the following continuous changes (Figure 6-6).

1. *Temperature change within a heating unit.* The weight of air, and therefore the pressure it exerts, varies directly with the temperature. Air at 30° has two times the weight of air at 400° F. Thus, the pressure of air

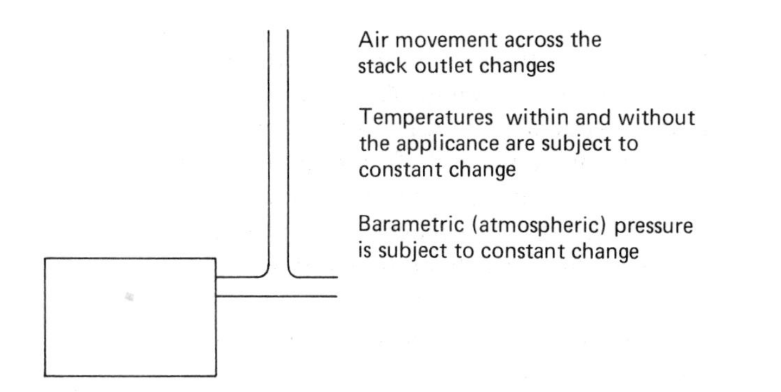

Figure 6-6. Representative changing of vent conditions.

within a heating unit changes as the temperature of the gases change. The vent system should be designed to maintain a temperature of at least 360° F. Under no condition should the inside vent temperature drop below 225° F above ambient temperature.

2. *Temperature change outside a heating unit.* The pressure exerted by the air around a heating unit changes as the ambient air temperatures change. Vent piping and fittings are designed for a 60° F outside ambient temperature.

3. *Barometric pressure changes.* As the barometric pressure rises and falls, the draft produced by a vent system of any given height, at any given temperature differential, also rises and falls proportionately.

4. *Air turbulence.* The capturing and retarding action of the wind gusts that alternate with calm periods at the stack outlet causes constant, abrupt, and severe changes in the draft action of the vent system.

It should be noted that four out of the five forces governing draft production are subject to constant changes. Only the stack height remains fixed, while air turbulence changes, ambient air temperatures change, stack temperatures change, and the barometric pressure changes.

As stated before, the capacities given in the tables apply specifically to UL listed double-wall type B gas vents constructed entirely of listed materials from the draft diverter outlet to the vent cap. Recommendations for venting system design and capacity in gas vent tables are based on the insulating qualities and fluid flow characteristics of gas vent pipe and fittings. These tables do not apply to materials having lower values such as masonry, asbestos-cement, or single-wall metal pipe. Nor do they consider vents less than 6 feet in height.

VENT DESIGN

In the study of gas vent design and installation, we will first consider the individual vent system. An *individual vent* may be defined as a single and independent vent for a single heating unit. The total vent height is the vertical from the draft diverter outlet and the vent cap. The position or rise of a connector, or the location of an offset has no effect on the total vent height. The length of lateral is the horizontal distance or offset between the draft diverter outlet and the final vertical portion of the vent (Figure 6–7).

When determining the vent size of an individual vent, use Table 6–1 with the following instructions.

1. Determine the total vent height and the length of lateral, based

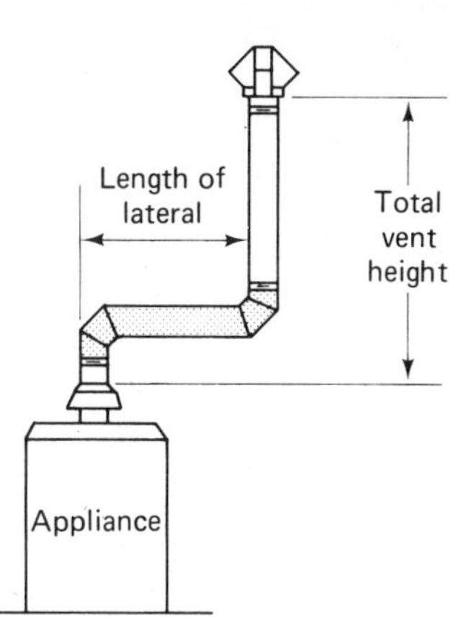

Figure 6-7. Total vent height and length of lateral. (*Courtesy William Wallace Division, Wallace-Murray Corporation*)

on the location of the unit and the vent, and the height to the top of the vent.

2. Follow the total height column, at the left side, down to a height equal to or less than the total vent height.

3. Select the horizontal row for the appropriate length of lateral, this would be zero for straight vents.

4. Next, read to the right to the first column which indicates a capacity equal to or greater than the unit name plate Btu input rating.

5. If the vent size indicated at the top of the column containing the correct capacity is equal to or larger than the unit draft diverter, use the vent shown by the table.

6. However, if the vent size indicated is smaller than the draft diverter connection, the vent size indicated by the table may be used. Even though a draft diverter-to-vent reduction takes place, it is permissible as long as the vent is at least 10 feet high. When a vent is shorter than 10 feet, it should be at least as large as the draft diverter connection.

A typical individual venting project is as follows (Figure 6-8). The furnace has a 150,000 Btu input rating and a 6-inch draft diverter connection.

The procedure would be to follow the total vent height column down to the 20 foot height, then right on the line for 10-foot lateral. The 150,000 Btu input rating is located in the 5-inch column. A capacity of 228,000 Btu is shown in the 6-inch column. However, the 5-inch vent is adequate because the total height exceeds 10 feet.

Table 6-1
Individual Vent Table
(FOR ONE APPLIANCE ONLY)

TOTAL VENT HEIGHT	LENGTH OF LATERAL	MAXIMUM CAPACITY, Thousands of BTU Per Hour *													
		METALBESTOS VENT SIZE, INCHES													
		3″	4″	5″	6″	7″	8″	10″	12″	14″	16″	18″	20″	22″	24″
4′	0	45	80	125	180	245	320	500	720	970	1280	1630	2000	2430	2900
	2′	30	59	94	137	190	250	405	580	740	960	1220	1520	1830	2180
	4′	29	56	90	133	185	243	395	570	730	950	1215	1510	1820	2170
	8′	27	53	84	125	175	232	377	525	700	930	1190	1505	1810	2160
6′	0	46	86	141	205	285	370	570	850	1170	1530	1960	2430	2950	3520
	2′	36	67	105	157	217	285	455	650	890	1170	1480	1850	2220	2670
	6′	32	61	100	149	205	273	435	630	870	1150	1470	1820	2210	2650
	12′	28	55	91	137	190	255	406	610	840	1110	1430	1795	2180	2600
8′	0	50	94	155	235	320	415	660	970	1320	1740	2220	2750	3360	4010
	2′	40	75	120	180	247	322	515	745	1020	1340	1700	2110	2560	3050
	8′	35	66	109	165	227	303	490	720	1000	1320	1670	2070	2530	3030
	16′	28	58	96	148	206	281	458	685	950	1260	1600	2035	2470	2960
10′	0	53	100	166	255	345	450	720	1060	1450	1925	2450	3050	3710	4450
	2′	42	81	129	195	273	355	560	850	1130	1480	1890	2340	2840	3390
	10′	36	70	115	175	245	330	525	795	1080	1430	1840	2280	2780	3340
	20′	—	60	100	154	217	300	486	735	1030	1360	1780	2230	2720	3250
15′	0	58	112	187	285	390	525	840	1240	1720	2270	2900	3620	4410	5300
	2′	48	93	150	225	316	414	675	985	1350	1770	2260	2800	3410	4080
	15′	37	76	128	198	275	373	610	905	1250	1675	2150	2700	3300	3980
	30′	—	60	107	169	243	328	553	845	1180	1550	2050	2620	3210	3840
20′	0	61	119	202	307	430	575	930	1350	1900	2520	3250	4060	4980	6000
	2′	51	100	166	249	346	470	755	1100	1520	2000	2570	3200	3910	4700
	10′	44	89	150	228	321	443	710	1045	1460	1940	2500	3130	3830	4600
	20′	35	78	134	206	295	410	665	990	1390	1880	2430	3050	3760	4550
	30′	—	68	120	186	273	380	626	945	1270	1700	2330	2980	3650	4390
30′	0	64	128	220	336	475	650	1060	1550	2170	2920	3770	4750	5850	7060
	2′	56	112	185	280	394	535	865	1310	1800	2380	3050	3810	4650	5600
	20′	—	90	154	237	343	473	784	1185	1650	2200	2870	3650	4480	5310
	40′	—	—	—	200	298	415	705	1075	1520	2060	2700	3480	4270	5140
40′	0	66	132	228	353	500	685	1140	1730	2400	3230	4180	5270	6500	7860
	2′	59	118	198	298	420	579	960	1420	2000	2660	3420	4300	5260	6320
	20′	—	96	167	261	377	516	860	1310	1830	2460	3200	4050	5000	6070
	40′	—	—	—	223	333	460	785	1205	1710	2310	3020	3840	4780	5820
60′	0	—	136	236	373	535	730	1250	1920	2700	3650	4740	6000	7380	9000
	2′	—	125	213	330	470	650	1060	1605	2250	3020	3920	4960	6130	7400
	30′	—	—	170	275	397	555	930	1440	2050	2780	3640	4700	5730	7000
	60′	—	—	—	—	334	475	830	1285	1870	2560	3380	4330	5420	6660
80′	0	—	—	239	384	550	755	1290	2020	2880	3900	5100	6450	8000	9750
	2′	—	—	217	350	495	683	1145	1740	2460	3320	4310	5450	6740	8200
	40′	—	—	—	275	404	570	980	1515	2180	2980	3920	5000	6270	7650
	80′	—	—	—	—	—	—	850	1420	2000	2750	3640	4680	5850	7200
100′	0	—	—	—	400	560	770	1310	2050	2950	4050	5300	6700	8600	10,300
	2′	—	—	—	375	510	700	1170	1820	2550	3500	4600	5800	7200	8800
	50′	—	—	—	—	405	575	1000	1550	2250	3100	4050	5300	6600	8100
	100′	—	—	—	—	—	—	870	1430	2050	2850	3750	4900	6100	7500

NOTES:

A "*" See step 4 of paragraph 1-2.

B "—" Where no capacity is given, vent may be liable to both spillage and condensation.

C Regardless of altitude, always design vent for sea level nameplate appliance input.

D "0" lateral applies only to a vertical vent attached to a top outlet draft hood.

(Courtesy William Wallace Division, Wallace-Murray Corporation)

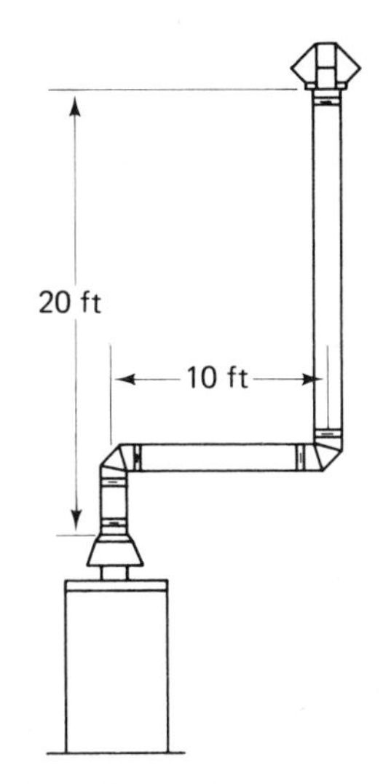

Figure 6-8. Typical individual vent estimation. (*Courtesy William Wallace Division, Wallace-Murray Corporation*)

Oversized or excessively long vents that may allow excessive cooling and condensation should be avoided. A rule of thumb: Be sure that the total length of the vent, including the connector and any offset, does not exceed 15 feet for every inch of vent diameter. As an example, see Figure 6-9. For a 4-inch vent, 60 feet (4 × 15) of vent pipe is permissible. When longer total lengths are shown in vent tables, they are based on maximum capacity, not on condensation factors.

Individual vent sizing is somewhat easier than combined vent sizing. However, when the proper tables are used, combined vent sizing should present no problems.

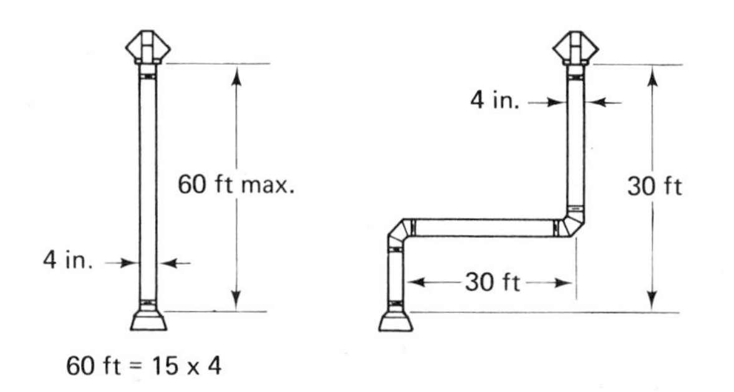

Figure 6-9. Maximum vent height. (*Courtesy William Wallace Division, Wallace-Murray Corporation*)

A *combined vent* may be defined as a vent system for two or more appliances at one level connected to a common vent. The least total vent height is the vertical distance from the highest draft diverter connection in the vent system to the vent top (Figure 6-10). This is one fixed dimension for any one system regardless of the number or placement of heating units in the system. The connector rise for any unit is the vertical distance from its draft diverter connection to the point where the next connector joins the system. A common vent is that portion of the venting system above the lowest interconnection. When the common vent is entirely vertical, it is called a vertical or V-type. Otherwise the common vent is a lateral or L-type.

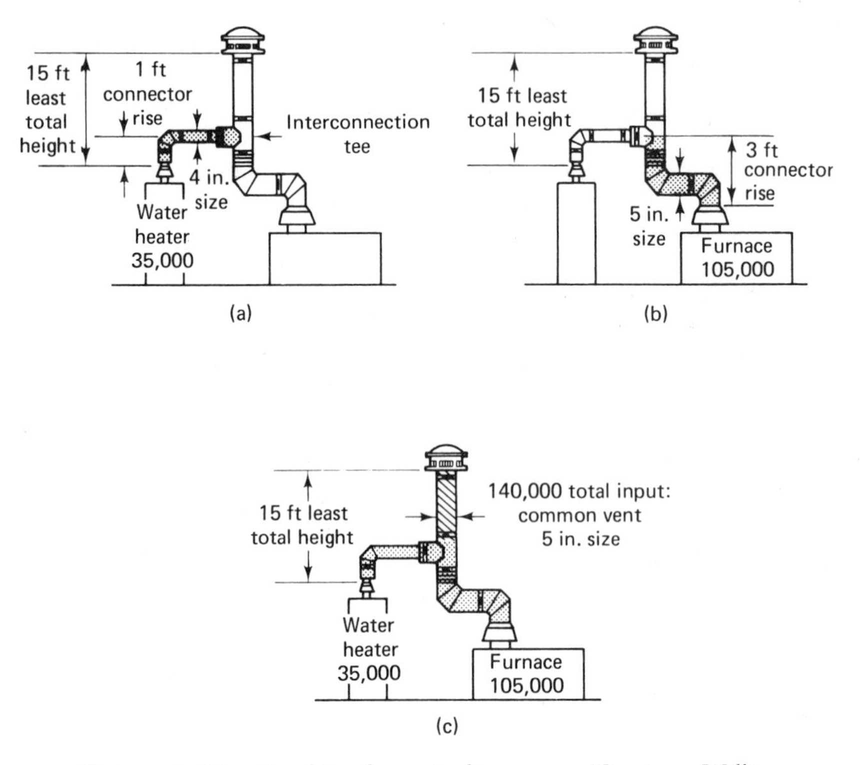

Figure 6-10. Combined vent diagrams. (*Courtesy William Wallace Division, Wallace-Murray Corporation*)

The procedure for determining each vent connector size is outlined below and is used in conjunction with Table 6-2.

1. Determine the least total height for the vent system.
2. Determine the connector rise for each unit.

3. Find the appropriate least total height in the vent connector table. Read across on the proper line for the first unit connector rise to the name plate Btu rating for that unit. Read the connector vent size for that appliance at the top of this column.

4. Repeat this procedure for the connector rise and Btu rating for each appliance at the same least total height.

The following procedure is recommended for determining the size of the common vent.

1. Add together all the unit Btu ratings to determine the total Btu input.

2. Use the common vent table at the same least height used above.

3. Read across to the right on either the L line or the V line as outlined earlier.

4. Stop at the first value which is equal to or greater than the total Btu rating.

5. Read the size of the common vent indicated at the top of this column.

Regardless of the table indications, the common vent must always be at least as large as the largest connector. If both connectors are the same size, the common vent must be at least one size larger.

To determine a common vent and the individual connector sizes, use the following procedure (Figure 6-11 and Table 6-2).

1. Locate on the vent connector table the least total height of 15 feet and a connector rise of 1 foot. Read to the right across to the next higher Btu rating than the water heater rating. As indicated at the top of the 53,000 Btu column, the vent size is a 4-inch connector size for the water heater [Figure 6-11 (a)].

2. Locate on the vent connector table the same least total height of 15 feet. However, this time for the furnace connector rise use 3 feet. Read across to the right and find 110,000 Btu column, the next higher rating above the 105,000 Btu input rating of the furnace. As indicated at the top of this column, the vent connector size for the furnace is 5 inches [Figure 6-11 (b)].

3. Total the two Btu ratings and find it to be 140,000 Btu. Read to the right, on the common vent table, at the same least total height of 15

Table 6-2
Combined Vent Tables

VENT CONNECTOR TABLE

LEAST TOTAL HEIGHT	CONNECTOR RISE	MAXIMUM INPUT TO VENT CONNECTOR, Thousands of BTU Per Hour													
		METALBESTOS VENT SIZE, INCHES													
		3"	4"	5"	6"	7"	8"	10"	12"	14"	16"	18"	20"	22"	24"
4'	1'	24	43	68	98	133	175	272	392	534	696	880	1090	1320	1570
	2'	29	51	80	116	158	206	322	464	600	782	990	1220	1480	1760
	3'	32	58	90	131	178	232	364	524	680	890	1125	1390	1680	2000
6'	1'	26	46	72	104	142	185	289	416	577	755	955	1180	1425	1700
	2'	31	55	86	124	168	220	345	496	653	853	1080	1335	1610	1920
	3'	35	62	96	139	189	248	386	556	740	967	1225	1510	1830	2180
8'	1'	27	48	76	109	148	194	303	439	601	805	1015	1255	1520	1810
	2'	32	57	90	129	175	230	358	516	696	910	1150	1420	1720	2050
	3'	36	64	101	145	198	258	402	580	790	1030	1305	1610	1950	2320
10'	1'	28	50	78	113	154	200	314	452	642	840	1060	1310	1585	1890
	2'	33	59	93	134	182	238	372	536	730	955	1205	1490	1800	2150
	3'	37	67	104	150	205	268	417	600	827	1080	1370	1690	2040	2430
15'	1'	30	53	83	120	163	214	333	480	697	910	1150	1420	1720	2050
	2'	35	63	99	142	193	253	394	568	790	1030	1305	1610	1950	2320
	3'	40	71	111	160	218	286	444	640	898	1175	1485	1835	2220	2640
20'	1'	31	56	87	125	171	224	347	500	740	965	1225	1510	1830	2190
	2'	37	66	104	149	202	265	414	596	840	1095	1385	1710	2070	2470
	3'	42	74	116	168	228	300	466	672	952	1245	1575	1945	2350	2800
30'	1'	33	59	93	134	182	238	372	536	805	1050	1330	1645	1990	2370
	2'	39	70	110	158	215	282	439	632	910	1190	1500	1855	2240	2670
	3'	44	79	124	178	242	317	494	712	1035	1350	1710	2110	2550	3040
40'	1'	35	62	97	140	190	248	389	560	850	1110	1405	1735	2100	2500
	2'	41	73	115	166	225	295	461	665	964	1260	1590	1965	2380	2830
	3'	46	83	129	187	253	331	520	748	1100	1435	1820	2240	2710	3230
60'	1'	37	66	104	150	204	266	417	600	926	1210	1530	1890	2280	2720
	2'	44	79	123	178	242	316	494	712	1050	1370	1740	2150	2590	3090
	3'	50	89	138	200	272	355	555	800	1198	1565	1980	2450	2960	3520

COMMON VENT TABLE

LEAST TOTAL HEIGHT	VENT TYPE*	MAXIMUM COMBINED INPUT TO EACH SECTION OF COMMON VENT, Thousands of BTU Per Hour													
		METALBESTOS VENT SIZE, INCHES													
		3"	4"	5"	6"	7"	8"	10"	12"	14"	16"	18"	20"	22"	24"
4'	L	—	45	70	101	138	180	280	404	600	780	990	1220	1445	1760
	V	—	56	88	127	173	225	355	508	688	900	1135	1400	1660	2020
6'	L	—	52	82	117	160	210	325	468	708	925	1170	1445	1680	2080
	V	—	65	103	147	200	260	410	588	815	1065	1345	1660	1970	2390
8'	L	—	58	91	130	178	230	365	520	793	1035	1310	1620	1920	2330
	V	—	73	114	163	223	290	465	652	912	1190	1510	1860	2200	2680
10'	L	—	63	98	142	193	250	395	568	865	1130	1430	1765	2090	2540
	V	—	79	124	178	242	315	495	712	995	1300	1645	2030	2400	2920
15'	L	—	73	114	164	224	290	460	656	1008	1315	1665	2060	2430	2960
	V	—	91	144	206	280	365	565	825	1158	1510	1910	2360	2790	3400
20'	L	—	81	127	182	250	325	510	728	1126	1470	1860	2300	2720	3310
	V	—	102	160	229	310	405	640	916	1290	1690	2140	2640	3120	3800
30'	L	—	94	147	211	290	375	590	844	1327	1735	2190	2710	3210	3900
	V	—	118	185	266	360	470	740	1025	1525	1990	2520	3110	3680	4480
40'	L	—	105	164	236	320	420	660	945	1492	1950	2470	3050	3610	4390
	V	—	131	203	295	405	525	820	1180	1715	2240	2830	3500	4150	5050
60'	L	—	—	178	259	352	460	720	1100	1750	2280	2890	3570	4230	5050
	V	—	—	224	324	440	575	900	1380	2010	2620	3320	4100	4850	5900
80'	L	—	—	—	275	374	488	765	1232	1950	2550	3230	3980	4720	5750
	V	—	—	—	344	468	610	955	1540	2250	2930	3710	4590	5420	6600
100'	L	—	—	—	—	383	500	780	1335	2140	2790	3530	4360	5160	6290
	V	—	—	—	—	479	625	975	1670	2450	3200	4050	5000	5920	7200

(Courtesy William Wallace Division, Wallace-Murray Corporation)

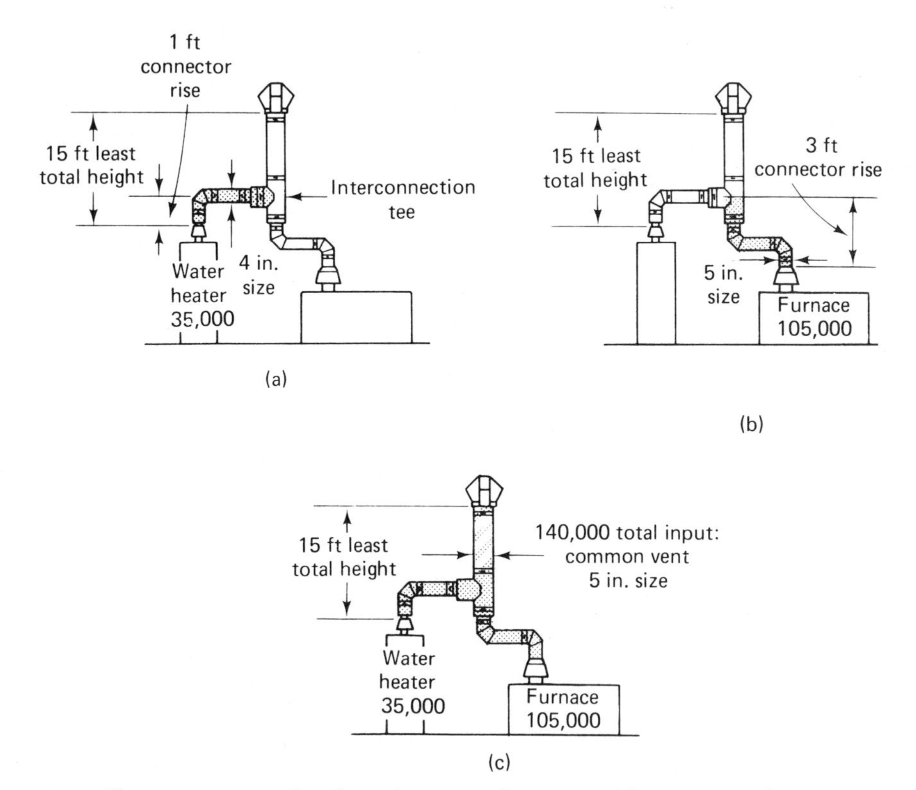

Figure 6–11. Combined vent diagram. (*Courtesy William Wallace Division, Wallace-Murray Corporation*)

feet and on the V line. Read the next higher Btu rating of 144,000 Btu. At the top of this column is indicated a 5-inch common vent for this system [Figure 6–11 (c)].

Vent heights of less than 4 feet are not recommended. When using short vents, with or without laterals, a size increase is often necessary and is indicated in the tables.

Short vents for horizontal furnaces located close to combustible materials should be of adequate height and size to agree with the tables and to avoid the hazard of draft diverter spillage caused by poor venting.

Vented recessed wall heaters or wall furnaces must terminate 12 feet or more above the bottom of the heater. For example, a 5-foot-high recessed wall heater normally requires 7 feet of vent height (Figure 6–12).

The recommendations given here may occasionally conflict with some local building codes and construction practices. If a particular installation given by the tables is in conflict with your local building code,

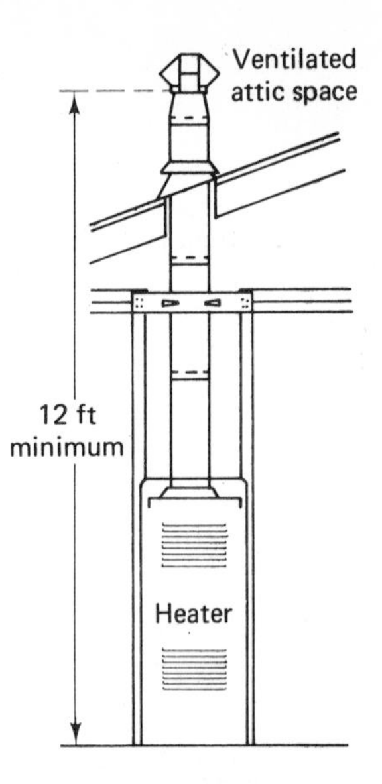

Figure 6–12. Minimum vent height. (*Courtesy William Wallace Division, Wallace-Murray Corporation*)

your local building inspector or other local administrative authority should be consulted. However, it must be stressed that these tables have been developed from the latest research and that their use will undoubtedly result in a better, safer venting system than the practices required by many local codes.

VENT DRAFT

Vent draft is extremely susceptible to the influence of the atmosphere. While it is general knowledge that outside atmospheric conditions, especially winds, bring about enormous changes in drafts, it is often disregarded that drafts change whenever the flue gas temperatures change. What this means is that drafts are changing almost constantly during the on and off cycles of the burner. Also, when two or more units are connected to a single vent, they affect each other, increasing or decreasing the draft as they start and stop individually.

Without some means for accurately controlling drafts, the fuel losses in a heating plant may be between 4% and 8%, and in some cases may reach 15%. This loss is occurring today in hundreds of thousands of existing heating plants.

If the heating plant is to operate without this waste, a definite and exact amount of air must be supplied to the fuel. This is the prime function of the barometric draft control.

When the correct draft is maintained automatically:

1. The weight or volume of air supplied to the fire is held to a minimum, reducing the weight and volume of the flue gas produced.

2. Combustion chamber temperatures are higher because the vent gases are not diluted with excess air.

3. A smaller volume of flue gases results in lower velocities through the heating unit, which increases the amount of time that the gases remain in contact with the heating surfaces.

4. More heat is absorbed and the flue gases are cooler when they enter the venting system than when the drafts are higher than needed.

5. Automatic draft control will also guard against incomplete combustion that will occur when manual dampers are closed too much. This is important with oil and coal fuels, but is tremendously important with gas as a fuel.

Because the draft needs of a heating unit are rigidly fixed, and because the draft forces being exerted are constantly varying, it is essential that a precisely accurate control be installed between the vent producing the draft and the unit using the draft. Note, however, that such a control must join the vent to its heating unit (Figure 6–13). Any device that simply isolates a vent system from its heating unit does not control draft. It simply eliminates it as an oxygen source.

Figure 6–13. A barometric draft control. (*Courtesy Conco, Field Control Division*)

When the barometric draft control is installed on gas fired equipment, the preferred location is part of the bull head tee [Figure 6–14 (a), (b), and (c)]. During normal operation the flue gases make a right angle turn behind the control but do not impinge on it. Should a downdraft occur, the air flowing in the opposite direction strikes the control directly, causing it to open outwardly, and venting the air into the room with a minimum of resistance. Entrained products of combustion are thus provided a greater relief.

With oil or solid fuels, locate the draft control as shown in Figure 6–14 (d)–(j). The locations are recommended for normal updraft operation. The bull head tee is not recommended, except for gas fired furnaces and boilers.

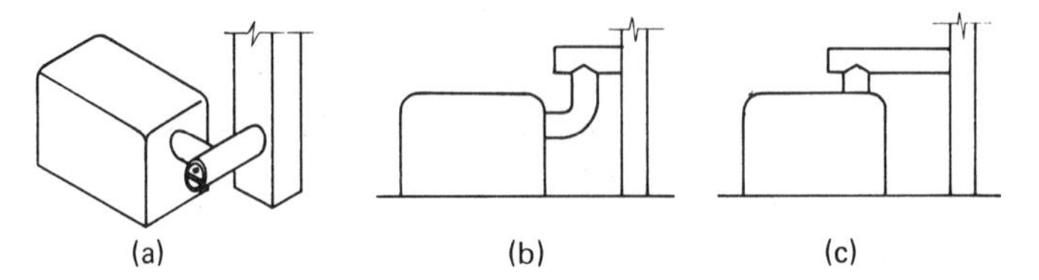

(a) Best locations for gas

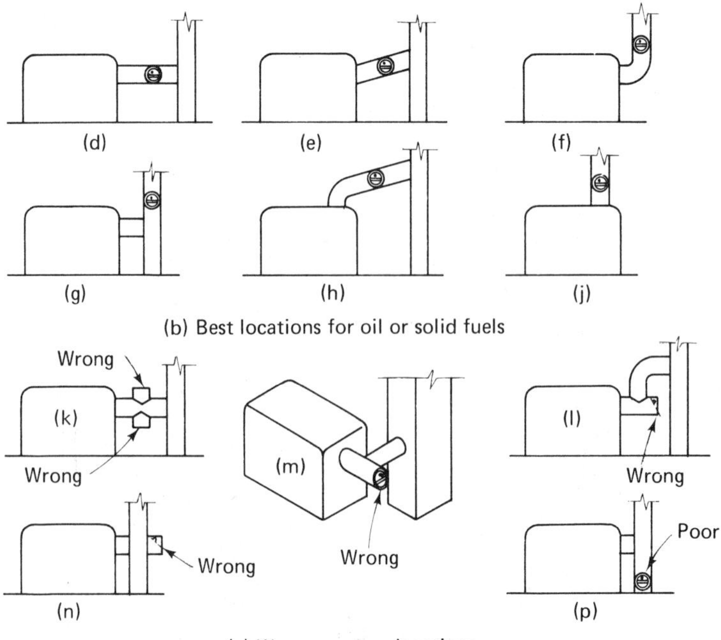

(b) Best locations for oil or solid fuels

(c) Wrong or poor locations

Figure 6–14. Draft control location. (*Courtesy Conco, Field Control Division*)

Parts (d)–(j) are acceptable for gas units where equipment room configuration does not permit the use of a bull head tee.

Parts (k)–(p) show the wrong way or poor locations. Even though (l), (m), and (n) appear to be bull head tees, they are incorrect. On correct installations, gases are directed toward the gate when the normal direction of flow is reversed. Part (k) shows the draft controls on top and bottom of a vent pipe, a placement that would cause the control to be inoperative. A draft control may be placed in a vent system at some point higher than the unit outlet. Part (p) would result in poor draft control operation.

The draft control should be placed as close as possible to the heating equipment except on forced draft systems. The draft control must be in the same room as the heating unit and at least 12 inches beyond the stack switch on oil-fired units and a minimum of 18 inches from any combustible materials.

POWER VENTING

A power supplied pressure increase can add to gravity draft or deliver all the energy required for venting. Many different types of heating equipment have forced or induced draft burner components capable of producing enough pressure for satisfactory power venting. For any common gas equipment depending on gravity for combustion air flow, power venting can be obtained with a fan or booster at the vent outlet, inlet, or anywhere in between.

Power venting provides safe, reliable equipment under a number of adverse conditions. It can also solve a number of common problems. For example:

1. Long horizontal vents for which access to the outside was prohibited by the building construction.

2. Too small a vent size due to space limitations or because of increasing the requirements of existing venting systems.

3. An excessive number of turns that create high pressure losses and reduce gravity flow.

4. A need for more draft than a short vent can supply, such as on roof mount units.

5. Erratic or inadequate venting caused by winds, adverse internal pressures, restricted air supply, or indoor-outdoor temperature changes.

6. A need for more dilution air to lower vent gas temperature thus reducing vent heat losses to surroundings.

7. An improper design of a combined vent system, such as with insufficient connector rise.

8. Roof location problems brought about by penthouses, nearby mechanical installations, smoke dissipation, etc.

The location recommended for any power vent device or draft booster is near the vent outlet, preferably outdoors above the roof. Thus, the entire venting system is kept under a negative pressure, and whether the system is an individual or combined vent, draft is insured to all interconnecting draft diverters.

GAS BURNER COMBUSTION TESTING

Automatically controlled gas heating represents the ultimate in trouble-free, economical heating comfort. The types of gas burners used in domestic heating units are termed *atmospheric injection burners.* This term is derived from the fact that gas pressure is used at only 2–5 inches of water column pressure and the air is furnished by atmospheric pressure. The following discussion will cover only the atmospheric injection burner and will not apply to power gas burners.

The atmospheric injection burner is tremendously flexible under adverse conditions. However, full advantage of this diversity can be obtained only if proper use is made of the adjustment means on the burner. Easy to follow, well-proven procedures of adjusting and testing gas heating equipment have been organized by the American Gas Association in cooperation with a large group of authorities, including the American Standards Association. Furthermore, many cities and towns have already legislated, or are considering legislating ordinances governing the installation and adjustment of gas fired heating equipment. All these codes include requirements for a combustion test. The principal objective of the combustion test on gas burning equipment is to analyze the flue gases for (1) safe combustion, and (2) maximum operating of the heating equipment.

Air for Combustion

In an atmospheric injection-type burner, there are two sources of combustion air—primary and secondary (Figure 6–15). Primary air is the air that is mixed with the gas before it burns. The air is drawn into the burner by the flow of gas through the venturi. The air starts moving into the burner the instant the gas is turned on because of the action of the gas flow through the venturi. As combustion of this gas-air mixture starts burning, additional air is sucked into the furnace by the draft through the furnace. This additional air is called secondary air. It is supplied around the burner head and mixes with the gas after it has been ignited.

None of the atmospheric injection-type burners used at present induce enough primary air for complete combustion. Therefore secondary air must be provided in sufficient quantity to complete combustion and thus eliminate the possibility of CO gas in the products of combustion. Since CO is an extremely dangerous gas, the basic requirement in installing and servicing any gas burning equipment is that no unit will be left with any CO in the flue products.

Air Adjustments and Excess Air

The primary air is adjusted by means of the primary air shutter for the proper flame characteristic. If, even with the air shutter completely open, the flame characteristics indicate insufficient primary air, this lack may be corrected by reducing the size of the orifice and increasing the manifold gas pressure, thereby maintaining the correct gas input. It should be mentioned that for atmospheric burners with normal limits of adjustment, this primary air setting has little effect on carbon dioxide (CO_2). It is the secondary air setting that greatly establishes the percentage of CO_2 in the flue products.

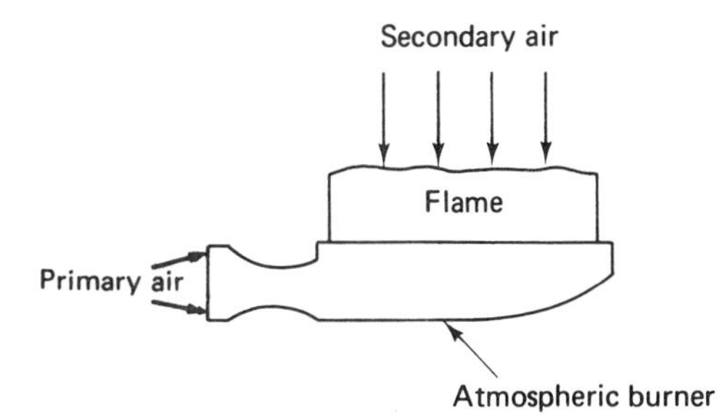

Figure 6-15. Atmospheric gas burner.

The secondary air should be adjusted to prevent excessive agitation of the flame and yet provide sufficient air for a CO free combustion. On gas-designed equipment, the amount of secondary air is fixed, and no secondary air adjustment can be made. However, many servicemen find it desirable to check the CO_2 content and flue gas temperature as an added check on the fuel input adjustment since these tests will indicate any error in metering or reading during this important adjustment.

The CO_2 in the combustion products is really an indication of the amount of excess air supplied for the combustion of the gas being burned. Excess air is the quantity of air admitted to the furnace in excess

Table 6-3
Ultimate CO_2 Percentage,
Characteristics of Fuel Gases Used Straight or Mixed
for Domestic Service

	Specific Gravity	Btu per Ft^3—Gross	Ultimate Percent CO_2
Natural Gas	0.57–0.70	900–1200	11.7–12.2
Carbureted Water Gas	0.65–0.71	500–600	15.8–17.3
Coke Oven Gas	0.35–0.49	500–600	10.3–12.1
Coal Gas (Retort Process)	0.41–0.49	500–600	11.5–12.1
Refinery Oil Gas	0.88–1.00	1470–1660	13.6–14.4
Butane Gas	1.95–2.04	3180–3260	14.0
Propane Gas	1.52–1.57	2500–2580	13.7

(Courtesy Bacharach Instrument Co.)

of that required for perfect combustion, which would produce the ultimate percentage of CO_2 in the flue products. Table 6-3 indicates the ultimate CO_2 percentage for common types of gaseous fuel. The addition of some excess air is required for all fuel gases to insure safe combustion; thus a gas burner is never adjusted to provide the ultimate percentage of CO_2. The generally accepted practice is to adjust the air supply for CO_2 levels from 70% of the ultimate to 80% of the ultimate CO_2.

It should be noted that if the ultimate CO_2 is known, the excess air may be determined by measuring either percentage of CO_2 or percentage of oxygen. It should be noted that the percentage of oxygen at a given excess air value varies less for different fuel gases than does the percentage of CO_2.

The amount of excess air required for safe heating unit operation cannot be stated specifically because most units have considerable tolerances. Table 6-4 lists generally accepted limits of excess air for flue connected heating units.

Table 6-4
Limits of Percent Excess Air for
Flue-Connected Gas Appliances

Central Heating Furnace or Boiler	25–50%
Gas Water Heater	25–50%
Space Heater	50–100%

(Courtesy Bacharach Instrument Co.)

Too much excess air is the principal reason for inefficient combustion. If the CO_2 test shows an abnormally high excess air, the secondary air adjustment means alone may not provide the desired results; it may be necessary also to restrict the flow of combustion products. This may be accomplished by either reducing the flue pipe size connecting the equipment to the draft diverter. The purpose is to restrict the discharge of combustion products from the equipment to such a point that there is a slight positive pressure in the bottom portion of the furnace.

CO_2 Check of Secondary Air Adjustment

When the neutral pressure point has been adjusted to the proper position, either by using a vent pipe of smaller diameter or by inserting into the flue pipe a neutral pressure point slide, the combustion air shutter should be adjusted so that the CO_2 percentage in the flue products is in accordance with the applicable codes or standards. The CO_2 test must be made at the inlet side of the draft hood (Figure 6-16). If the CO_2 reading is not within the recommended limits, the combustion air shutter should be readjusted and then the neutral pressure point should be rechecked and reset if necessary.

Figure 6-16. Checking CO_2 of furnace. (*Courtesy Bacharach Instrument Co.*)

Combustion Efficiency

The percentage of CO_2 and the temperature of the flue gases are the two measurements that establish the percentage of combustion efficiency. These two measurements should be taken on the inlet side of the draft diverter. The flue gas thermometer is inserted into the same hole for the CO_2 test.

Any heat contained in the flue gases at this point will be mostly wasted to the vent. Obviously, the higher the temperature of these flue gases, the greater will be the loss of heat that has not been absorbed by the heat exchanger. Furthermore, these flue gases cannot be cooled below the room air temperature, so that the difference between the actual flue gas temperature and room air temperature (known as the *net flue gas temperature*) represents a measure of the useable heat loss in the vent. The relationship between net flue gas temperature, percent CO_2, and combustion efficiency can be shown in a curve or table. However, the use of a slide rule calculator for obtaining the combustion efficiency is usually preferred. By means of such a calculator (Figure 6–17), setting one slide to the net stack temperature and the other to the percent CO_2 will permit direct readings of percent combustion efficiency. The percent combustion efficiency is an indication of the useful heat obtained from the gas being burned. It is expressed in a percentage of the total heat produced by the burning of the gas, assuming complete combustion with no excess air. For example, a combustion efficiency of 60% means that 60% of the total heat input has been absorbed by the furnace. The percentage of stack loss is an indication of the heat that is wasted in the hot vent gases instead of being absorbed by the furnace. The percentage of stack loss is obtained by subtracting the percentage of combustion efficiency from 100%.

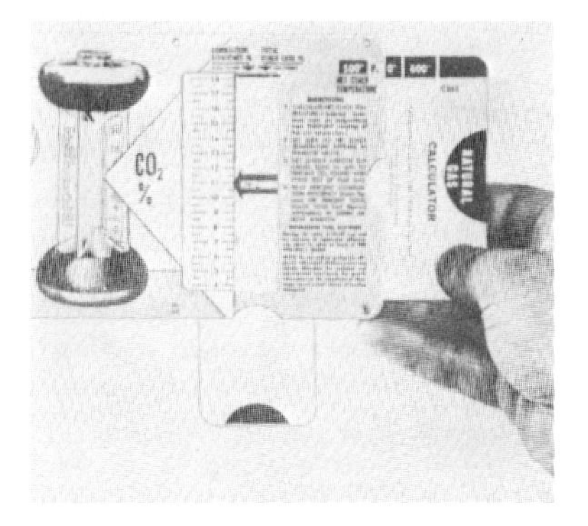

Figure 6–17. Combustion efficiency and stack loss calculator. (*Courtesy Bacharach Instrument Co.*)

A word of caution. The stack loss percentage is not a measure of what can be saved by readjustment of the equipment since the flue gases must still be allowed to escape. Furthermore, combustion efficiency must not be confused with overall efficiency of the furnace which must include allowance for radiation and miscellaneous heat losses.

As the percentage of CO_2 decreases and stack temperature increases, the heat wasted to the vent increases and combustion efficiency decreases. With ultimate CO_2 (no excess air) and flue gas temperature at room temperature, all available heat has been extracted and combustion efficiency is 100%. Gas burning equipment should always be capable of 75% combustion efficiency. Anything higher than 80% may adversely affect vent draft. In some cases it may cause a corrosive moisture to condense in the flue and at an extreme may result in incomplete combustion.

CO-Free Combustion

As has already been stated, it is important to make sure that the products of combustion do not contain any CO. It is possible that overfired equipment could have CO in the flue gases, especially with flame impingement on cold surfaces even with excess air, CO_2 and O_2 within acceptable limits. The flue gas sample for the carbon monoxide check must be taken at the inlet side of the draft diverter to make sure that the sample is not diluted by air being drawn in at the draft diverter.

The monoxor (Figure 6–18) is a practical, reliable instrument for testing CO concentrations in flue gas samples. According to gas industry standards CO-free combustion is defined as that which produces less than 0.04% CO in an air-free sample of the flue gas.

The CO test is similar to the CO_2 test but, of course, is made with the monoxor instead of the "firite." A flue gas sample is taken through the same hole used for the CO_2 test. The flue gas sample is first drawn into the collecting bladder of the sampling assembly and subsequently

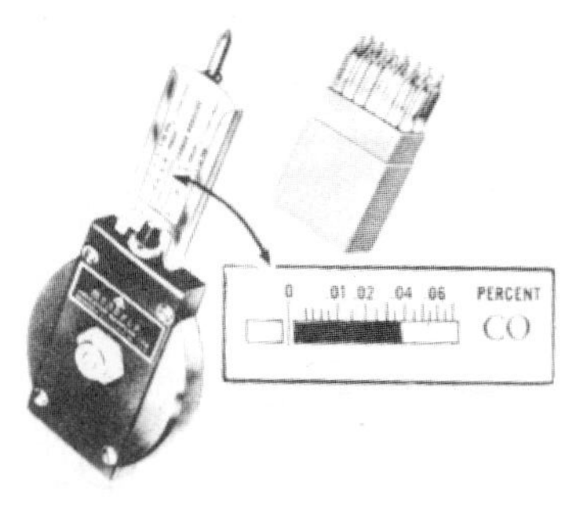

Figure 6–18. Monoxor indicator. (*Courtesy Bacharach Instrument Co.*)

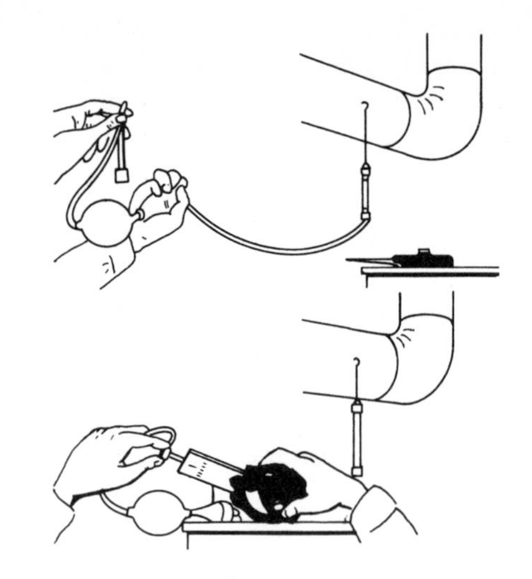

Figure 6–19. Sampling flue gas for monoxor CO test. (*Courtesy Bacharach Instrument Co.*)

drawn through the glass indicator tube of the monoxor. This indicator tube is inserted into the rubber connector of the monoxor sampler that contains a suction pump operated by a push button on the outside of the sampler housing. These two steps are the flue gas sampling procedure shown in Figure 6–19.

The monoxor indicator tube is filled with a yellow-colored CO sensitive chemical. The CO content of the sample is absorbed by the yellow-colored chemical causing the production of a dark-brownish stain. The length of this stain is directly proportional to the CO concentration in the sample, and the stain length is measured on the instrument's etched metal scale. The scale is calibrated directly in CO percentage. The monoxor indicator tube can be used for two tests if stain from the first test penetrates less than one-half the length of the yellow chemical. The unstained end of the tube may be used for the second test by reversing the tube's position in the monoxor sampler.

Draft Diverter and Flue

After the air adjustments and the flue gas tests have been completed, the draft diverter should be checked with the neutral pressure point gauge. This can be done by holding the fish tail tip of the instrument just inside the edge of the diverter opening (Figure 6–20). A negative pressure (draft) should be indicated. It is also sound practice to check the flue draft on the vent side of the diverter to be certain that the vent is adequate for moving the products of combustion from the draft

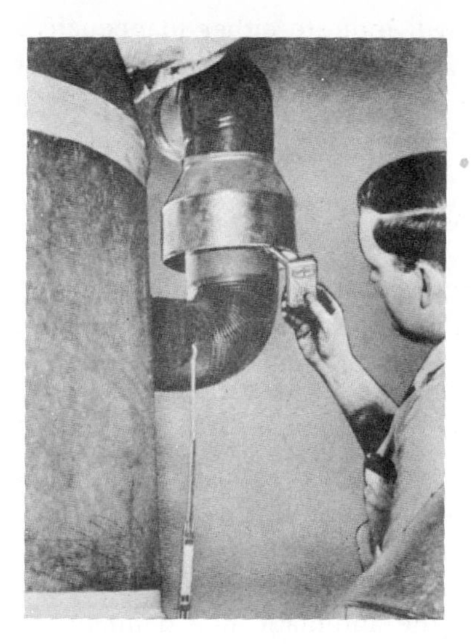

Figure 6-20. Using N.P.P. indicator. (*Courtesy Bacharach Instrument Co.*)

diverter to the outside atmosphere under all weather conditions. Otherwise, there will be a smothering of the flame that will produce CO. The location of the check point is above the draft diverter (Figure 6-21). The gauge illustrated has a range of 0.14 inch water column updraft, which is the preferred range for this test. The draft check must be made with the burner operating. Inadequate venting capacity of the vent will be indicated by (1) very low updraft, (2) temporary downdraft, or (3) at an extreme, a steady downdraft reading.

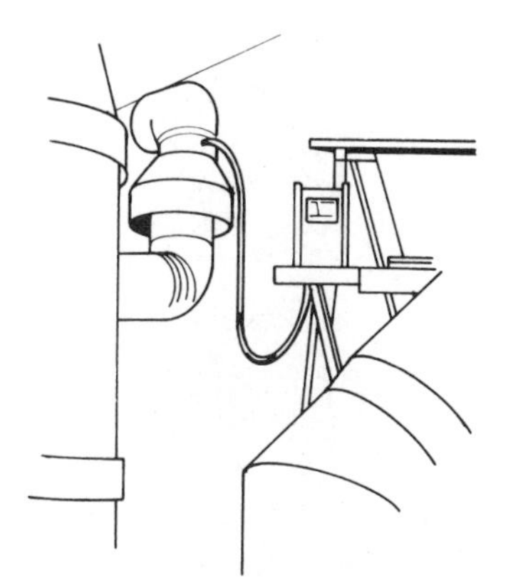

Figure 6-21. Using draft gauge. (*Courtesy Bacharach Instrument Co.*)

All of these will indicate either inadequate vent size or height, a blocked vent, or vent downdraft caused by roof obstructions or nearby structures. Appropriate steps must be taken to remedy these defective conditions.

AUTOMATIC FLUE DAMPER

When the furnace burner is operating, air must be drawn from inside the home and be allowed to pass up through the venting system to the atmosphere with the products of combustion. This flow of air is not required when the burner is not functioning. The automatic flue damper prevents this wasteful flow of heated household air by opening and closing on demand from the thermostat. The automatic flue damper can reduce fuel consumption as much as 16%.

The automatic flue damper is installed on top of the draft diverter before the venting system is connected (Figure 6-22). When the thermostat calls for heat, the automatic flue damper blade moves to the full open position before the burner can ignite (Figure 6-23). While the burner is on, the damper blade stays in the full open position. A spring loaded mechanism will open the damper fully in case of a power failure.

When the burner shuts off, an electric motor automatically closes the automatic flue damper to stop the flow of heated air out the venting system (Figure 6-24), thus providing a savings during the heating season.

Figure 6-22. Automatic flue damper location. (*Courtesy Carrier Corporation*)

Figure 6–23. Automatic flue damper open. (*Courtesy Carrier Corporation*)

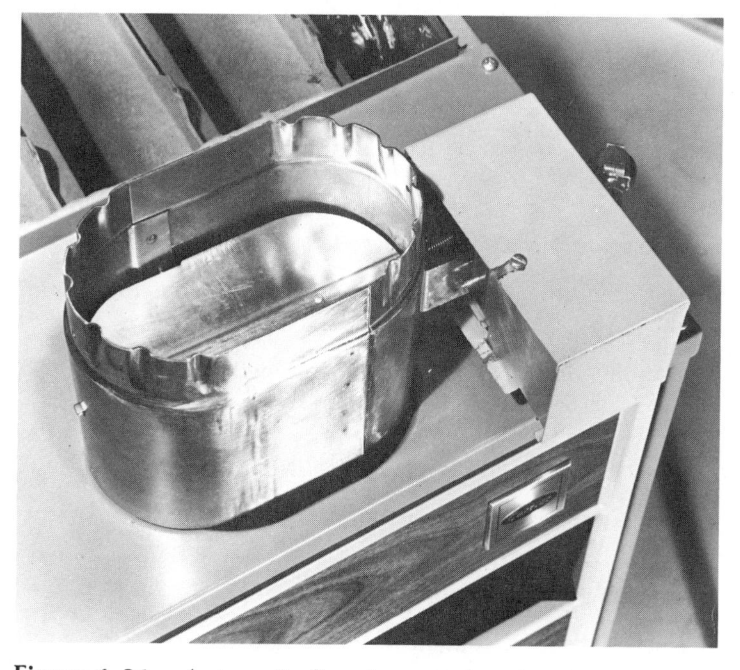

Figure 6–24. Automatic flue damper closed. (*Courtesy Carrier Corporation*)

REVIEW QUESTIONS

1. What is the purpose of the venting system?
2. Outline the flow of air through a heating unit, from combustion air to the vent termination.
3. What is dilution air?
4. What percentage of combustion air is dilution air?
5. What is the purpose of a draft diverter?
6. Is it good practice to alter a draft diverter?
7. What could happen if the venting system were insufficient?
8. What route is the most desirable when installing a vent pipe?
9. What institute makes recommendations involving vent construction?
10. Which type of vent pipe is most desirable, single wall or double wall?
11. What two factors are involved in determining vent pipe size?
12. Name three functions of air to heating equipment.
13. Define draft as it applies to heating equipment.
14. Can natural draft be produced without a venting system?
15. Do draft conditions on a natural draft system vary?
16. What force does not change in a venting system?
17. What may be the result of oversized or excessively long vents?
18. What is the rule of thumb for determining the maximum vent length?
19. What is the vent connector on a combined venting system?
20. What is the common vent on a combined venting system?
21. How does a barometric draft control function?
22. How does a power vent operate?
23. On gas burner combustion testing, why is it desirable to test the CO_2 and the flue gas temperature?
24. Of what is the CO_2 content in the combustion products an indication on gas burner testing?

25. Why can the amount of excess air required for safe heating unit operation not be stated specifically?
26. What is the principal reason for inefficient combustion?
27. What is the minimum temperature allowed of flue gases?
28. Of what is combustion efficiency an indication?
29. What happens to the gas temperature when the percentage of CO_2 increases?
30. What happens to combustion efficiency when the vent gas temperature increases?
31. For what is the monoxor used?
32. Why is it important to maintain CO-free combustion?
33. Define net stack temperature.

7

Electric Heating

The objectives of this chapter are:

- To provide you with the operating principles of electric heating.
- To describe heating element construction to you.
- To cause you to become aware of the different methods of using electricity for heating purposes.
- To provide you with the principles of heat pump operation.
- To introduce you to the method of repairing broken heating elements.

Electric heating is beneficial from the standpoint of ecology because the fuel burned in producing electricity is consumed in one remote area. At the electricity consumer level, there is no waste to be removed. There are no vent pipes to be installed or ashes to be removed. The homeowner realizes clean, efficient heat.

However certain adjustments must be made for electric heating. The user must be willing to pay higher electric bills. There is also the increased initial cost of construction. This cost is consumed in the added electrical service and wiring and the extra insulation required when electric heating is used. The contractor must be more knowledgeable about construction codes and service required.

Electric heating is probably the most efficient form of heating known today. Theoretically, when 1 Btu of electrical energy is used, 1 Btu of heat energy is released. In other words, the Btu input equals the Btu output. When one kW of electricity is converted to heat, 3,412 Btu are released.

TYPES OF ELECTRIC HEATING EQUIPMENT

There are several operations that must be considered in a discussion of electric heating. These areas are: resistance heating, heat pump, and a combination of the two.

Resistance Heating

Resistance heating involves a large number of operations, all of which are merely variations but are considered different in use and application. Under resistance heating, we may list resistance elements, baseboard elements, heating cable, portable and unit type heaters, and duct heaters. However, regardless of the use or application, resistance heating wire is selected for high resistance per unit of length, and for the stability of this resistance at high temperatures. The type of wire used for this purpose is a high nickel chrome alloy resistance wire. Resistance heating does not use a centrally heated medium. The energy is expended directly into the air stream.

Durability also must be considered when determining the wire used for resistance heating. The high nickel content in this wire retards oxidation and separation of the wire. If, however, the resistance wire does become broken, it may be repaired by cleaning the two ends, overlapping them, and applying borax liberally to the two pieces. Apply heat to the cleaned area until the two metals fuse together. After the wire has cooled to room temperature, remove the excess borax with a wet cloth. If this borax is allowed to remain on the wire, an immediate burnout will occur because the borax causes excessive spot heating.

Heating elements are staggered for more uniform heat transfer to eliminate hot spots and to insure black heat. Each element is protected by an over-temperature disc in one end and a fuse link in the other.

Figure 7–1. Resistance element installed in a furnace frame. (*Courtesy Electric Products Manufacturing Corp.*)

Resistance Elements

Resistance elements are more commonly used in forced air central heating systems. They are termed such to differentiate them from duct heater strips, and are mounted directly in the air stream inside the furnace (Figure 7-1). The elements are staggered for more uniform heat transfer, to eliminate hotspots, and to insure maximum heat transfer. Each element is protected by a temperature limit switch and a fuse link. These units are located inside the furnace proper (Figure 7-2) in an element compartment. These units emit approximately 3,412 Btu/kW input.

Figure 7-2. Electric element compartment. (*Courtesy Electric Products Manufacturing Corp.*)

Baseboard Elements

Baseboard elements may be used for the whole house or for supplementary heating. These elements are usually enclosed in a housing that provides safety as well as efficient use of the available heat (Figure 7-3).

These units are available in a wide variety of sizes and may be joined together, usually end to end, to provide the Btu required for any given installation (Figure 7-4). As shown, the ideal location of installation is at the point of greatest heat loss. The electrical element in these units is normally of the enclosed type. That is, the electric resistance has fins added for increasing the surface area of the heating element and increasing efficiency. Baseboard units have a direct Btu output per W input. An example of this may be seen in Table 7-1, where we have the

Figure 7–3. Electric baseboard heating units. (*Courtesy Chromalox Comfort Conditioning Division, Emerson Electric Co.*)

Figure 7–4. Baseboard unit location, units joined for capacity. (*Courtesy Chromolox Comfort Conditioning Division, Emerson Electric Co.*)

applied voltage, the element length, the wattage rating, and the corresponding Btu output rating. A similar table should be consulted when baseboard heating is planned. All manufacturers have such tables representing these values for their equipment.

Duct Heaters

Electric duct heaters are factory assembled units consisting of a steel frame, open coil heating elements, and an integral control compartment. These matched combinations are fabricated to order in a wide

Table 7-1
Wattage and BTU/hr Rating (Typical)

Volts	Length	Watts	Btu/hr.	Cat. No.
120	28 in.	500	1707	BB-SC-281
	40 in.	750	2360	BB-SC-401
	48 in.	1000	3413	BB-SC-481
	60 in.	1250	4266	BB-SC-601
208/240	28 in.	375/500	1280/1707	BB-SC-284
	40 in.	560/750	1927/2360	BB-SC-404
	48 in.	750/1000	2360/3413	BB-SC-484
	60 in.	935/1250	3108/4266	BB-SC-604
	72 in.	1125/1500	3840/5118	BB-SC-724
	96 in.	1500/2000	5118/6824	BB-SC-964
	120 in.	1875/2500	6399/8547	BB-SC-1204
240/277	28 in.	375/500	1280/1707	BB-SC-287
	40 in.	560/750	1927/2360	BB-SC-407
	48 in.	750/1000	2360/3413	BB-SC-487
	60 in.	935/1250	3108/4266	BB-SC-607
	72 in.	1125/1500	3840/5118	BB-SC-727
	96 in.	1500/2000	5118/6824	BB-SC-967
	120 in.	1875/2500	6399/8547	BB-SC-1207

(Courtesy of Chromalox Comfort Conditioning Division, Emerson Electric Co.)

range of standard and custom sizes, heating capacities, and control modes. They are prewired at the factory, inspected, and ready for installation at the job site.

The duct heater frame is made in two basic configurations—slip-in (Figure 7-5) or flanged (Figure 7-6). The slip-in model is normally used in ducts up to 72 inches wide by 36 inches high, while the flanged model is used in larger ducts or where duct layout would make it impossible or impractical to use the slip-in type.

The slip-in heater, which is standard, is designed so that the heater frame is slightly smaller than the duct dimensions. The frame is inserted through a rectangular opening cut in the side of the duct, with the face of the heater at a right angle to the air stream. It is secured in place by sheet metal screws running through the control compartment.

The flanged-type heater is designed so that the frame matches the duct dimensions. The frame is then attached directly to the external flanges of the duct.

Open coil heating elements (Figure 7-7) are a high nickel chrome alloy wire (usually 80% nickel and 20% chromium) physically designed

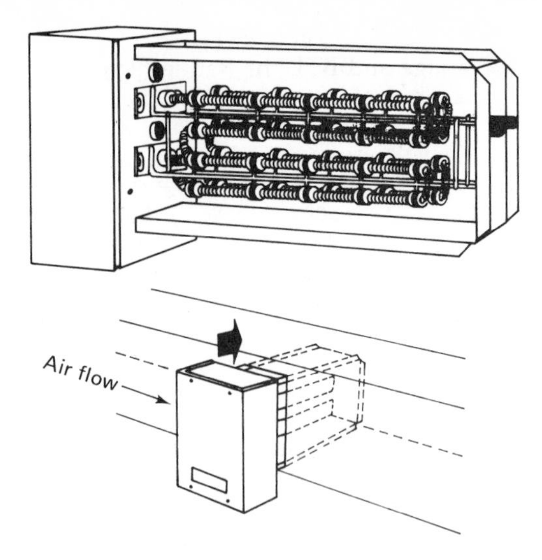

Figure 7-5. Slip-in duct heater. (*Courtesy Gould Inc., Heating Element Division.*)

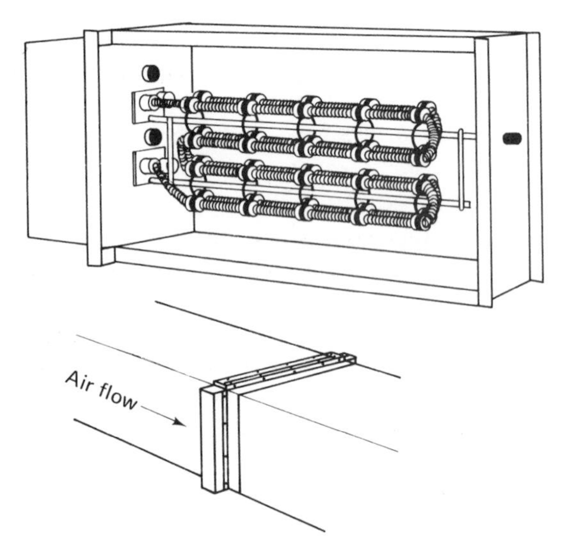

Figure 7-6. Flange type duct heater. (*Courtesy Gould Inc., Heating Element Division*)

to meet the application requirements of duct heaters. Essentially 100% of the electrical energy input to the coil is converted to heat energy, regardless of the temperature of the surrounding air or the velocity with which it passes over the heating element.

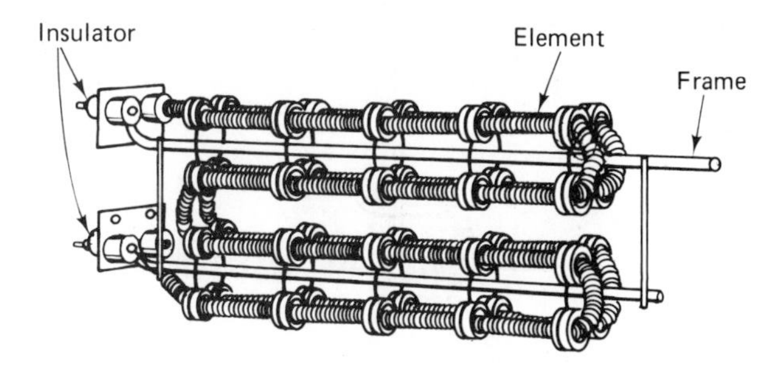

Figure 7-7. Heater coils and rack. (*Courtesy Gould Inc., Heating Element Division*)

An iron-chrome-aluminum alloy is also used for coils in special low wattage applications. Regardless of the type of wire used, ceramic bushings (Figure 7-8) are used that insulate the coil from surrounding sheet metal, float freely in embossed openings, and prevent binding and cracking as the heater cycles. The bushings are held in place by curved sheet metal tabs. Their extra-heavy body enables the bushings to withstand high humidity conditions and a 2,000 V dielectric test.

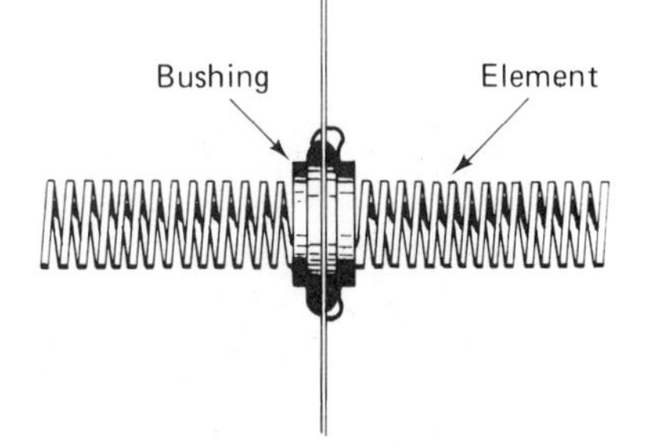

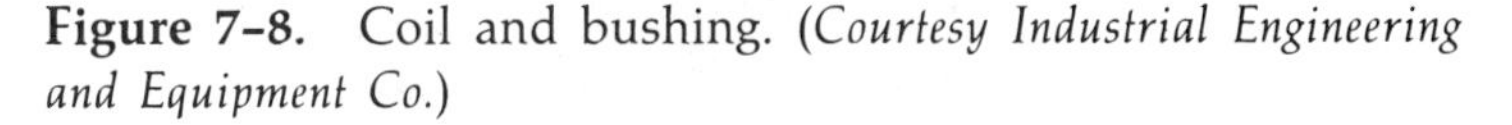

Figure 7-8. Coil and bushing. (*Courtesy Industrial Engineering and Equipment Co.*)

Stainless steel terminals (Figure 7-9) have threads for securing electrical connections. The coil is mechanically crimped into its terminal with closely adjusted tools to insure cool, minimum resistance connections. High-temperature molded phenolic terminal insulators are used that will not crack or chip during normal use.

Aluminized steel brackets support the coils (Figure 7-10). They are spot welded in place and spaced to prevent coil sag. The brackets are reinforced two ways: (1) by ribbing the bends at each end and (2) by grooving the edges facing the air stream.

While UL listed heaters are available for virtually any kW capacity

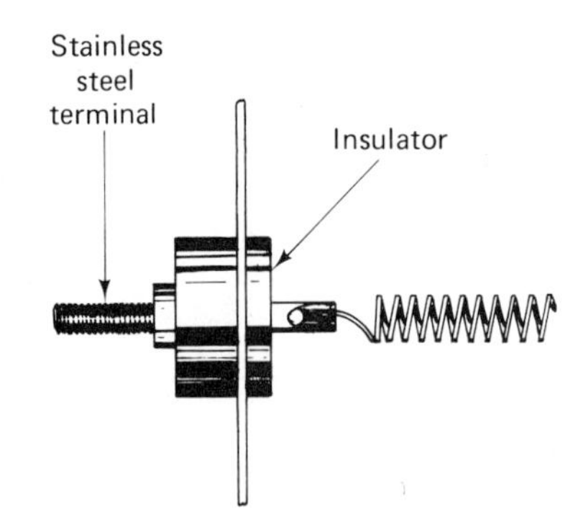

Figure 7–9. Heater element terminal. (*Courtesy Industrial Engineering and Equipment Co.*)

specified, they are restricted to a maximum watt density of 156 W/inch of duct face area for both slip-in and flange-type heaters. To determine the maximum kW for given duct size, the following formula (*courtesy of Gould Inc., Heating Element Division*) may be used.

$$\text{Maximum kW} = \frac{156 \text{ duct width (in.) height (in.)}}{1{,}000}$$

Figure 7–10. Element brackets. (*Courtesy Industrial Engineering and Equipment Co.*)

AIR FLOW

Electric heaters differ from steam or hot water coils in that the Btu/hr output is constant as long as the heater is energized. It is, therefore, necessary that sufficient and uniform air flow be provided to carry away this heat and to prevent overheating and nuisance tripping of the thermal cutouts. The minimum velocity required is determined from Figure 7-11 on the basis of entering air temperature and W/ft^2 of cross-sectional duct area.

Example: To find the minimum air velocity required for a 10,000 W heater installed in a duct 12 inches high and 24 inches wide and operating in a minimum inlet air temperature of 65° F.

1. Use the top curve (below 80° inlet air).

2. W/ft^2 of duct area $= \dfrac{10{,}000}{1 \text{ ft } 2 \text{ ft}} = 5{,}000$ w/ft^2.

3. This point on the curve corresponds to 310 fpm; thus, the minimum velocity required is 310 fpm. Since the duct area is 2 ft^2, the minimum cfm required is 620 cfm.

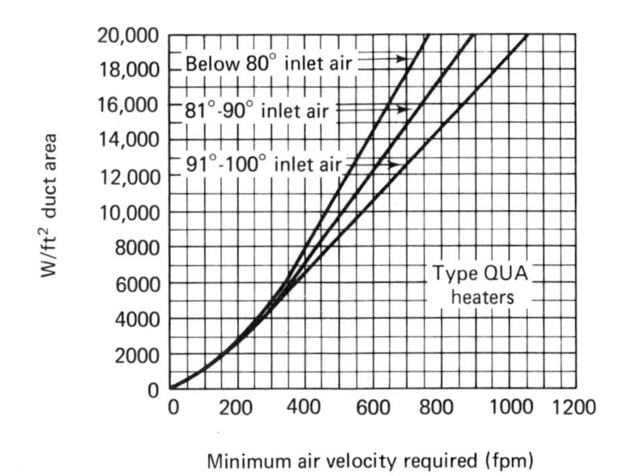

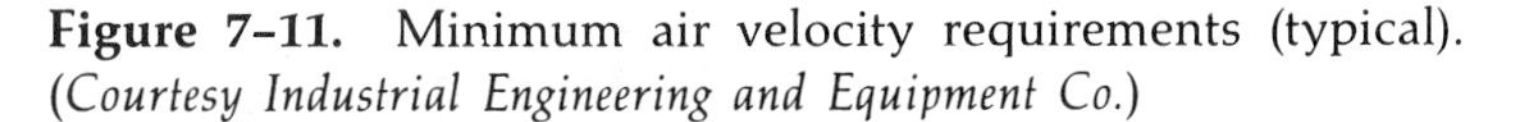

Figure 7-11. Minimum air velocity requirements (typical). (*Courtesy Industrial Engineering and Equipment Co.*)

The minimum velocity should be uniformly distributed across the duct at the point of insertion of the heater. See Figure 7-12 for typical heater misapplications that result in nonuniform airflow. It is suggested that the heater be installed at least 48 inches from any change in duct direction, any abrupt changes in duct size, and any air moving equipment.

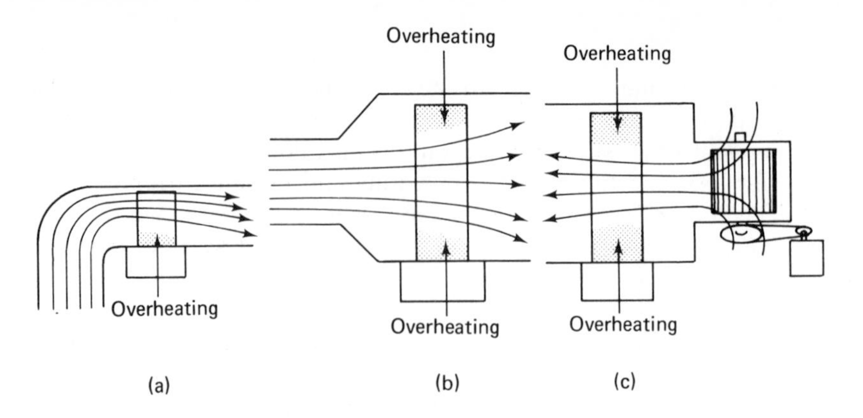

Figure 7-12. Typical heater misapplications. (*Courtesy Industrial Engineering and Equipment Co.*)

KILOWATT CALCULATION

For space heating, the required kilowatt input may be calculated as outlined in the *ASHRAE Guide or NEMA Publication Number HE 1-1966* to arrive at the actual heat loss.

(typical equation)

$$kW = \frac{\text{Btu/hr heat loss}}{3.412}$$

(*Courtesy Gould Inc., Heating Element Division*)

ELEMENT COMPARISON

Although it is less expensive, the *open coil* is far superior to the enclosed design for most space heating applications. Only for special applications, such as exposed heaters and hazardous areas, is the finned tubular construction recommended.

The reasons for recommending the open coil are as follows:

Longer Heater Life: The open coil releases its heat directly to the airstream. As a result, the open coil runs cooler than the coil in a finned tubular element, where it is isolated from the air by insulation and a metal sheath. Low coil temperatures mean long life.

Low Pressure Drop: Large open spaces between coils result in free air flow and negligible pressure drop (Figure 7-13). This important factor can result in a reduced fan size for a given cfm.

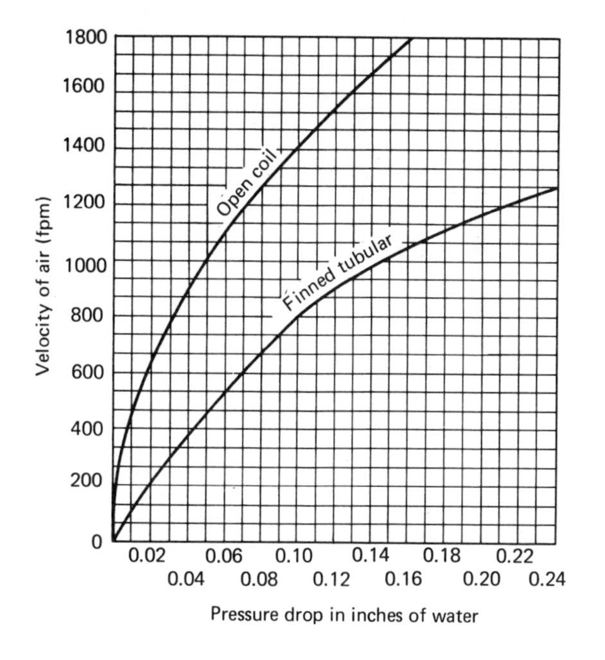

Figure 7-13. Pressure drop comparison of an open coil and a finned tubular duct heater (typical). (*Courtesy Industrial Engineering and Equipment Co.*)

Accurate Control: The lightweight coil reaches operating temperature very quickly when it is energized. This low thermal inertia results in more precise temperature control, as the air temperature responds rapidly to a call for heat. Comparing the open coil to the finned tubular construction (Figure 7-14), note that the air heated with an open coil reaches steady-state temperature in less than half the time required for an enclosed heater.

Easy Handling: Their sturdy, lightweight construction makes open coils easy to mount and handle. For example, a 30,000 W heater to fit a 30-inch 16-inch duct weighs only 30 pounds. Such a heater can be installed quickly without the use of elaborate hoisting equipment.

Large Electrical Clearances: Clearances between the coil and frame enable open coils to withstand severe applications such as vibration and high voltages.

Duct heaters are advantageous because they are adaptable to cooling, air cleaning, ventilation, and humidification. They are also easily applied to zone control.

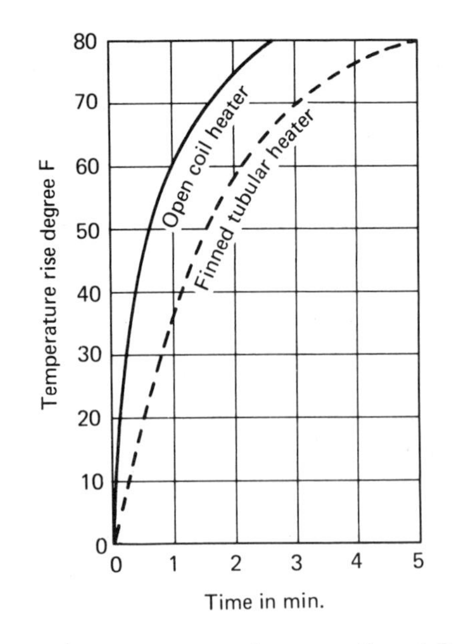

Figure 7–14. Response times of open coil and finned tubular duct heaters (typical). (*Courtesy Industrial Engineering and Equipment Co.*)

On the other hand, a tempering heater may be required at the fan location to prevent delivery of cold air to rooms when the primary heater is off, thus increasing the overall cost of the equipment and its operation.

HEAT PUMP

A heat pump may be defined as an air conditioner used to move heat to and from a conditioned area. Its operation is based on the normal refrigeration cycle. In winter it employs a reversed freon flow through the use of valves. Heat is extracted from the outside air and released into the home.

Heat is energy; it is contained in all substances. Although heat cannot be picked up by hand like a rock or poured into a container like water, it can be moved from one place to another. For heat to be moved, it must be absorbed by a substance, except in the case of radiant energy. The substance is then relocated to where the heat is needed and the heat is released. Thus, the heat is moved by the intermediate substance. The intermediate substance used in a heat pump is known as a refrigerant.

The fundamentals of the refrigeration cycle must be known before a clear understanding of the operation of a heat pump can be obtained.

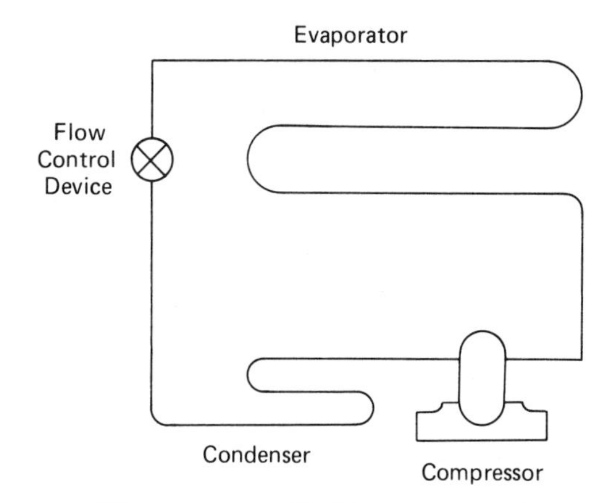

Figure 7-15. Refrigeration cycle.

The same basic components are used in both systems (Figure 7-15). The difference is that the flow of refrigerant is reversed.

The purpose of a standard refrigeration unit is to pick up heat from a place where it is not wanted and pump it to some place where it is unobjectionable. Heat is absorbed by the refrigerant as it changes from a liquid to a vapor in the evaporator (Figure 7-16). The gaseous refrigerant is then pumped from the evaporator to the condenser by the compressor. To accomplish this, work is done on the refrigerant, resulting in heat being added to the refrigerant. As the refrigerant passes through the condenser, the total heat contained in the gaseous refrigerant is given up and the gaseous refrigerant turns to a liquid refrigerant. In a normal refrigeration cycle, the rejection of heat from the condenser is not so important as the heat absorption process in the evaporator.

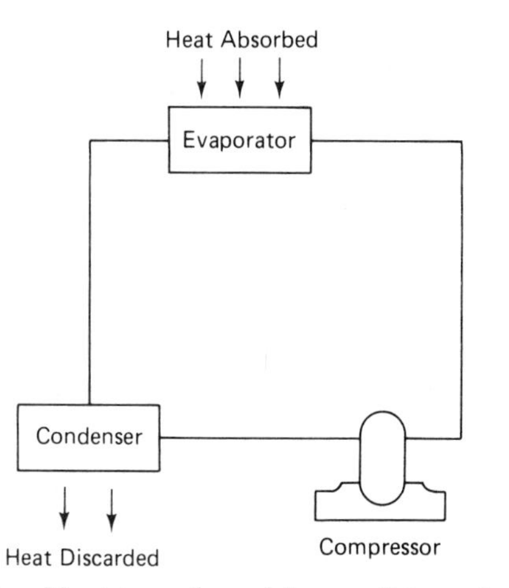

Figure 7-16. Heat transferred in a refrigeration system.

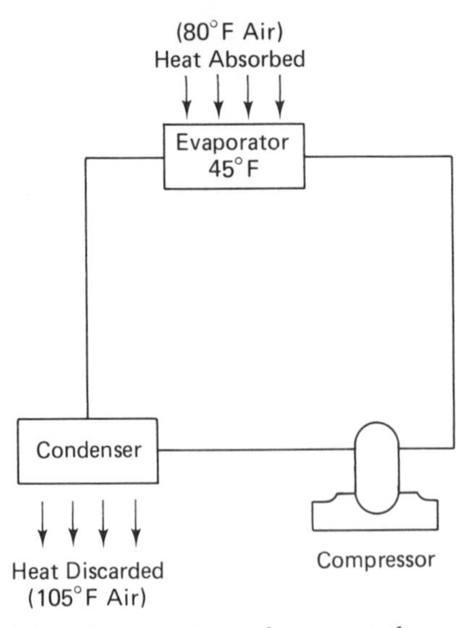

Figure 7–17. Temperature change at the evaporator.

In a heat pump system, the heat rejected by the condenser is just as important, if not more important, than the refrigeration effect provided by the evaporator in a typical refrigeration system. A heat pump system operating on the cooling cycle absorbs heat into the evaporator (Figure 7–17). If the evaporator temperature is maintained at 45° F, as shown, air or water can flow over the outside of the evaporator coil and be cooled as the air or water gives up heat to the refrigerant inside the evaporator. The refrigerant is then pumped to the condenser where it gives up its heat of condensation to some cooling medium, such as air or water, which carries away the heat. When the heat pump is in the cooling cycle, its operation is identical to a refrigeration system. Therefore, a refrigeration system is a heat pump. However, it is customary to use the term heat pump only for those refrigeration systems that are provided with special devices to use the heat which is rejected from the condenser.

Heat pump units are classified by their heat source, that is the medium from which the outside coil obtains its heat. These heat sources are: air to air; water to air; water to water; and ground to air. For the most part we will be concerned with the air to air unit because of the expense involved in installation and maintenance.

Operation

The name heat pump is descriptive of its operation. If this term is divided and the meaning of each word studied, the true meaning of the term will be more clear. The first word, heat, is a physical property which has two physical factors-level and amount. Both factors are important in providing comfort. The level or intensity of heat is measured by the temperature of a substance. The amount or quantity is measured by the British Thermal Unit (Btu).

The second word, pump, indicates a device for making something go in a direction that it ordinarily would not go; for example, making water run uphill or making a boat go upstream against the current. The pump, or compressor, in a heat pump serves a similar function because it forces heat to move in a way it would not ordinarily move.

Some outside force is required to lift heat or energy from a lower temperature to a higher temperature. This force is provided by the compressor. The amount of force required is determined by how far the heat must be raised. For example, it is easier to remove heat from air at 40° F and release it at 75° F than it is to remove the heat from 40° F air and release it at 90° F. The compressor provides the force required to lift the heat—not only in the heat pump but in a refrigeration or air conditioning unit.

Cooling Cycle: A typical air conditioning system has two coils, or heat exchangers, one inside the conditioned space and one outside. The coil inside the home serves as the evaporator where liquid refrigerant evaporates and absorbs heat from the air passing over it. It is termed the indoor coil (Figure 7–18).

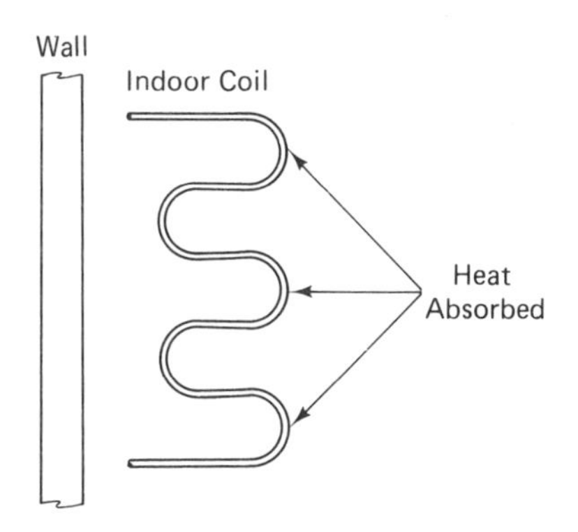

Figure 7–18. Indoor coil.

The evaporated refrigerant leaves the evaporator and goes to the compressor where its temperature and pressure are both raised to a higher level. This high pressure vapor then passes through the outside coil, termed the outdoor coil, where heat is removed and the vapor is changed back into a liquid (Figure 7–19).

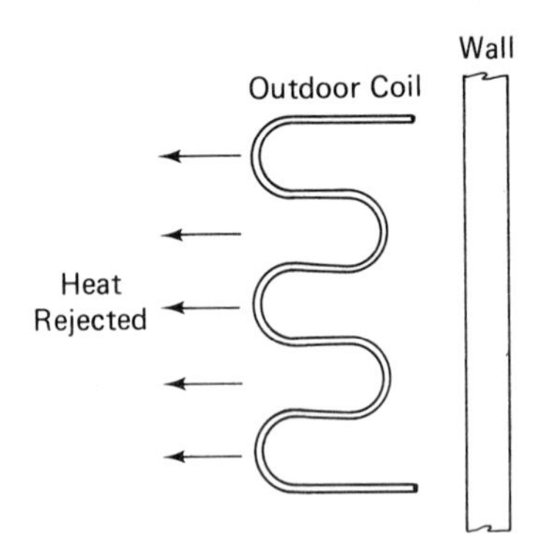

Figure 7-19. Outdoor coil.

The liquid refrigerant then flows back inside the home and is metered into the indoor coil by the flow control device. These devices are usually an expansion valve or a capillary tube (Figure 7-20). When the unit is in the cooling cycle, the evaporation of the refrigerant takes place at a fairly constant temperature, usually between 30° F. and 40° F. The condensing temperature will vary depending on the outdoor air temperature.

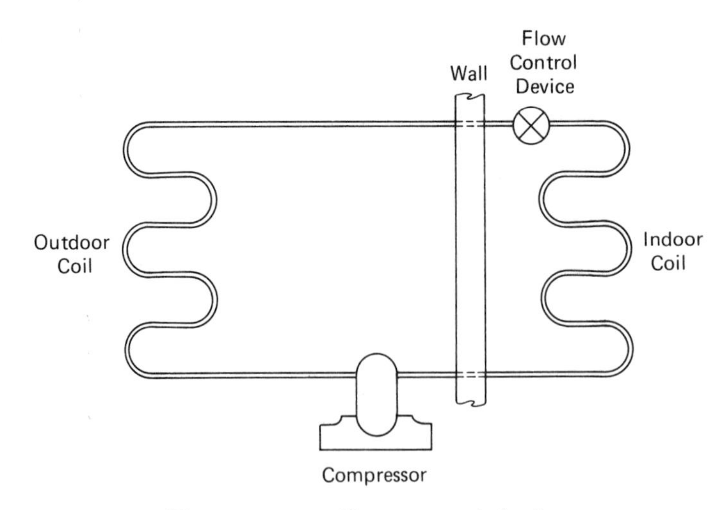

Figure 7-20. Flow control device.

Heating Cycle: The basic components used on a heat pump are similar to those used on a normal cooling system with some additional controls. Both the indoor and outdoor coils on a heat pump system will serve as either the evaporator or the condenser, depending on the direction of flow of the refrigerant.

HEAT PUMP REFRIGERANT SYSTEM COMPONENTS

For the heat pump system to operate automatically, certain other refrigeration system components are required. These extra components are required to reverse the flow of refrigerant or to protect the compressor.

Reversing Valve

A reversing valve, or four-way valve, is used to change the direction of the refrigerant flow. The direction of flow will depend on whether heating or cooling is needed indoors (Figure 7-21). The reversing valve is an important part of the heat pump cycle and it must operate smoothly, reliably, and efficiently. Valve manufacturers have designed reversing valves which meet these requirements (Figure 7-22).

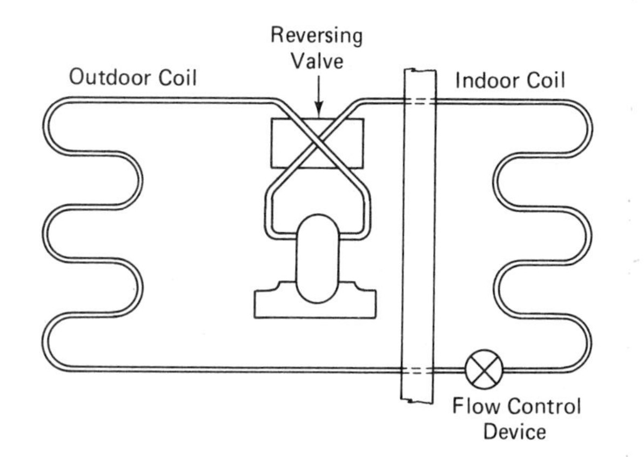

Figure 7-21. Reversing valve directing refrigerant.

Two types of reversing valves are in use on modern equipment. In the direct acting valve, the valve plunger is connected directly to the solenoid plunger. The second type is pilot operated. That is, a solenoid valve controls a flow of discharge gas which operates the valve plunger. When the solenoid is energized, the discharge gas is directed to one end of the plunger and forces it to the other end of the reversing valve. When the solenoid is de-energized, the discharge gas is directed to the other end of the plunger and forces it to the opposite end of the reversing valve (Figure 7-23). The reversing valve may be in either the heating or cooling position when de-energized; different manufacturers use different positions. The trend, however, is toward having the valve in the heating position when de-energized.

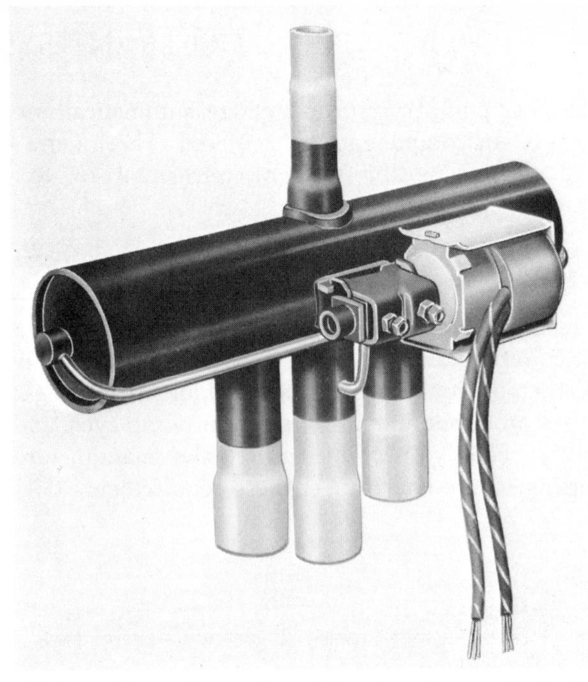

Figure 7-22. Reversing valve. (*Courtesy Ranco Controls Division*)

Suction Line Accumulator

A suction line accumulator is normally used to catch any liquid refrigerant and prevent it from entering the compressor (Figure 7-24). These devices are located in the suction line between the reversing valve and the compressor. If liquid refrigerant is allowed to enter the compressor, damage to the bearings, valves, and all wearing surfaces may occur.

Flow Control Devices

There are two flow control devices used on heat pump systems. They may be either expansion valves or capillary tubes. These devices are used to control the flow of refrigerant to the coil. One is needed for each coil so that the desired function may be obtained (Figure 7-25).

Check Valves

There are two check valves used to direct the flow of refrigerant through the proper metering device (Figure 7-26). They are installed so that they will open and allow the refrigerant to pass or close and prevent passage of the refrigerant during a particular function of the unit.

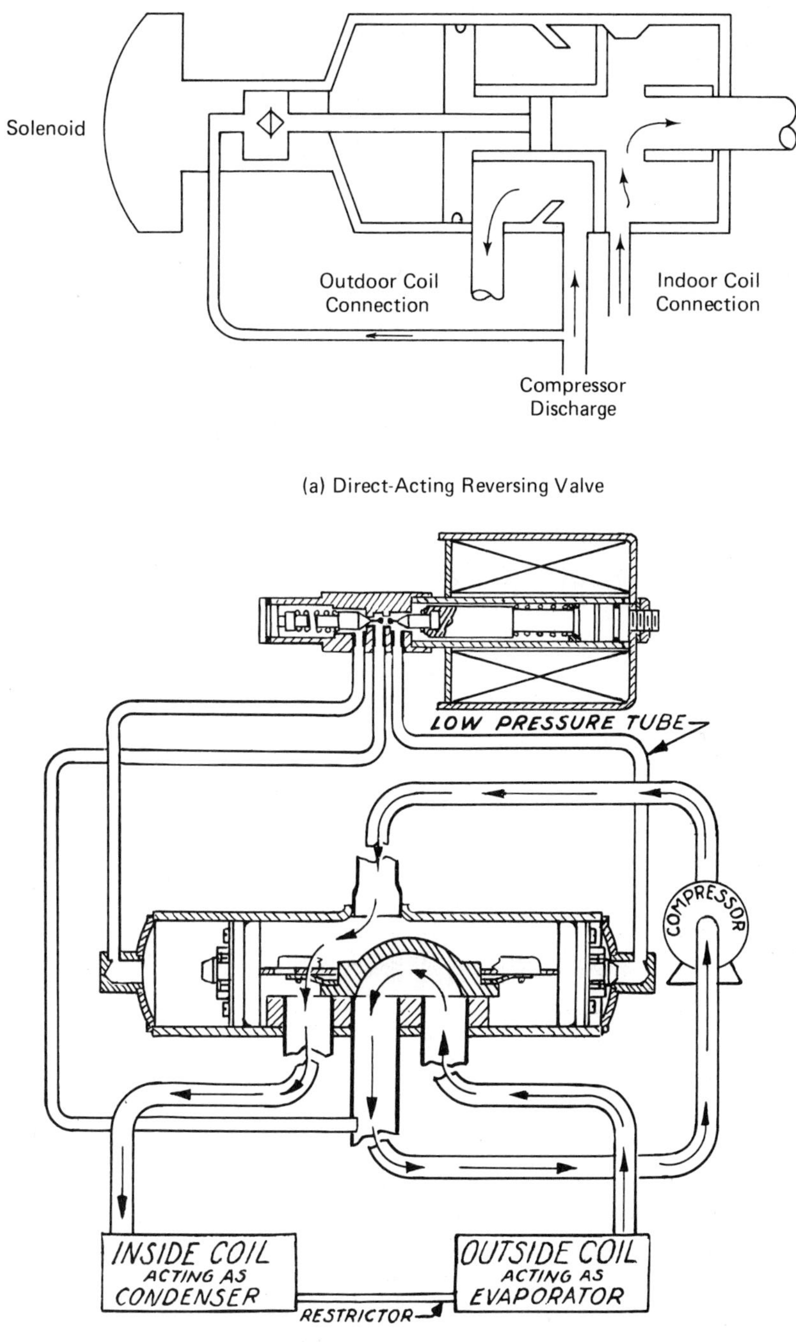

(a) Direct-Acting Reversing Valve

(b) Pilot-operated reversing valve

Figure 7-23. Reversing valve operating methods. (*Courtesy Ranco Controls Division*)

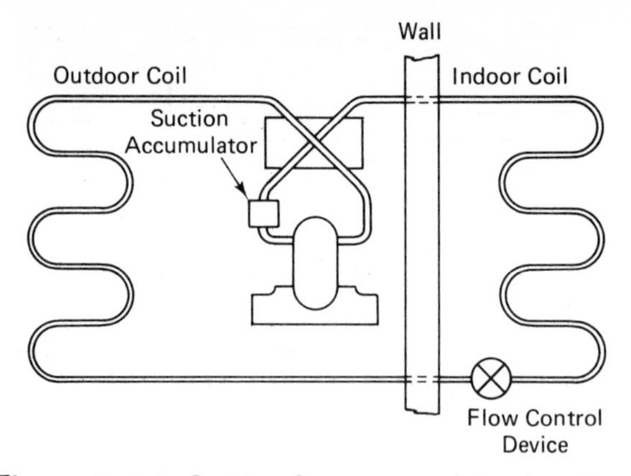

Figure 7-24. Suction line accumulator location.

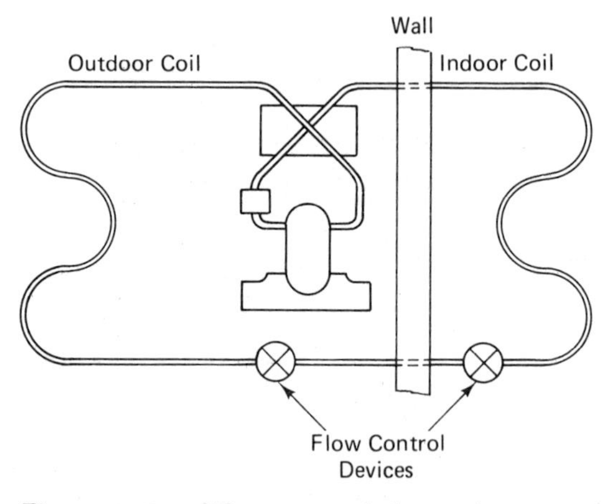

Figure 7-25. Flow control device location.

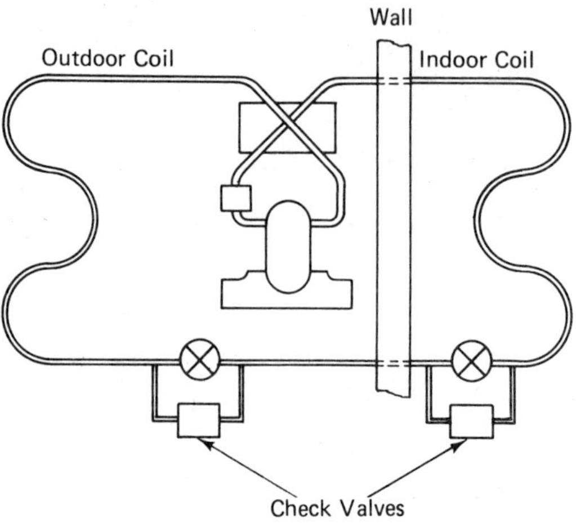

Figure 7-26. Check valve locations.

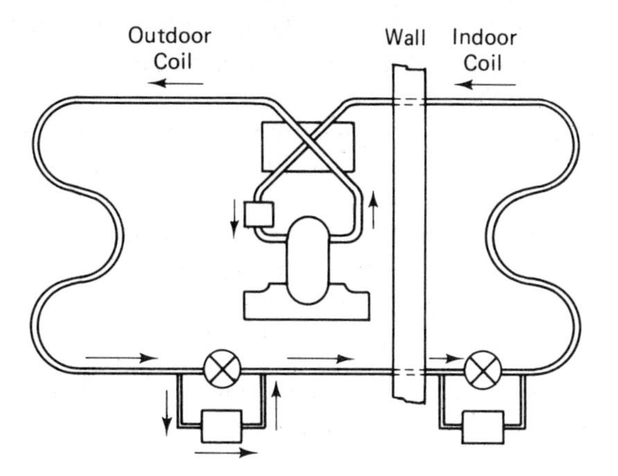

Figure 7-27. Heat pump in cooling cycle.

When the thermostat calls for cooling, the flow of refrigerant is in the same direction as during the air conditioning cycle. Liquid refrigerant is evaporated inside the indoor coil and the vapor flows to the compressor (Figure 7-27). The high-pressure, high-temperature discharge gas then goes to the outdoor coil where it is again condensed to a liquid. The liquid is caused to bypass the first flow control device by the check valves and is controlled by the second flow control device.

When the thermostat calls for heating, the reversing valve is actuated. The hot discharge gas from the compressor is directed to the indoor coil (Figure 7-28). The condensed liquid leaves the indoor coil and flows to the outdoor coil. It is again caused to by-pass the first flow control device because of the check valve operation and is controlled by the second flow control device. The liquid is evaporated in the outdoor coil

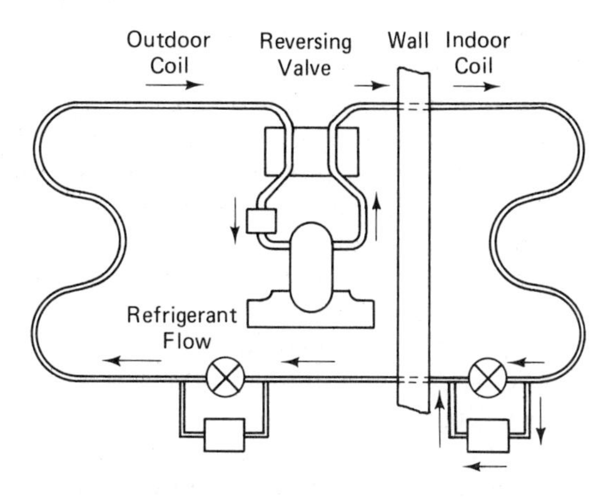

Figure 7-28. Heat pump in heating cycle.

where it absorbs heat from the outside air. This heat is then transferred into the home by the refrigerant when it gives up its latent heat of condensation in the indoor coil.

For these units to be accepted by the public, they must be as economical as possible. Every possible Btu must be squeezed from every kW input. For each kW put into a heat pump, more than 3,412 Btu are released. A properly designed heat pump can produce as much as 3 to 4 times the amount of heat released by resistance heating equipment. This is known as the coeffcient of performance (COP). The COP is determined by dividing the Btu output by the Btu paid for.

Example: If we have a 10 ton heat pump that has been found to be delivering 86,000 Btu, the COP would be:

$$\text{COP} = \frac{86{,}000 \text{ Btu}}{\text{09Bhp} \quad 2{,}545 \text{ Btu/hp}} = 3.3 \text{ total heat available.}$$

That is, the COP of this particular unit is 3.3. Or the Btu output is 3.3 times the Btu input. It should be noted here that the higher the COP, the greater the amount of heat available. In comparison, direct resistance heating has a COP of only 1.

When properly designed, installed and maintained, a heat pump can be superior to a gas heating unit. This can be attributed to a lower discharge air temperature that provides a more even temperature during the heating cycle.

However, these units do not operate satisfactorily in cold climates. Some manufacturers introduce an automatic means of stopping the compressor completely when the outdoor temperature drops to around 20° F. Below this temperature the COP has fallen below the economical point. Also, the outdoor coil must operate below the ambient air temperature in order to absorb heat in the evaporation process. At these low temperatures the suction pressure is low enough to cause damage to the compressor. Some units are equipped with an electric heater on the air inlet side of the outdoor coil to provide heat for this purpose (Figure 7-29). Because of the added expense in manufacturing and operation, this is done only in special cases.

Solar energy is also being used to assist heat pump equipment in performing its job. Three major methods are presently used for this solar assist of heat pump units. One method utilizes a second coil, or an added feature to the outdoor coil, through which water heated by solar collectors is pumped (Figure 7-30).

Another method uses the solar heated water pumped through a coil which is mounted downstream from the indoor coil (Figure 7-31). When

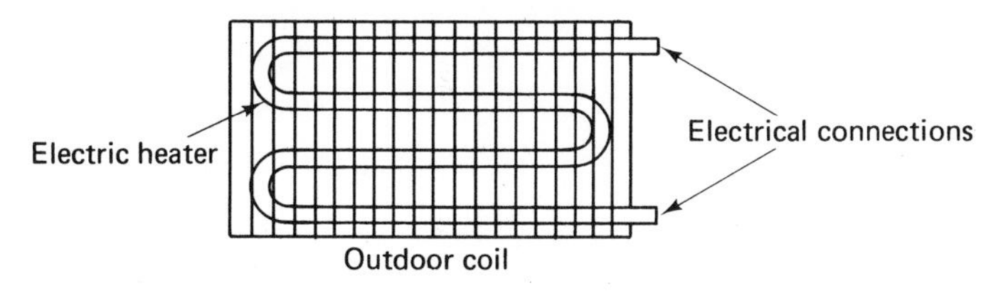

Figure 7-29. Supplemental outdoor heat for a heat pump.

the outdoor temperature falls below the balance point, the heat pump compressor is stopped by a thermostat and the solar heated water is pumped through the coil to provide heating for the home.

Still another method of using solar energy to aid in heat pump operation, uses the solar heated water to heat the home directly as in Figure 7-31, until the solar heated water temperatures falls to approximately 90° F. When this temperature is reached, the heat pump is brought into service. In this system the outdoor coil is immersed in the solar heated water (Figure 7-32). This is probably the most efficient method because heat can be extracted with water temperatures as low as 40° F. Also, most of the required heat is provided by the sun practically free of operational costs.

Defrost Cycle

When the heat pump is operating in the heating cycle, the refrigerant is being evaporated in the outdoor coil. When the temperature of the outdoor coil falls below 32° F, frost will appear on the coil surface. If

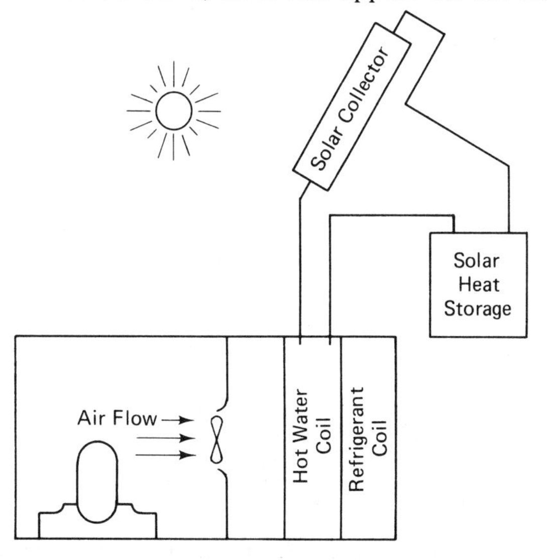

Figure 7-30. Solar assist heat pump (outdoor hot water coil).

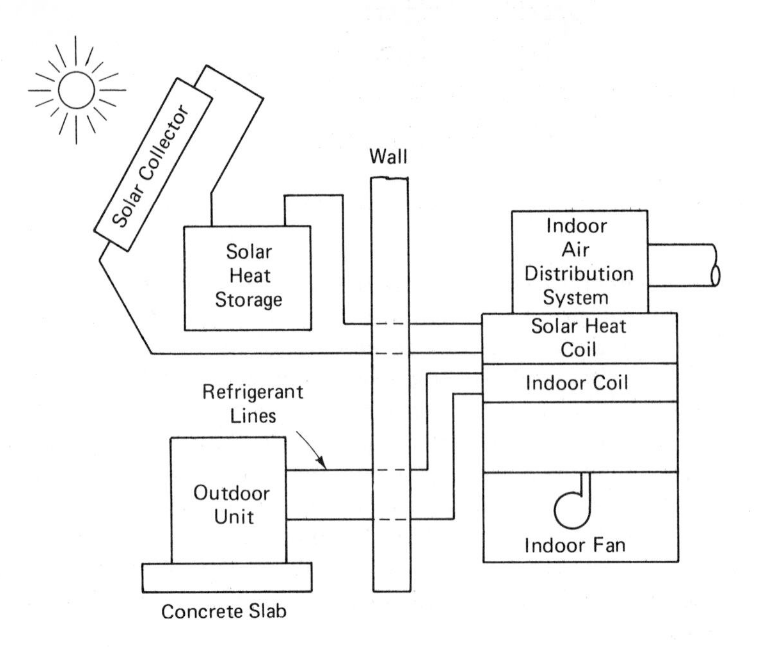

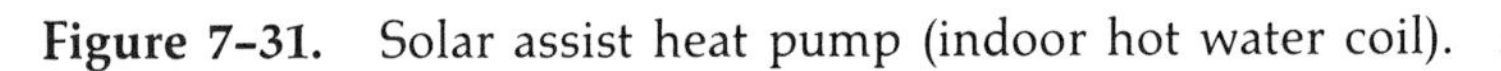

Figure 7-31. Solar assist heat pump (indoor hot water coil).

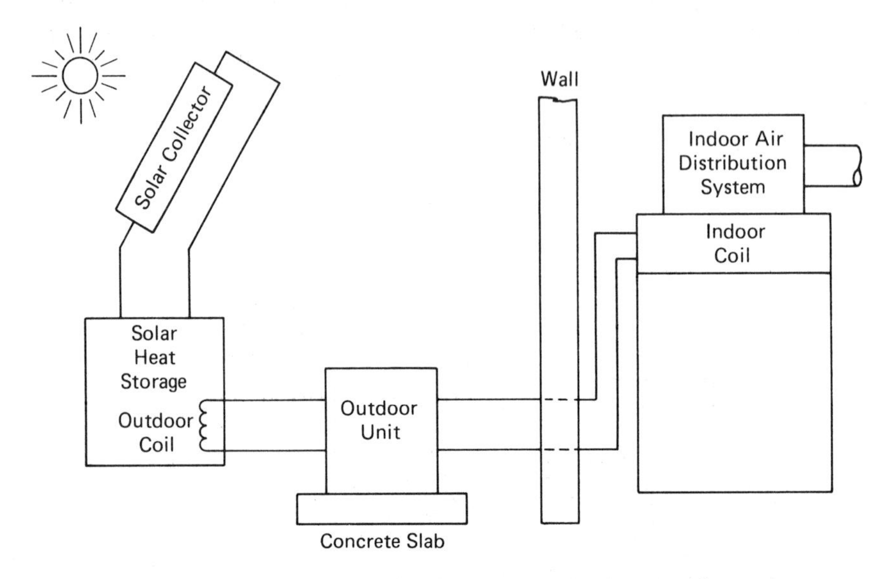

Figure 7-32. Solar assist heat pump (immersed outdoor coil).

this frosting is allowed to continue, the deposit of ice will gradually build up until the flow of air through the coil is restricted. This restriction of air will decrease the heat transfer and seriously affect the efficiency of the unit. Modern heat pump units, however, are designed to handle this frosting problem.

The most popular and effective method of removing the frost is to direct the hot gas from the compressor discharge through the outdoor coil for a period of time long enough to melt the accumulated ice. This change in the direction of the refrigerant flow is caused by operation of the reversing valve.

At the same time the reversing valve is actuated, the outdoor blower is stopped in order to reduce the flow of cold, outside air over the outdoor coil, thus allowing the frost to melt much faster. Since no heat is provided for the home during the defrost cycle, supplementary heat is supplied by an electric resistance heater during this period.

The methods used for frost detection are many and varied. The following are some of the more common methods.

Air Pressure: This method uses the air pressure differential across the outdoor coil (Figure 7-33). This control has a time delay to prevent false initiation. When the pressure approaches a given drop, the defrost cycle is initiated.

Time-Temperature: This control uses a temperature bulb attached to the outdoor coil to initiate the defrost cycle at the prescribed temperature. Sometimes incorporated in this control is a time clock that will only allow the defrost cycle to be initiated at given intervals. This method often has a temperature termination control with a temperature sensing device attached to the outdoor coil. At a present temperature, the defrost cycle will be terminated and the regular heating cycle maintianed (Figure 7-34).

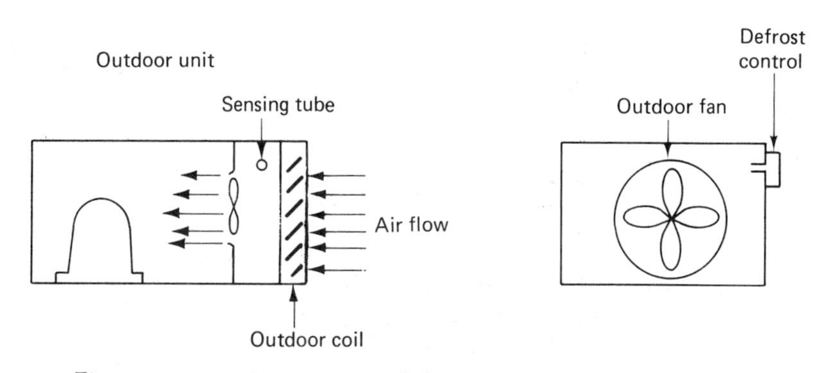

Figure 7-33. Air pressure defrost initiation.

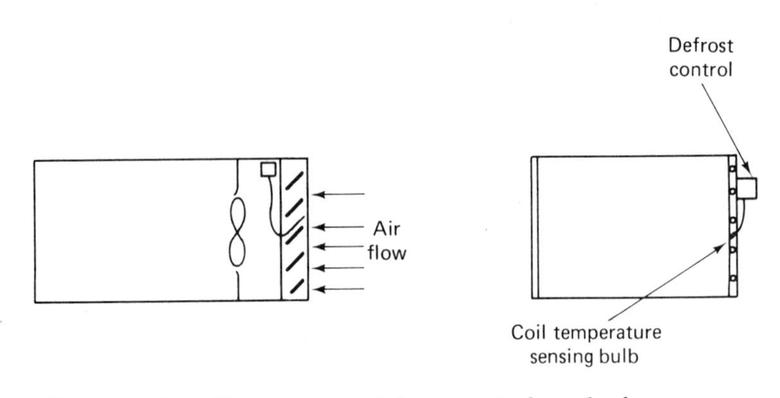

Figure 7-34. Temperature defrost control method.

Time Termination: This is usually used in conjunction with any of the previously mentioned initiation methods.

Pressure Termination: This type of termination uses the fact that the refrigerant pressure will increase when the frost is gone.

Some manufacturers incorporate electric strip heaters attached to the outdoor coil to speed up the defrosting of the coil. This heater is energized by the defrost initiation control (Figure 7-35).

A heat pump should not be used for reheating purposes, especially in the summer. The compressors used in these units have a higher compression ratio than normal refrigeration compressors. It is not difficult to obtain a 300° F discharge gas temperature. This is the maximum allowable temperature and is caused by a reduction in air over the indoor coil. A high indoor air temperature (90° F or higher) will require more air to be moved across the indoor coil.

Balance Point

The ability of a heat pump to provide heat decreases as the outdoor temperature goes down. At the same time, the need for heating or the heating load becomes greater when the outside temperature goes down.

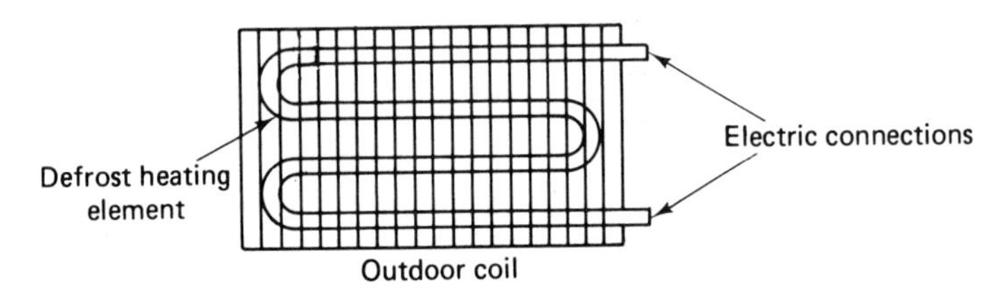

Figure 7-35. Electric defrost heater.

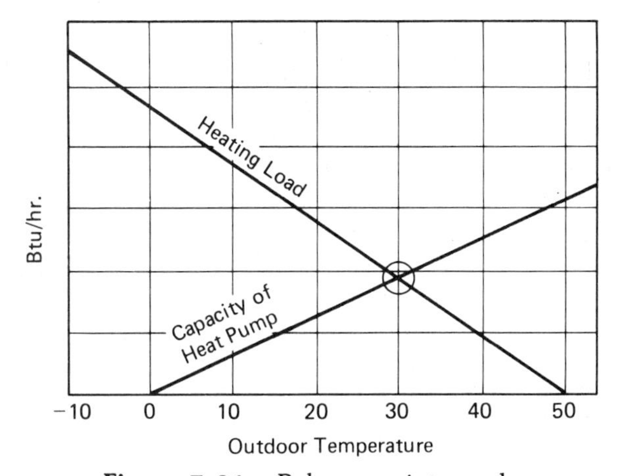

Figure 7-36. Balance point graph.

When these two changes—heat pump capacity and the heating load—are plotted on the same graph, the point of intersection is called the *balance point* (Figure 7-36). When the outdoor temperature falls below the balance point, the heat pump compressor can not supply all of the required heating and some form of additional heat is necessary. When the outdoor temperature is above the balance point, the heat pump can supply all of the required heating without help. Generally, the balance point temperature is between 25° F and 35° F. However, the rate of change in capacity will depend on each individual heat pump.

The balance point of a unit is affected by other sources of heat inside the conditioned space, such as electric lights, toasters, other appliances and the occupants. Solar heat is also a major factor. All of these sources of heat tend to lower the balance point.

For example, consider a heating season in Chicago (Figure 7-37). Since heating is not generally required in most buildings until the outside temperature falls below 65° F, Figure 7-37 shows a curve that starts at about 65° F at the bottom right. The highest percentage of days at each temperature is at the top center at about 32° F and drops rapidly to the bottom left at −20° F. This is a representation of the percentage of days at each temperature between these two extremes during the heating season in Chicago. The vertical dotted line in the figure is located at 25° F. Only 18% of the heating season in Chicago is below 25° F, while 82% is above 25° F. Thus, if the balance point of the heat pump is 25° F, the unit could supply the heating requirements for a major part of the cold weather for this location without any supplementary heat.

These units are sized to satisfy the cooling load, and supplementary heat is added to bring the unit to the desired heating capacity. Each manufacturer will provide charts indicating the cooling and heating capacity

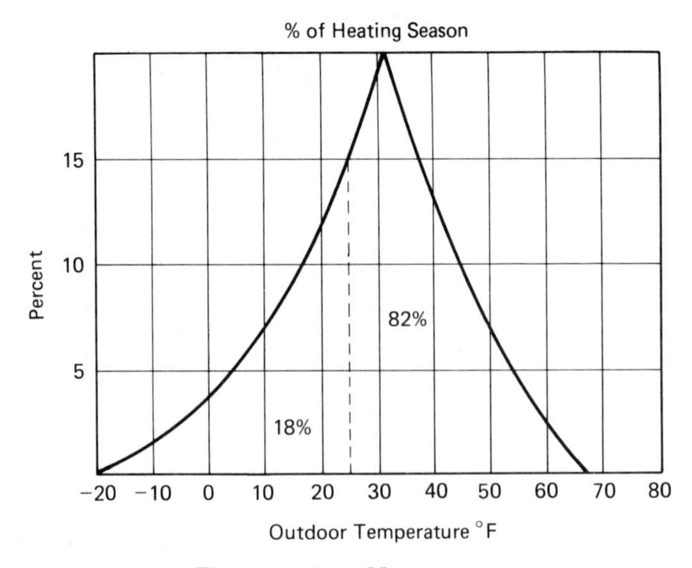

Figure 7–37. Heating curve.

in Btu for any given unit. From these charts and the calculated heat load, the amount of supplementary heat may be determined. This supplementary heat usually is in the form of duct heaters installed in the supply duct of the equipment (Figure 7–38).

Supplementary heat is brought to use only when the heat pump cannot supply the desired amount of heat. It is controlled from two points: the second stage on the indoor thermostat and the outdoor thermostats mounted under the eaves of the house. Only when both of these thermostats are demanding will the supplementary heat be energized.

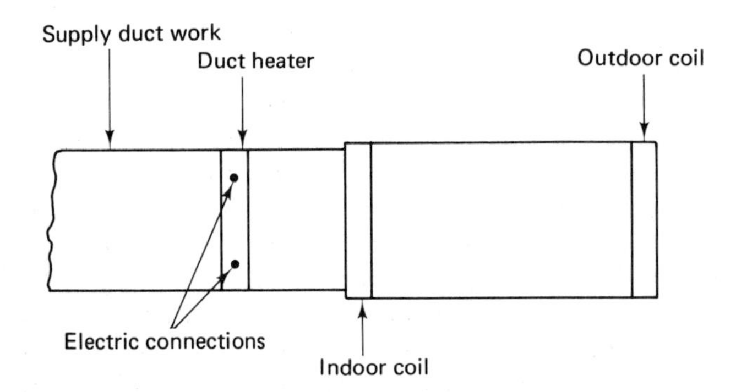

Figure 7–38. Location of supplementary duct heaters.

The design engineer determines the temperature setting of the outdoor thermostats. If these thermostats are set too high, the electricity bill will be excessive. On the other hand, if they are set too low, enough heat may not be supplied to the conditioned area. There must be a point reached where comfort and economy are maintained.

When adding refrigerant to these units or making any type of adjustment, the manufacturers' specifications must be followed. The trial and error method is too time-consuming and the desired conditions may never be reached because of the numerous overlaps and interlocks incorporated in these units. Therefore, for the sake of comfort and economy follow the proper specifications.

The foregoing explanation discussed the fundamentals of using heat pump systems. Since there is always a demand for more economical operation of heating equipment, some manufacturers are developing equipment to meet this requirement. Such a system, developed by Westinghouse Central Air Conditioning Division, is known as the Hi/Re/Li heat pump. The following discussion compares the Hi/Re/Li system and the conventional heat pump system in greater detail.

HI/RE/LI HEAT PUMP

Like any system, the Hi/Re/Li system uses a compressor, an indoor coil, an outdoor coil, and a reversing valve. However, there are no thermostatic expansion valves or capillary tubes used to meter the refrigerant to the coils (Figure 7-39). Instead, a single liquid subcooling control valve and a single manifold check valve is used to control and direct the flow of refrigerant on both cycles. Also, the accumulator has been replaced with an insulated combination accumulator-heat exchanger designed especially for use in the Hi/Re/Li circuit.

Hi/Re/Li Cooling Cycle

The performance of the Hi/Re/Li system during the cooling cycle is similar to that of a conventional air conditioning system, except that the Hi/Re/Li provides positive compressor protection against liquid slugging, even while operating at the extremely low suction pressures often caused by dirty filters, a common maintenance problem in most air conditioning systems (Figure 7-40). Notice that the reversing valve and the check valve are positioned to direct the flow of refrigerant so that the indoor coil functions as the evaporator and the outdoor coil functions as the condenser.

However, the subcooling control valve functions to control the flow of refrigerant "from" the condenser instead of "into" the evaporator.

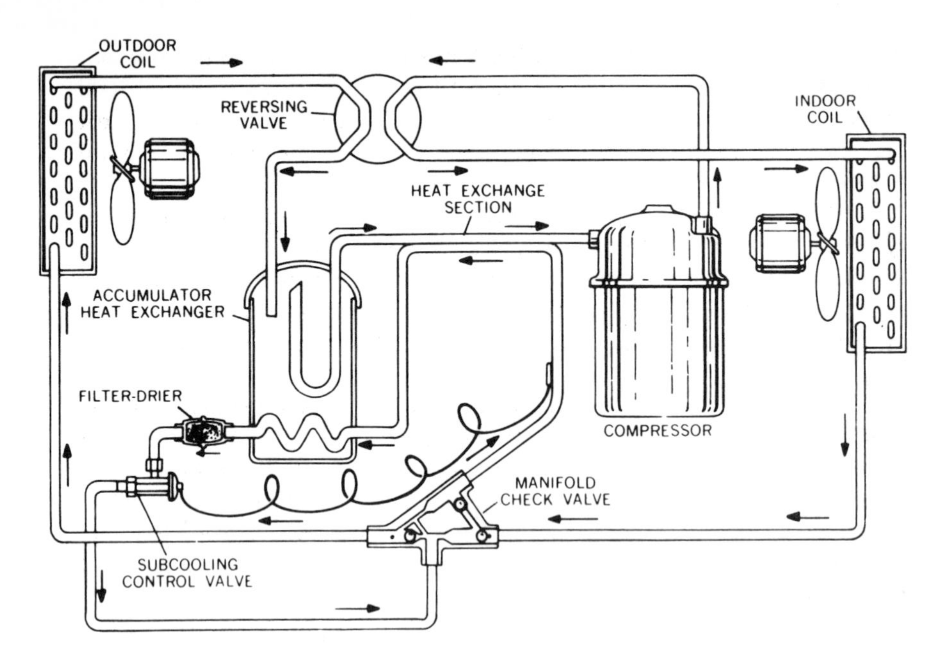

Figure 7-39. Hi/Re/Li system. (*Courtesy Westinghouse Central Residential Air Conditioning Division*)

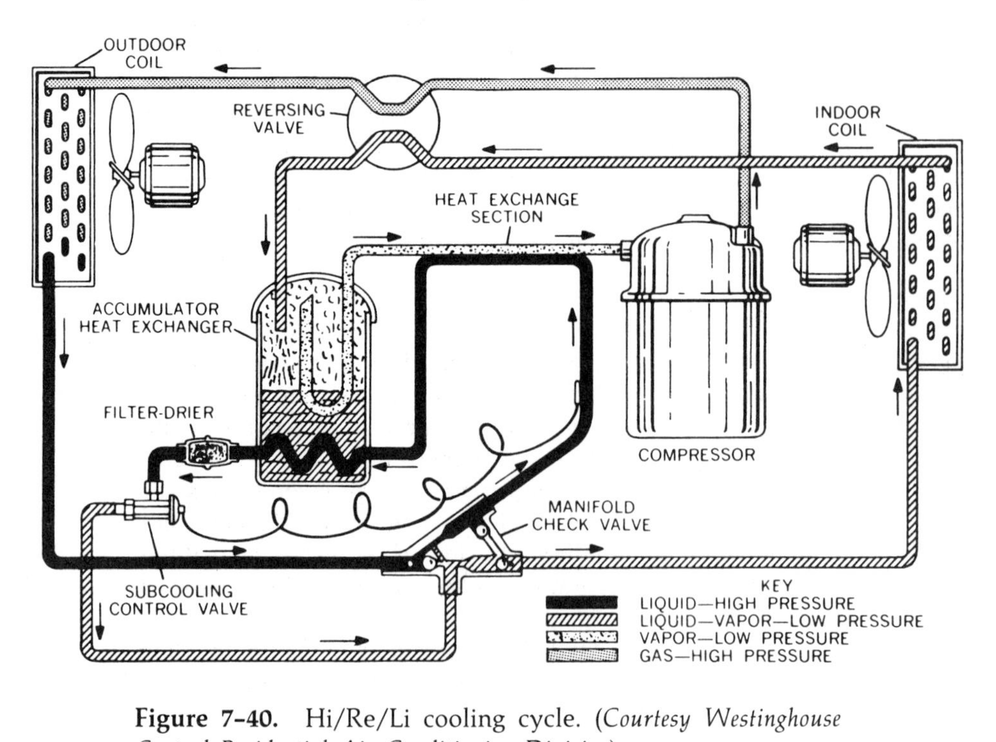

Figure 7-40. Hi/Re/Li cooling cycle. (*Courtesy Westinghouse Central Residential Air Conditioning Division*)

This feature assures practically complete drainage of the condenser, except for the required liquid seal, and provides a flooded evaporator for best system efficiency at all times during operation. The accumulator-heat exchanger provides additional subcooling, assures quick oil return to the compressor, and gives positive protection against liquid slugging. Although the cooling cycle performance of the Hi/Re/Li system is similar to a conventional system, the big differences show up most during the heating cycle.

Hi/Re/Li Heating Cycle

On the heating cycle, the reversing valve and the check valve change positions (Figure 7-41). The refrigerant flow is now directed so that the outdoor coil becomes the evaporator and the indoor coil is the condenser. The subcooling control valve is factory set to maintain 10° to 15° F of subcooling in the high pressure liquid leaving the condenser. From this point, the subcooled liquid refrigerant from the condenser flows through the manifold check valve to a heat exchange section of piping, or a section where the liquid and suction piping are in contact with each other. In this section, heat flows from the warm, subcooled liquid into the cold suction gas returning to the compressor. This heat exchange subcools the high pressure liquid another 10° to 15° F for a total

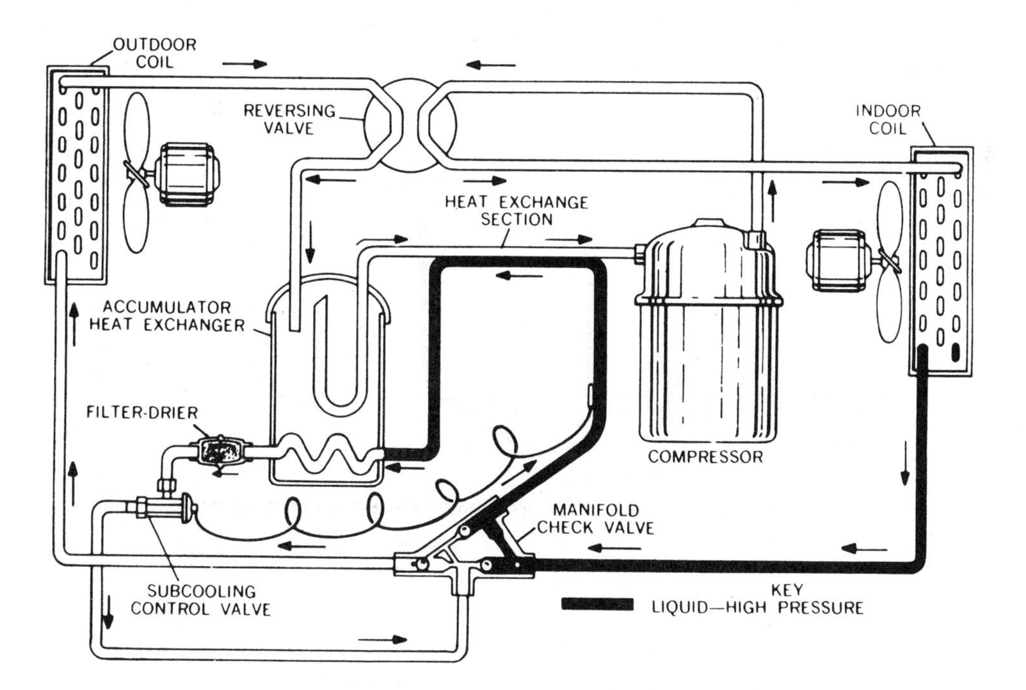

Figure 7-41. Hi/Re/Li heating cycle. (*Courtesy Westinghouse Central Residential Air Conditioning Division*)

of 20° to 30° F of subcooling at this point. At the same time, the cold suction gas is superheated 2° or 3° F for final assurance that the refrigerant is absolutely dry before it enters the compressor. (That is, no liquid refrigerant enters the compressor.) Although the suction gas is essentially dry before it reaches this point, superheating it a few degrees is simply an extra step to assure protection against liquid slugging of the compressor. Compare the difference of 10° to 25° F of superheat in a conventional heat pump to 2° to 3° F in the Hi/Re/Li system.

The 20° F subcooled liquid next flows through the coil of the accumulator heat exchanger where the high pressure liquid is further subcooled another 50° to 60° F by the cold low pressure refrigerant surrounding the coil. The high pressure liquid leaves the accumulator-heat exchanger coil, now subcooled a total of 70° to 80° F, and flows through a filter drier to the subcooling control valve (Figure 7-42). As the liquid passes through the subcooling control valve, its pressure is reduced to evaporator pressure and it flows into the evaporator. Remember, in contrast to the thermal expansion valve or the capillary tube of conventional heat pump systems, the subcooling control valve in the Hi/Re/Li system controls the liquid refrigerant flow "from" the condenser, not the liquid flow "into" the evaporator. As a result, the evaporator is flooded with

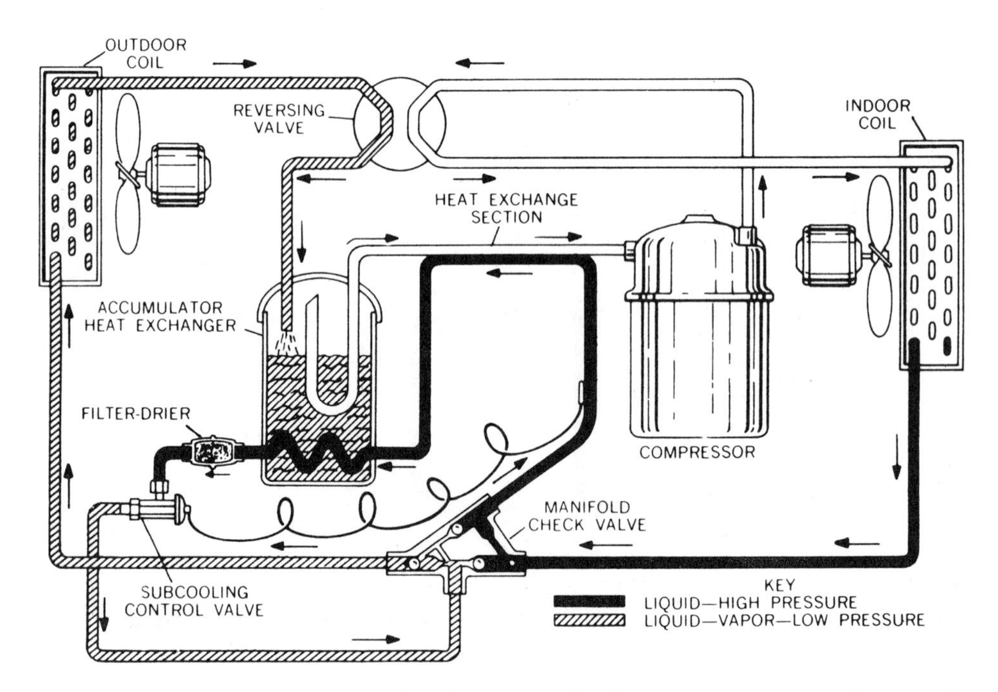

Figure 7-42. Subcooling control valve. (*Courtesy Westinghouse Central Residential Air Conditioning Division*)

subcooled liquid at all times during operation. This means that every square inch of coil surface is used for effective cooling, flash gas is practically eliminated and the subcooled liquid provides faster, more efficient heat transfer. Since the subcooled liquid can absorb more heat than the evaporator coil is capable of transferring, all of the liquid will not be evaporated and the mixture of saturated gas and liquid flows on into the shell of the accumulator-heat exchanger. The elimination of flash gas, the flooded coil, the subcooled liquid, and no evaporator superheat all mean more heat pickup and better heat transfer efficiency than any other heat pump available.

In the accumulator-heat exchanger, the mixture of vapor and unused liquid from the evaporator is separated. The saturated gas is drawn into the compressor suction line where it is superheated about 2° to 3° F before entering the compressor. The excess liquid is stored in the accumulator shell surrounding the coil (Figure 7-43). Some of this liquid evaporates as it subcools the warm liquid from the condenser; the remainder stays in the shell until it is needed in the system. Only part of the total charge required for cooling is used during the heating cycle, and the accumulator is sized to hold the extra charge of refrigerant. The

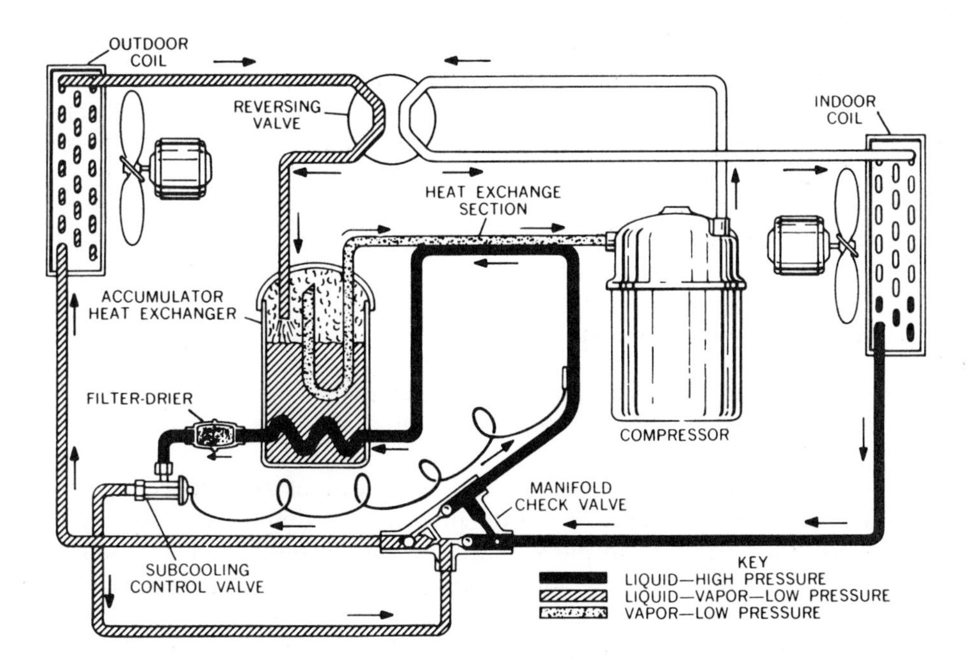

Figure 7-43. Hi/Re/Li accumulator heat exchanger operation. (*Courtesy Westinghouse Central Residential Air Conditioning Division*)

U-tube of the suction line inside the accumulator shell is designed to assure positive return of the oil which normally circulates during operation of the system. The bottom of the U-tube is equipped with a filtered orifice through which a mixture of oil and refrigerant is drawn into the compressor suction line by the high velocity suction gas, which creates a lower pressure as it passes over the orifice. The refrigerant in the mixture is dried up as it passes through the heat exchanger section, and the oil flows into the compressor oil reservoir. The unique design of the accumulator-heat exchanger is a major factor in the superior reliability of the Hi/Re/Li system. It assures positive compressor protection against liquid slugging by permitting only dry, but essentially saturated, gas to enter the compressor. This promotes a cooler running compressor and considerably lower discharge temperatures than can be found in any conventional heat pump. In addition, positive return of undiluted oil provides proper compressor lubrication under any operating condition. The result is at least three times more reliability.

The compressor picks up the dry, but essentially saturated, refrigerant vapor from the top of the accumulator-heat exchanger shell and discharges the compressed gas into the condenser. As the high pressure refrigerant vapor passes through the condenser coil, it gives up its latent heat of condensation to the atmosphere and condenses to a high pressure liquid ready to begin another cycle through the system (Figure 7-44). In the condenser another important advantage is gained over the conventional system. The use of the subcooling control valve to maintain a constant 10° to 15° F of subcooling in the liquid leaving the condenser keeps the condenser essentially drained of liquid at all times during system operation. This permits full use of the entire condensing surface for effective condensing. More efficiency and more condensing capacity is obtained for every square foot of condenser surface.

Hi/Re/Li System Components

An examination of the individual components of the Hi/Re/Li system shows how they make this system efficient.

Compressor: The compressor is the heart of any refrigeration or air conditioning system. It creates the pressure difference necessary to cause the refrigerant to move from one part of the system to another. This is done by raising the low pressure of the vapor in the evaporator to the point where it will condense to a liquid at ordinary atmospheric temperatures. The Westinghouse Model CD compressor is used in the Hi/Re/Li system (Figure 7-45). It is equipped with special low lift, high capacity valves designed for higher volumetric efficiency under low temperature operating conditions. The oversized oil reservoir is combined

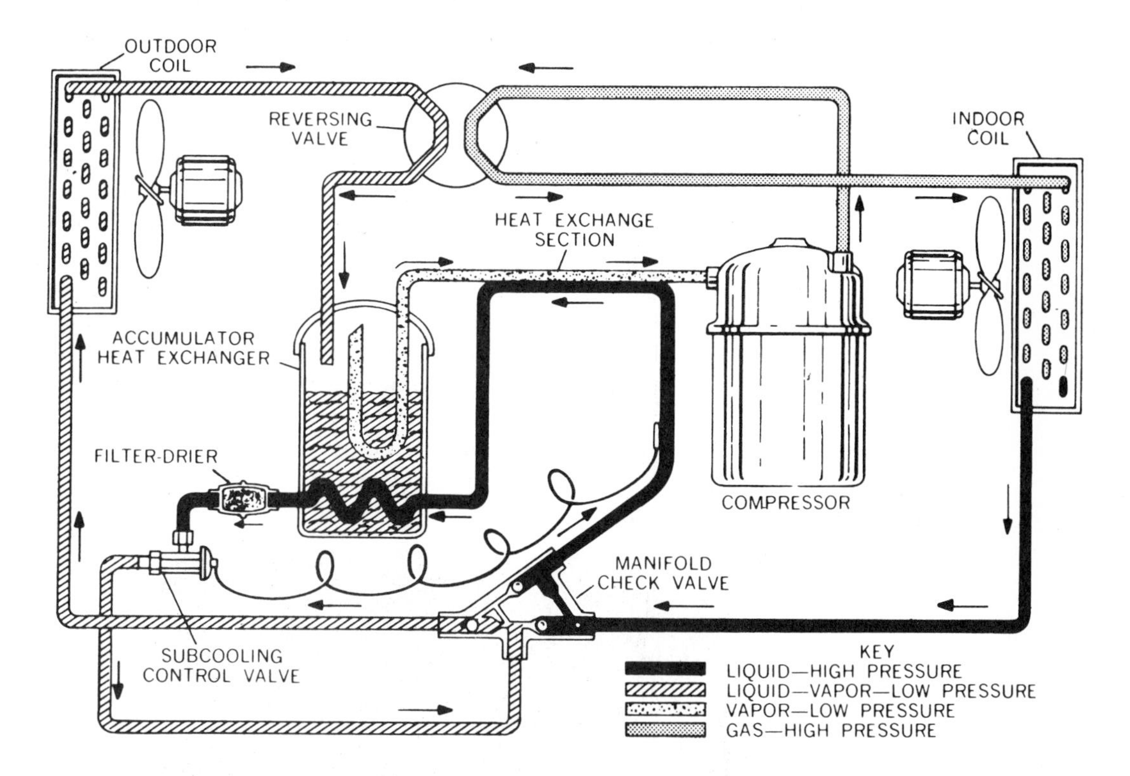

Figure 7-44. Hi/Re/Li condensing cycle. (*Courtesy Westinghouse Central Residential Air Conditioning Division*)

Figure 7-45. Westinghouse Hi/Re/Li compressor. (*Courtesy Westinghouse Central Residential Air Conditioning Division*)

with the positive action, self-priming oil pump and a unique gas venting system all designed to assure better lubrication during low temperature operation.

Coils: The coils have a large face area and are engineered for both condensing and evaporating functions (Figure 7–46). The staggered tube arrangements as well as the fin configuration and spacing are precisely engineered for uniform distribution of air flow through the coil for best heat transfer characteristics and high capacity. Both coils are carefully sized to provide balanced system performance and to obtain maximum efficiency during both the heating and cooling cycles. In the conventional heat pump system, refrigerant distribution to the indoor coil presents serious difficulties, which all kinds of elaborate distributors have been developed to overcome. This problem does not exist in the Hi/Re/Li system. In fact, there are no refrigerant distributors. In the Hi/Re/Li system, the evaporator coil is flooded with refrigerant and each coil pass is fed from a common liquid header.

Figure 7–46. Hi/Re/Li coils. (*Courtesy Westinghouse Central Residential Air Conditioning Division*)

Subcooling Control Valve: The liquid subcooling control valve controls the refrigerant flow in the Hi/Re/Li system. The same valve controls the flow during both the heating and cooling cycles (Figure 7-47) taking the place of both flow control devices required on a conventional heat pump. Although the subcooling control valve resembles a thermostatic expansion valve, its location, function, and operation are entirely different. To understand the principle of the Hi/Re/Li system, it is important to have a thorough knowledge of this valve. The valve assembly consists of a spring-balanced valve and a remote sensing element which is pressure-sensitive to the temperature sensed by the bulb. The

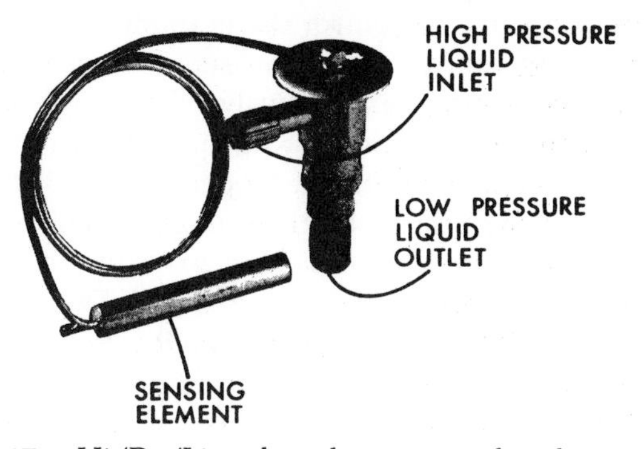

Figure 7-47. Hi/Re/Li subcooling control valve. (*Courtesy Westinghouse Central Residential Air Conditioning Division*)

valve is installed in the high pressure liquid line with its sensing bulb clamped to the liquid line leaving the condenser (Figure 7-48) instead of to the suction line, as is a thermostatic expansion valve. To control the liquid subcooling, in this location the subcooling control valve is actuated by the temperature of the liquid refrigerant leaving the condenser

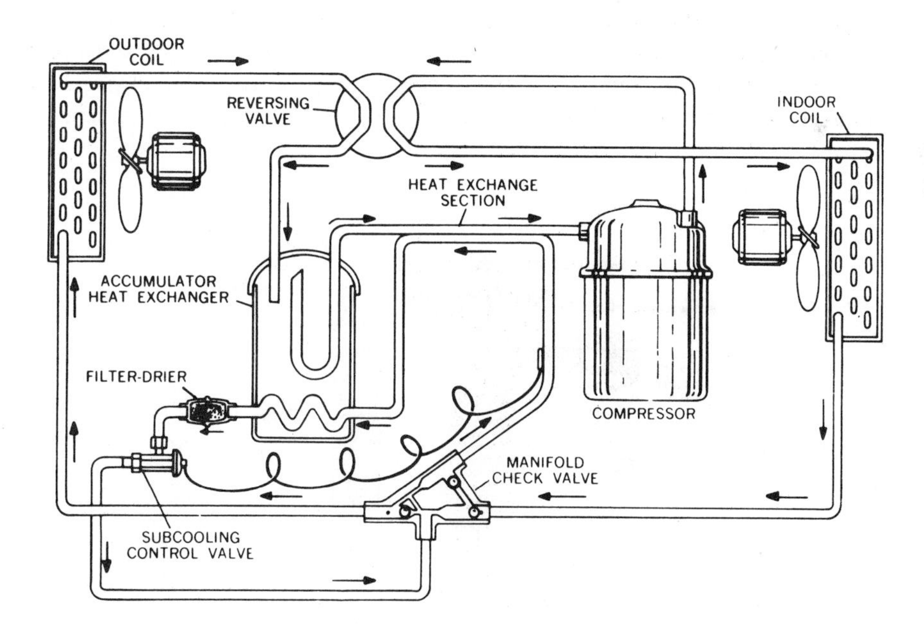

Figure 7-48. Subcooling control valve location. (*Courtesy Westinghouse Central Residential Air Conditioning Division*)

instead of by the temperature of the suction gas which actuates the thermostatic expansion valve to control the evaporator superheat.

The subcooling control valve consists of a power element and the internally equalized, spring-balanced valve. The power element contains a diaphragm which flexes to close or open the control valve as more or less pressure is exerted upon it. The upper surface of the diaphragm is exposed to the pressure exerted by the remote sensing element while the underside is exposed to the condenser pressure through an internal equalizing port in the valve body (Figure 7-49). The valve is factory set to maintain 10° to 15° F of subcooling in the liquid leaving the condenser. During operation at any condensing pressure, if the liquid temperature rises, the control valve bulb will sense the warmer temperature, increase the pressure on the upper side of the diaphragm and cause the valve to move toward the closed position. As the valve closes, the liquid refrigerant backs up slightly in the condenser until the liquid temperature is reduced to correspond to 10° to 15° F subcooling or 10° to 15° F degrees below saturation. On the other hand, if the liquid refrigerant leaving the condenser is subcooled more than 10° to 15° F, the sensing bulb of the control valve senses the drop in temperature and exerts less pressure on top of the diaphragm. Thus the condenser pressure under the diaphragm, assisted by the balance spring, is allowed to move the valve toward the open position, thereby increasing the flow of liquid refrigerant from the condenser and reducing the degree of subcooling. Unlike the thermostatic expansion valve which opens when its sensing bulb feels a warmer temperature, the subcooling control valve closes as its bulb temperature rises. If the diaphragm should rupture or the sens-

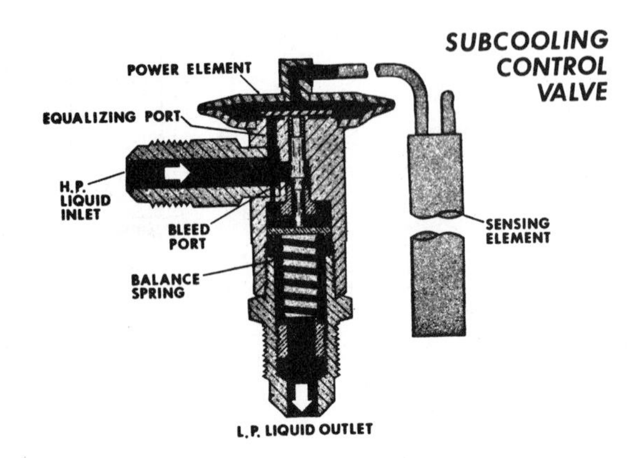

Figure 7-49. Subcooling control valve cutaway. (*Courtesy Westinghouse Central Residential Air Conditioning Division*)

ing element loses its charge, liquid refrigerant simply flows through the evaporator into the accumulator shell, which is large enough to hold the refrigerant charge and protect the compressor from liquid flood-back.

The subcooling control valve contains a small bleed port which by-passes the valve between the liquid inlet from the condenser and the outlet to the evaporator. This tiny port is not only essential to valve operation; it is extremely important to the success of the overall system performance in three different ways (Figure 7-50). First, the bleed port assures quick equalization of the suction and discharge pressures after shutdown on the cooling cycle. On the heating cycle, pressure equalization after shutdown is accomplished by simply de-energizing the reversing valve and allowing it to return to the cooling position. Pressure equalization is important because the compressor is not required to start against a high pressure differential.

Second, the bleed port provides additional assurance of proper oil return during all types of operating conditions (Figure 7-51). Even when the liquid subcooling valve may be closed, a small but sufficient amount of refrigerant is allowed to by-pass the control valve to assure adequate oil return during both the heating and cooling cycles.

Third, the bleed port assures that the valve quickly responds to changing system conditions. For example, at the instant the system is reversed from heating to cooling for defrosting the outdoor coil, the high side pressure tends to build up slowly under the valve diaphragm, and the valve would be slow in opening if it were not for the bleed port (Figure 7-52). The bleed port permits enough refrigerant to by-pass the valve and flow into the system during this period, causing the pressure to build up much more rapidly and allowing the control valve to recover full control in a very short time.

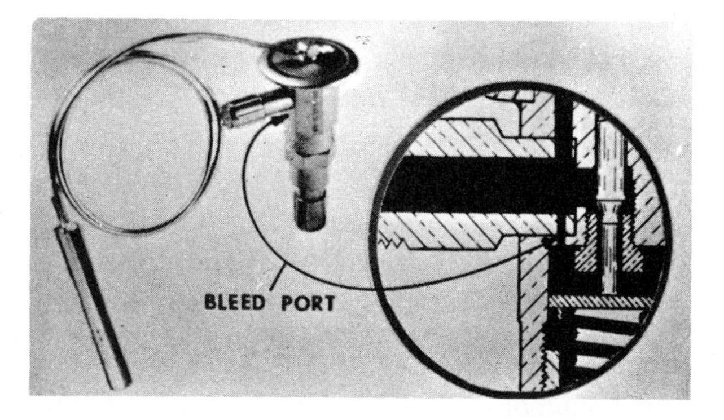

Figure 7-50. Subcooling control valve bleed port for pressure equalization. (*Courtesy Westinghouse Central Residential Air Conditioning Division*)

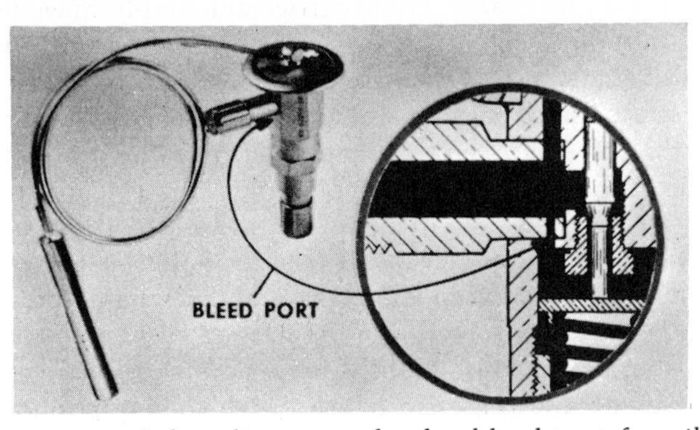

Figure 7-51. Subcooling control valve bleed port for oil return. (*Courtesy Westinghouse Central Residential Air Conditioning Division*)

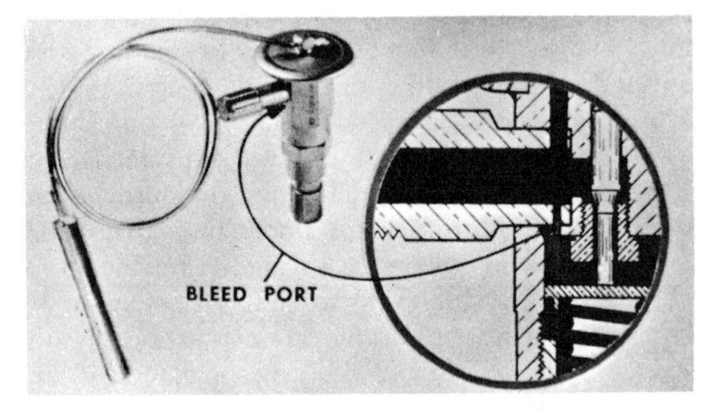

Figure 7-52. Subcooling control valve bleed port for quick recovery. (*Courtesy Westinghouse Central Residential Air Conditioning Division*)

Reversing Valve: The reversing valve is the same electrically actuated four-way solenoid valve which is used in many heat pump applications. It mechanically changes the flow of refrigerant to place the system in either the heating or cooling cycle (Figure 7-53). When the system calls for cooling, the solenoid is de-energized and the valve directs the compressor discharge gas to the outdoor coil, and the flow of low pressure liquid-vapor mixture from the indoor coil is directed into the shell of the accumulator-heat exchanger. In other words, during the cooling cycle the outdoor coil functions as the condenser; the indoor coil functions as the evaporator. When heating is required, the solenoid is energized and the valve reverses its position. The compressor discharge gas is directed to the indoor coil, and the liquid-vapor mixture from the

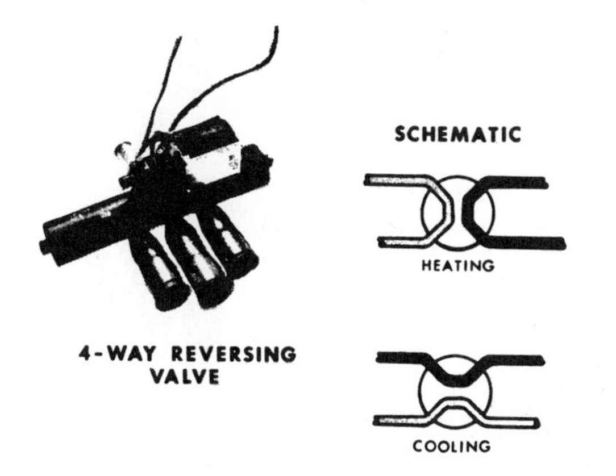

Figure 7-53. Reversing valve. (*Courtesy Westinghouse Central Residential Air Conditioning Division*)

outdoor coil flows through the accumulator-heat exchanger shell. The outdoor coil now functions as the evaporator. The indoor coil functions as the condenser. In the Hi/Re/Li system, the reversing valve has no direct effect upon the flow of subcooled liquid refrigerant from the condenser to the evaporator. However, the changes in system pressure created when the reversing valve changes position cause the manifold check valve to direct the flow of subcooled liquid to the proper coil.

The reversing valve is made up of two pistons connected to a ported sliding block. It is actuated by a solenoid operated plunger which opens and closes gas ports to build up pressure and move the piston assembly either to the right or to the left, depending on whether the thermostat is calling for heating or cooling (Figure 7-54). During the cooling cycle, the solenoid is de-energized, allowing the plunger to drop and close the lower vent line to the right side of the assembly. High pressure discharge gas seeps through the bleed port in the right hand piston to build up pressure behind the piston. Since the left hand piston is equalized with suction pressure, the piston assembly moves to the left as shown. In this position, the compressor discharge gas is directed to the outdoor coil and the compressor reduces the pressure in the indoor coil through the accumulator-heat exchanger. The outdoor coil now serves as the condenser, and the indoor coil serves as the evaporator.

When the thermostat calls for heating, the solenoid is energized and the plunger is pulled up to cover the top vent line on the left side of the piston assembly (Figure 7-55). The discharge gas pressure now seeps through the small port in the left hand piston and the gas pressure builds up in back of the piston. With gas pressure on the left hand piston and the right hand piston vented to the suction pressure, the piston assembly

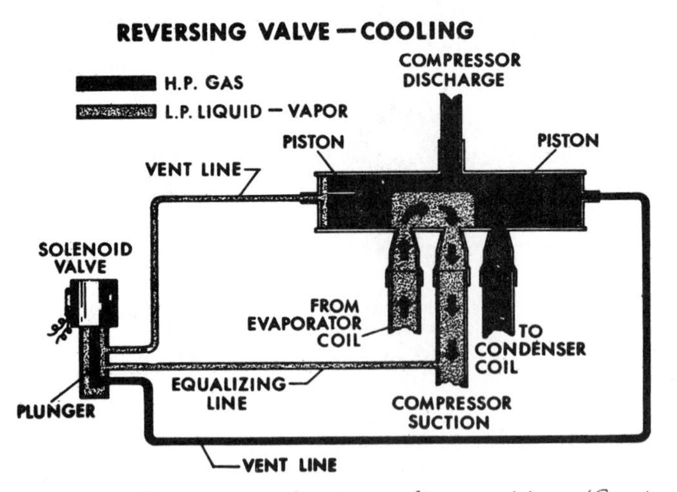

Figure 7-54. Reversing valve in cooling position. (*Courtesy Westinghouse Central Residential Air Conditioning Division*)

moves to the right. In this position, the valve directs the compressor discharge gas to the indoor coil, and the compressor removes the gaseous refrigerant from the outdoor coil through the accumulator shell. In this situation, the outdoor coil serves as the evaporator and the indoor coil serves as the condenser.

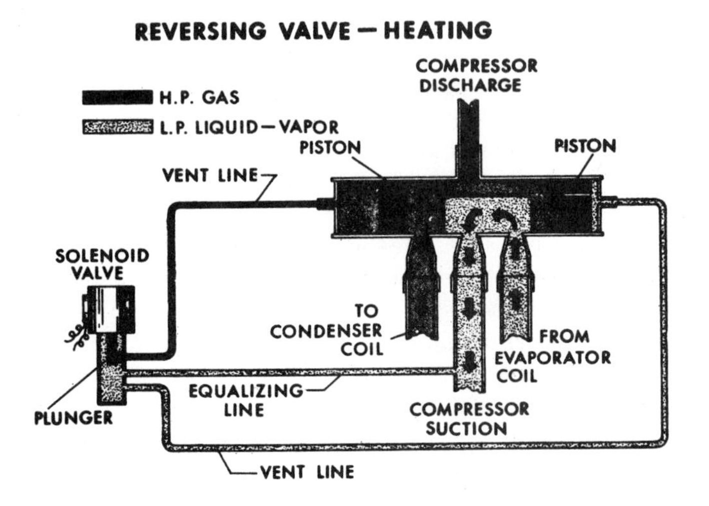

Figure 7-55. Reversing valve in heating position. (*Courtesy Westinghouse Central Residential Air Conditioning Division*)

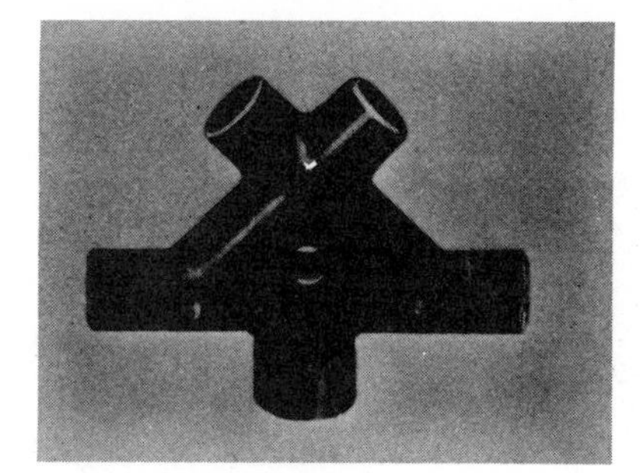

Figure 7-56. Manifold check valve. (*Courtesy Westinghouse Central Residential Air Conditioning Division*)

Manifold Check Valve: The manifold check valve is a simple device which serves a vital function in the Hi/Re/Li system. It is a combination of four separate ball check positions in a single housing (Figure 7-56). The balls are positioned by changes in the system pressure created when the electrically operated reversing valve changes its position from heating to cooling or cooling to heating. Its purpose is to direct the flow of liquid refrigerant from the condenser through the coil of the accumulator-heat exchanger and the subcooling control valve to the evaporator coil during both the heating and cooling cycles. The manifold check valve is simple and requires no adjustment.

When the reversing valve moves to the cooling position, the outdoor coil becomes the condenser and the indoor coil becomes the evaporator. The system pressures move the balls to cover the check valve ports (Figure 7-57). The high pressure liquid from the outdoor coil is directed to the coil of the accumulator-heat exchanger, and the low pressure liquid is directed to the indoor coil.

When the thermostat calls for heat, the reversing valve moves to reverse the flow of refrigerant vapor. The indoor coil is now the condenser; the outdoor coil is the evaporator. The system pressures are reversed, as are the positions of the balls in the manifold check valve. However, the flow of liquid remains the same with respect to the condenser and evaporator. The high pressure liquid from the condenser still flows to the coil of the accumulator-heat exchanger; the low pressure liquid from the subcooling valve still flows to the evaporator. Only the path of liquid through the check valve is changed. The actual liquid refrigerant flow remains the same.

The small bleed port between the check valve passages (Figure 7-58) is used only during the cooling cycle and serves two important purposes. First, it assists the bleed port in the liquid subcooling valve to quickly equalize the system pressures on shutdown. Second, it helps to provide a minimal flow of refrigerant during operation at extremely low

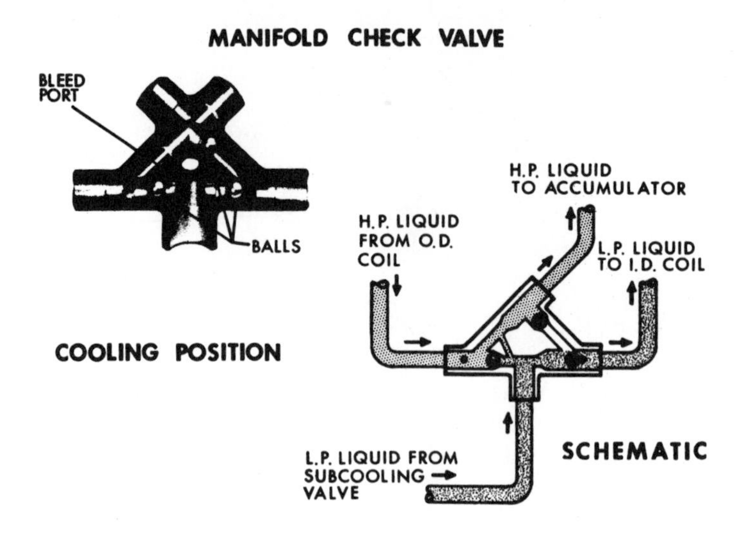

Figure 7-57. Manifold check valve in cooling position. *(Courtesy Westinghouse Central Residential Air Conditioning Division)*

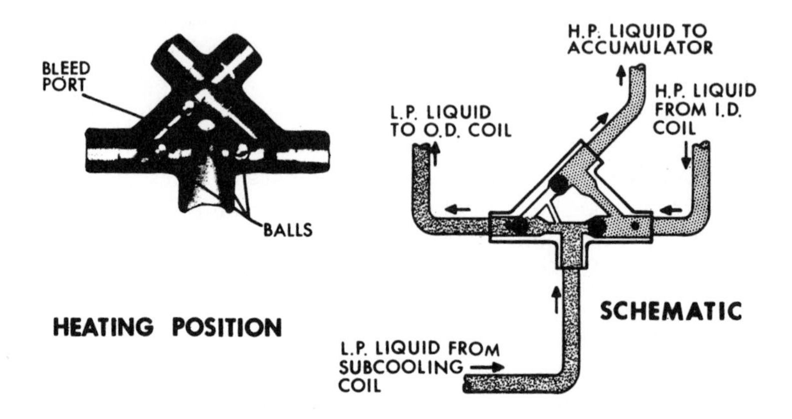

Figure 7-58. Manifold check valve in heating position. *(Courtesy Westinghouse Central Residential Air Conditioning Division)*

load conditions, an important factor in maintaining a cool running compressor and assuring adequate oil return from proper compressor lubrication.

Accumulator-Heat Exchanger: Especially designed for use in the Hi/Re/Li system (Figure 7-59), this component serves three important functions. First, it adds subcooling to the high pressure liquid on its way to the evaporator: approximately 35° to 45° F subcooling during the cooling cycle and approximately 50° to 60° F during the heating cycle. Second, it provides a positive separation of the low pressure liquid and vapor from the evaporator so that only dry, practically saturated, gas reaches the compressor suction. Third, it assures positive oil return to the compressor at all times during operation. The accumulator-heat exchanger assembly consists of a steel shell containing a flat, circular liquid subcooling coil and a U-shaped suction connection with a filtered oil return orifice. The entire shell is insulated.

During operation in either the cooling or heating cycle, the mixture of low pressure vapor and unevaporated liquid from the evaporator dumps into the accumulator-heat exchanger shell (Figure 7-60). The saturated vapor separates in the upper part of the shell ready for return to the compressor. The cold liquid refrigerant accumulates in the bottom of the shell around the liquid subcooling coil. The warm, high pressure liquid from the condenser flows through the subcooling coil on its way to the subcooling valve and the evaporator. As the refrigerant leaves the condenser, the high pressure liquid is already subcooled 10° to 15° F. It

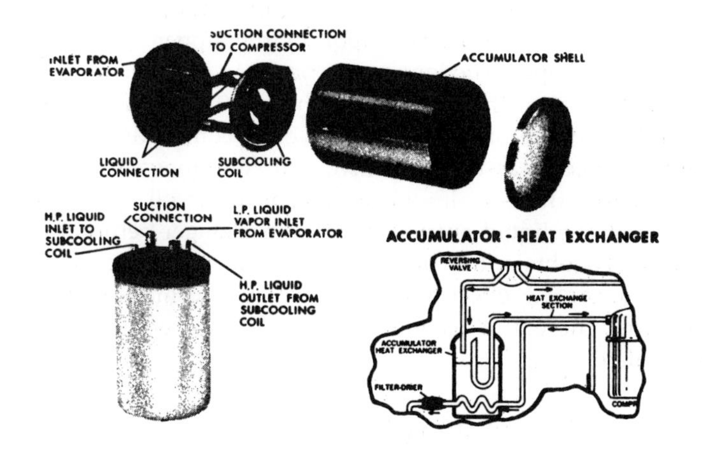

Figure 7-59. Accumulator-heat exchanger exploded view. *(Courtesy Westinghouse Central Residential Air Conditioning Division)*

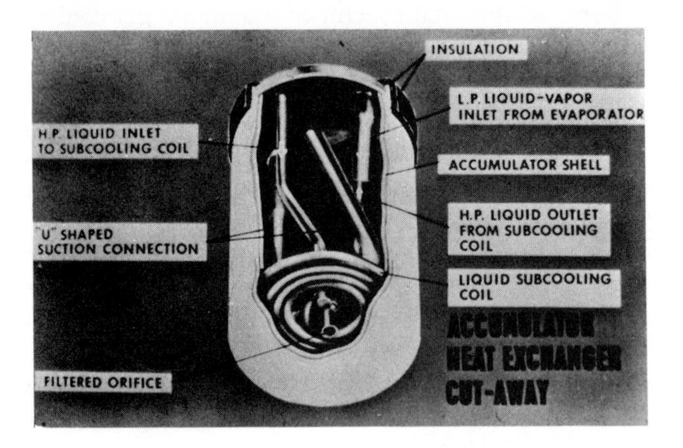

Figure 7-60. Accumulator-heat exchanger cutaway. (*Courtesy Westinghouse Central Residential Air Conditioning Division*)

is subcooled another 10° to 15° F as it passes through the liquid-suction line heat exchanger. Therefore, the refrigerant enters the subcooling coil with approximately 20° to 30° F of subcooling, or 20° to 30° F below its saturation temperature. The heat exchanged between the warm, high pressure liquid in the coil and the cold low pressure liquid in the shell further subcools the high pressure liquid another 35° to 45° during the cooling cycle and 50° to 60° during the heating cycle. In addition to subcooling the high pressure liquid, this heat exchange also boils off some of the low pressure liquid in the accumulator shell. The compressor removes the dry saturated vapor in the upper part of the shell through the U-shaped suction connection. As the suction gas passes through the bottom of the U-tube at high velocity, a low pressure area is created and a mixture of refrigerant and oil is constantly drawn through the filtered orifice to assure a positive and continuous oil return to the compressor. Any traces of liquid refrigerant in the oil are dried up as it passes through the liquid-suction line heat exchanger before it enters the compressor. To obtain the best possible heat exchange efficiency for maximum subcooling effect and to prevent the formation of condensation and frost, the entire accumulator-heat exchanger assembly is completely insulated.

REVIEW QUESTIONS

1. Why is electric heat beneficial from the ecology standpoint?
2. What are some of the adjustments which must be made when considering electric heat?

3. Why is electric heating so efficient?
4. What are the three facets of electric heating?
5. Does resistance heating use a centrally heated medium?
6. Of what material is resistance wire used for heating purposes made?
7. What two factors determine the wire selected for resistance heating?
8. Describe the process used in repairing broken resistance wire.
9. Where are resistance elements located?
10. How is each element protected from overheating?
11. Where is the ideal location for a baseboard heater?
12. What type element is used in baseboard heaters?
13. What are two types of duct heaters?
14. What is the nickel content of the resistance wire?
15. When is iron-chrome-aluminum alloy wire used?
16. What is the maximum watt density per square inch of duct area for duct heaters?
17. Why is a uniform air flow necessary in electric heating?
18. Which type of coil is recommended for most space heating applications?
19. Name four reasons for recommending open coil heaters.
20. Define a heat pump.
21. How are heat pumps classified?
22. How are the coils termed on heat pump installations?
23. What does COP of a unit indicate?
24. Give two reasons for a heat pump being superior to gas heating.
25. What is the maximum COP of a resistance element?
26. What may be used to combat low ambient temperatures when heating with a heat pump?
27. What are the three major methods of using solar energy to assist heat pump operation?

28. Why must the frost be removed from the outdoor coil of a heat pump?
29. Should a heat pump be used for reheat purposes?
30. Describe the procedure of operation of the supplementary heating unit.
31. What two general types of flow control devices are used on heat pump systems?
32. What is the balance point of a heat pump system?
33. What device controls the flow of refrigerant in the Hi/Re/Li system?
34. What are the three functions of the bleed port in the subcooling control valve?
35. What are the three functions of the accumulator-heat exchanger in the Hi/Re/Li system?

8

Heating Boilers and Piping

The objectives of this chapter are:

- To acquaint you with boiler construction.
- To introduce you to the water flow through a boiler.
- To acquaint you with the different methods of boiler firing.
- To provide you with advantages and disadvantages of hydronic heating.
- To introduce you to the problems encountered in heat transfer in a boiler.
- To provide you with some solutions to problems encountered in boiler heat transfer.
- To acquaint you with the one-pipe system.
- To acquaint you with the two-pipe system.

Hydronic heating maintains comfort in the home by circulating hot water to the secondary heat exchanger. This system basically consists of a boiler, a pump, a secondary heat exchanger, and a series of piping. The principle of hydronic heating dates back to ancient times, when the Romans used it by circulating warm water through the walls and floors. Today, however, hydronics bears no resemblance to ancient methods. These systems are as modern as can be installed, with refinements such as zoning, which allows the temperature to vary in the different areas of the home.

HEAT TRANSFER

The transmission or flow of heat ranks in the field of engineering second only to problems involving the strength of materials. Unfortunately, most engineers are probably more familiar with the calculations applying the material strengths and stresses and can recognize the limitations of the laws and methods of solution. In their approach to a problem involving heat transfer, their possible blind acceptance of published elementary laws governing heat flow may lead to gross error in the solution.

The heat transfer problem in heating water, no matter what the design of the heater, is the transmission of the required amount of heat from hot combustion gases through a metal wall into the water. As noted, the chief resistance to the free flow of heat from the source to the point of use is the insulating effect of the film of stagnant gas that always tends to cling to a heat transfer surface. The resistance of the metal wall is insignificant. Likewise the slight resistance of a thin, stationary film of water on the other side is also insignificant.

The relative importance of the three barriers to heat flow are dramatically illustrated in Figure 8–1. A combustion gas temperature of 820° F is assumed. In overcoming the resistance of the normal dead gas film, the temperature at the surface of the metal wall has dropped to 404° F. The slight resistance to conduction of heat through the steel tube wall or other heating surface causes a drop of only 3° F. The resistance

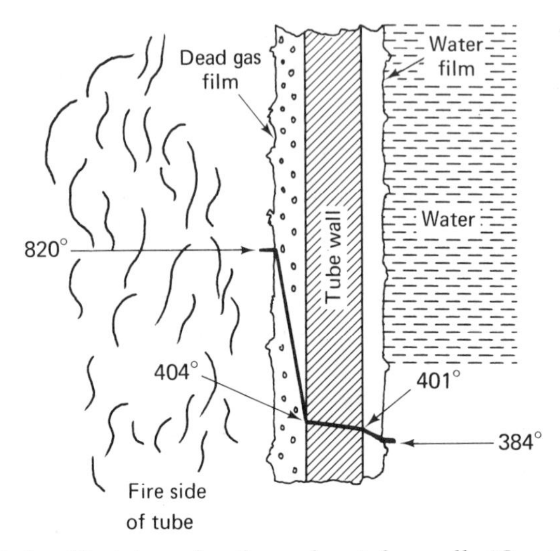

Figure 8–1. Heat transfer through a tube wall. (*Courtesy Sellers Engineering Co.*)

of the water film is equivalent to a drop of 17° F. Of the total heat loss of 436° F from the hot gases to the water side of the tube, 416° F or 95% represents the direct effect of the stagnant gas film. There, obviously, is the point of attack.

For effective heat transfer and all that it means in the way of heating surface allowances, safe temperature differentials, economical operation, and life of equipment, the stagnant gas films must be effectively scrubbed off the heating surfaces. It has long been known that no method of combustion accomplishes this as well as a correctly engineered system of hydronic heating.

The scouring effect is produced mainly by the high velocity turbulent flow of hot gases through the firing area and the rapid expansion of the gases on burning that increases both velocity and turbulence.

The more efficient boilers are designed with a water back combustion area with water circulated all around the firebox. The crown sheet and side walls in the firebox provide a maximum amount of radiant heat transfer surface.

BOILERS

A boiler maintains water temperatures between 120° F and 210° F. This water is pumped through piping to the secondary heat exchange. The term boiler may bring to mind large heating plants such as those found in schools and other large buildings. These are commercial boilers. Actually, a modern cast iron boiler used for comfort heating is very compact. Most units are the size of an automatic washing machine, and some are as small as a suitcase and can be hung on the wall.

Basic Cast Iron Boiler Designs

There are four basic section designs for cast iron boilers.

1. Sections that have two equally sized horizontal push nipples on the vertical center line of the section, and with horizontal and equally sized tappings for supply and return (Figure 8-2).

2. Sections with two horizontal push nipples of different sizes off the vertical center of the section, and with a horizontal lower return tapping and horizontal or vertical upper supply tapping [Figure 8-3 (a) and (b)].

3. Sections with three horizontal push nipples: two smaller nipples at the bottom and one larger nipple in the center at the top; one or two horizontal lower return tappings and one or more vertical top supply tappings [Figure 8-4 (a) and (b)].

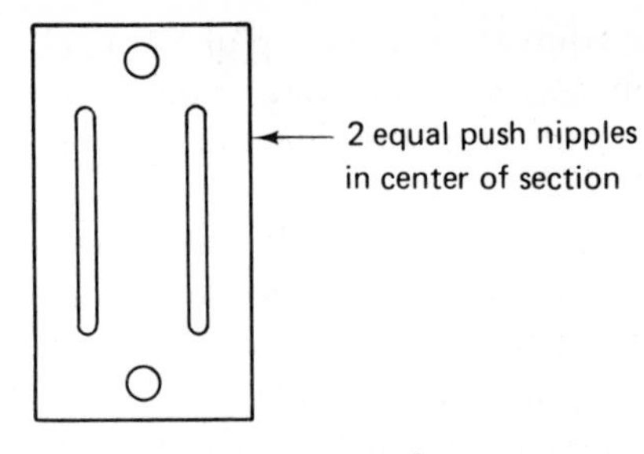

Figure 8-2. Two equal push nipples in center of section. *(Courtesy Weil-McLain Co., Inc., Hydronic Division)*

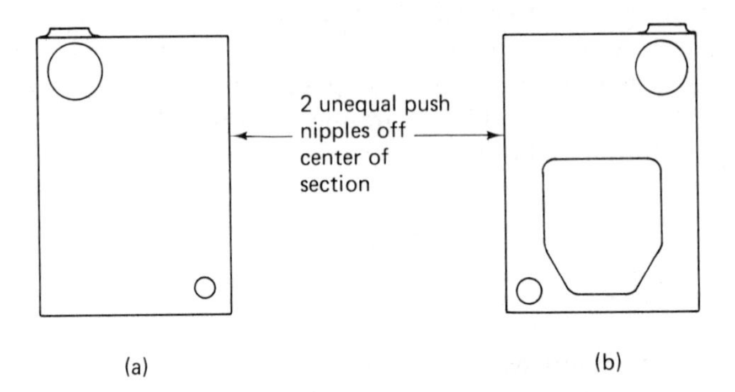

Figure 8-3. Two unequal push nipples off center of section. *(Courtesy Weil-McLain Co., Inc., Hydronic Division)*

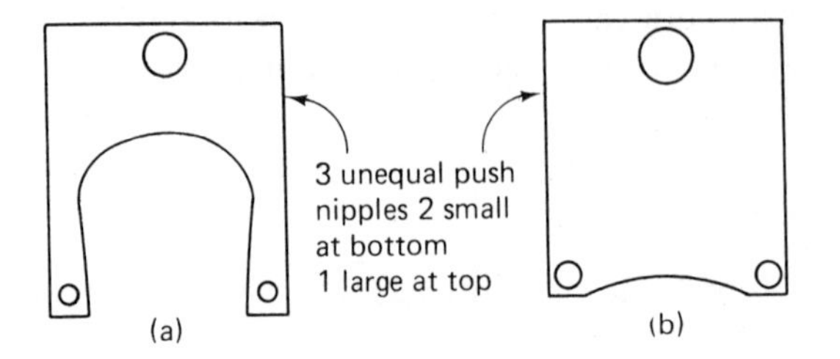

Figure 8-4. Three unequal push nipples. *(Courtesy Weil-McLain Co., Inc., Hydronic Division)*

4. Split sections with horizontal or vertical lower screw nipples and horizontal upper screw nipples; two horizontal lower return tappings and one or more vertical top supply tappings (Figure 8-5).

The variety of nipple arrangements and the variety of supply and return locations make it evident that each type of boiler has a particular internal flow pattern best suited to its design, and that each type will

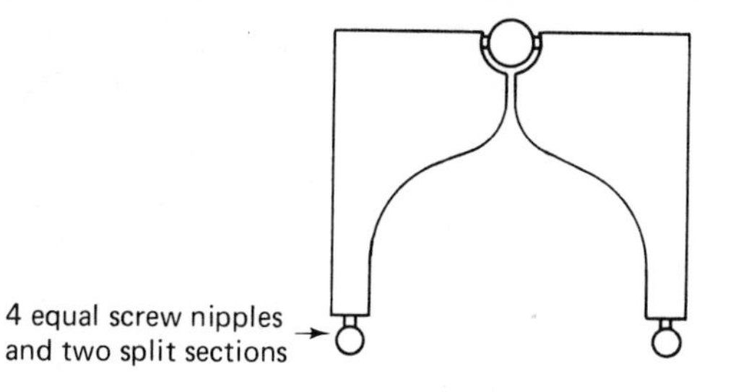

Figure 8-5. Split sections. *(Courtesy Weil-McLain Co., Inc., Hydronic Division)*

react differently to a change in its flow pattern. All these boilers are basically designed for a normal upward flow.

Basic Internal Flow Patterns: The four basic cast iron boiler designs have five basic flow patterns. Two of the basic flow patterns in the four basic boiler designs with normal upward flow are shown in Figures 8-6 and 8-7.

Figures 8-8, 8-9, and 8-10 illustrate a reverse flow and a combination of upward and reverse flow in the basic boiler designs. Keep in mind that these are internal flow patterns.

Almost all European cast iron boilers with two nipples in the center of the section have used the upward, U-shaped pattern [Figure 8-3 (a)] with good results on gravity system for decades. In fact, many European gravity boilers of this design have supply and return tappings only in the back section, thus offering no choice but the upward, U-shaped flow.

Laboratory tests conducted on gas-fired boilers with two nipples, both off center of the section [Figure 8-3 (b)] and on center of the section

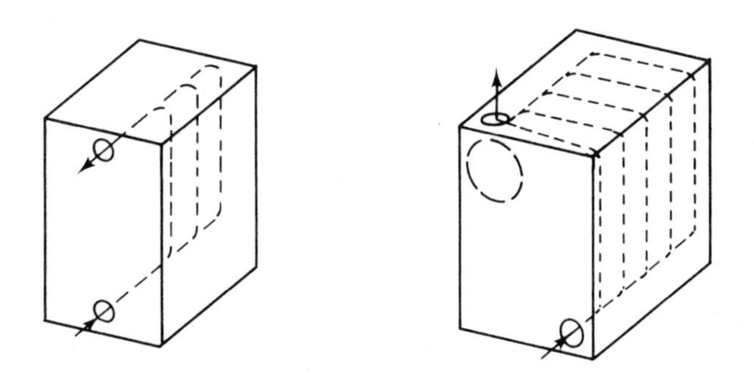

Figure 8-6. Upward U-shaped flow pattern in a two nipple boiler. *(Courtesy Weil-McLain Co., Inc., Hydronic Division)*

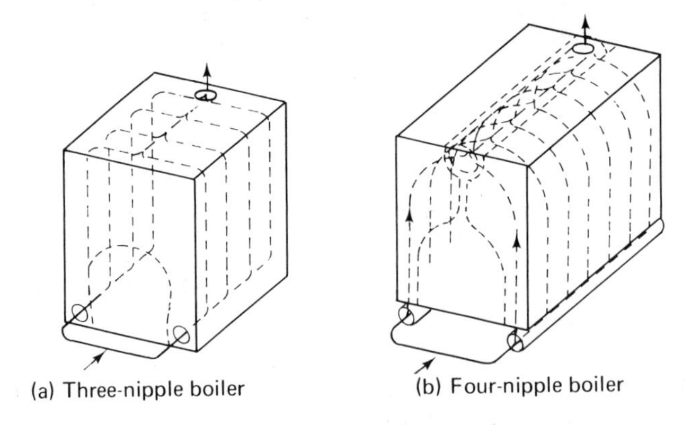

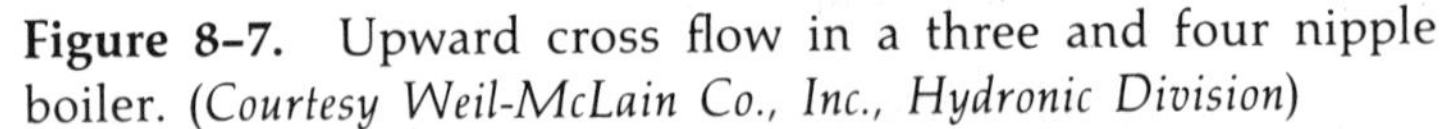

Figure 8-7. Upward cross flow in a three and four nipple boiler. (*Courtesy Weil-McLain Co., Inc., Hydronic Division*)

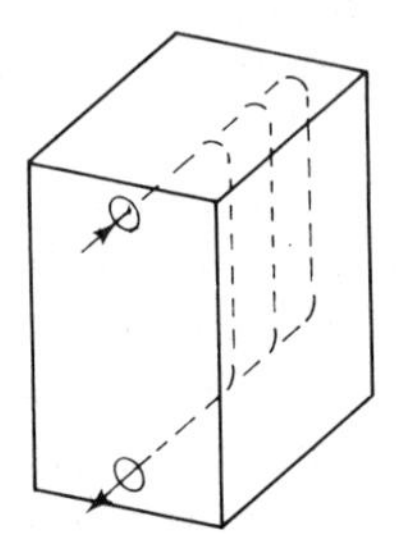

Figure 8-8. Reverse U-shaped flow pattern in a two nipple boiler. (*Courtesy Weil-McLain Co., Inc., Hydronic Division*)

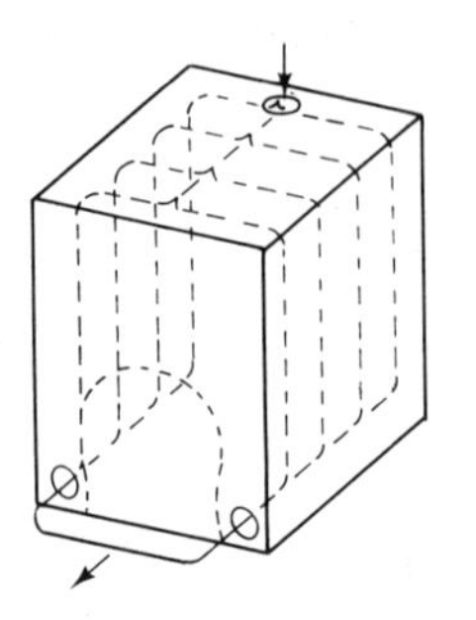

Figure 8-9. Reverse cross-flow pattern in a three or four nipple boiler. (*Courtesy Weil-McLain Co., Inc., Hydronic Division*)

[Figure 8-3 (a)], indicated a minimum temperature override and excellent temperature gradient with the upward, U-shaped flow for forced as well as gravity circulation. The key factors are that the return water is introduced into the axis of the lower nipple ports and that the stay arrangement in the waterways minimizes shortcuts.

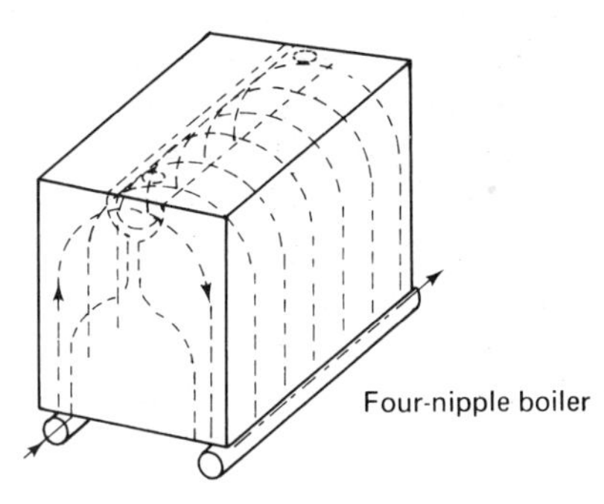

Figure 8-10. Combination upward and reverse cross-flow pattern. (*Courtesy Weil-McLain Co., Inc., Hydronic Division*)

This does not mean that the upward, U-shaped flow can be used in a sectional boiler of unlimited length. The limiting factor is the maximum allowable water velocity through the smallest-sized nipple port. It does mean, however, that the upward, U-shaped flow pattern, within this limitation, meets the requirements for internal flow—namely, equal water distribution, minimum temperature override, prevention of sudden great temperature change, and effective air elimination.

Similarly, field experience and laboratory tests have proven over the years that in boilers with three or more nipples per section (Figure 8-5), the upward cross flow meets the basic requirements for efficient internal flow better than other patterns. As in the two nipple designs, the limiting factor for the length of the boiler is the maximum allowable water velocity consistent with good design.

Effect of the Pump Head: The proper placement of the pump and compression tank in relation to the boiler has been established as an important factor in the functioning of high pump head systems. The pump should be located at the supply water outlet side of the boiler (discharging away from the boiler) with the compression tank at the pump suction side. The purpose of including this subject with internal boiler waterflow direction is to emphasize that the proper placement of the pump and compression tank, along with good system design in general, applies equally to systems where the boiler uses either an upward flow pattern, a reverse flow pattern, or a combination upward and reverse flow pattern. Regardless of the internal flow direction, the system design must always take into consideration the factors of pressure gradient, gpm, supply water temperature, temperature drop, and air control.

Regardless of the flow direction, a valve in the bypass line can provide a means of adjustment for the desired division of positive and negative pressure gradient in the system.

The proper placement of the pump and compression tank in relation to the boiler is the only complete answer to the elimination of undesirable effects of high positive pump head on the boiler and high negative pump head on other parts of the system. Reversing the internal boiler flow may help or hinder the pressure condition, depending on the location of the pump and compression tank, and may be of some advantage on existing systems with or without a divided pump head by changing the pressure on the boiler from a positive to a negative gradient. In new work, the best choice would be a negative gradient. In new work the best choice should be evident—begin with good system design.

It must be emphasized to the designer or installer of residential hot water systems that the location of the pump, either in the return or the supply line of a boiler, is not critical in heating systems with a low pump head (no more than 15 feet). The combination of the dynamic and static pressures in the system is, by definition, low enough to be disregarded.

Advantages of reverse flow:

1. Reverse flow may be useful in a complex piping layout, particularly on a replacement job, where a base mounted pump may need a simplified connection to the boiler.

2. Reverse flow can be a compromise solution to a negative pump head, or to a positive pump head on a boiler in a system in which the head is divided.

3. It can correct an internal temperature problem in certain boiler designs. For example, it is known that steel boilers with a large water volume in one drum sometimes develop high temperature gradients within the shell. This condition has been minimized with downward (reverse) flow through the boiler. Recognizing this is only a temporary solution, several manufacturers of steel boilers have conducted extensive tests with various flow patterns. They found a better solution to the problem by using natural upward flow and by producing a U-shaped internal flow with a special internal elbow at the return inlet. The supply outlet as well as the return inlet are located at the top of the boiler. A low temperature gradient for this type of boiler was achieved.

4. A reverse flow should only be used in boilers which are designed for it.

The question of whether reversing the flow in a cast iron boiler will increase the heat transfer from the flue gases in the boiler water may be answered in the affirmative, if the heat were transmitted by convection and/or conduction only and if a true counter-flow between water and flue gases existed. However, it is a well-known fact that the flow of flue gases along the secondary heating surface of a cast iron boiler is not a true counterflow or parallel flow or cross flow, but a combination of the three. It is also known that a modern cast iron boiler has up to 60% primary heating surface where the heat is transmitted mainly by radiation and where the water flow direction has no bearing on the heat absorption. In other words, reversing the water flow through the boiler will not appreciably change the heat transfer from the flue gases to the boiler water.

Disadvantages of reverse flow include the following:

1. As air must rise against water flow direction, its elimination from the boiler becomes a problem. Air released from the heating of the boiler water can be trapped and can build air pockets. The pockets will often create hot spots and, in turn, steam bubbles. Steam bubbles also can be caused by an incorrectly sized or poorly operating firing device. On their way to the top of the boiler, steam bubbles will come in contact with the cool return water. This causes them to collapse, producing noise which is usually transmitted throughout the entire piping system. In severe cases steam bubbles can create vibration of the whole boiler.

Air that will be drawn into the system can also cause noise and uneven heat distribution if the point of air elimination is not changed for reverse flow.

2. Since the water velocity within the boiler sections slows down considerably, no matter whether normal—upflow, or reversed—downflow, the reverse water flow direction within the boiler can be upset, particularly in a long boiler with many sections. In other words, it is possible that one part of the boiler can have reverse flow whereas in another part of the boiler the natural gravity flow is stronger than the reverse flow, thus upsetting the water circulation as well as the temperature gradients within the boiler. Actual tests have proven this fact. This condition will further aggravate air elimination and control functions.

3. Intermittent pump operation reverses the temperature pattern within the boiler each time the pump starts or stops and puts additional strain and stress on the boiler.

4. Controls in predetermined locations for normal upward flow generally do not provide satisfactory operation when used with reverse flow.

5. Reverse flow definitely decreases the output of a built-in tankless or storage heater because the cool return water comes in contact with the heater coil. The boiler manufacturer is responsible for trouble-

free operation of his product and therefore, he must recommend the waterflow best suited to each of his boiler designs.

Combustion Chamber: Because the combustion area is entirely surrounded by water and because flame retention burners do not require hot refractory for combustion, a refractory combustion chamber is not required for this type of boiler. The advantages are obvious—the cost of the combustion chamber material and the labor to install it are saved, and there is no future replacement cost.

Asbestos Rope Seal: For forced draft firing, it is absolutely necessary that a boiler be completely gas-tight. A boiler with a separate base can be made gas-tight for a short time only. But when the firebox is surrounded with water, a separate base is not necessary; therefore, the seal between adjacent sections can be made permanently gas-tight [Figure 8-11 (a)]. The seal between the platework and the boiler must also be permanently gas-tight.

The sections of these boilers have a grooved seal strip that receives the asbestos rope. When installed, the outer edge of the rope is accessible between sections, so that the boiler can be visually checked for tightness. The rope is compressible, allowing ample contraction and expansion of the boiler. Asbestos rope does not crack, lasts the life of the boiler, and assures a gas-tight assembly. The sections are not face ground, but retain the tough original skin which reduces the rate of rust

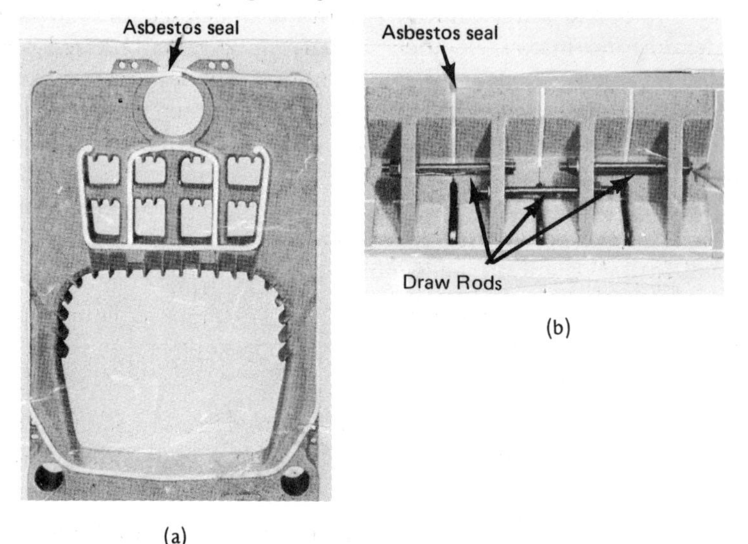

Figure 8-11. Boiler sections. (*Courtesy Weil-McLain Co., Inc., Hydronic Division*)

growth. The rope seal is impervious to the intense heat in the firebox and to floor moisture.

Draw Rods: Multiple short draw rods, instead of a single long rod and expansion washers, are desired to tie the sections together [see Figure 811(b)]. This feature is endorsed by leading insurance companies. The short draw rods permit faster, easier erection of the boiler and a strain-free assembly.

Flue Gas Travel: The flue gases rise from the combustion area into uptakes between each boiler section at the right and left side. The flue gases are directed through the two outside flue passages to the front of the boiler and then back through the center flueways to the flue connection (Figure 8-12).

The uptakes between each section expose more radiant heat-absorbing surface, and divide the hot gas volume into many small gas streams. These gas streams are individually directed to the cooler sides of the firebox, creating a wiping action that increases heat absorption. The turn-around flue-gas travel at the upper part of the uptakes uses the heating surface to the top of the sections. This feature aids in producing a dry steam in the steam boiler.

Multiple uptakes, combined with three pass design, for hot gases to wipe the entire flue area, assure balanced flue gas travel and prevent shortcuts to the chimney. Extra long flue gas travel and higher velocities increase heat absorption of the secondary heating surfaces.

Cleaning: Easy cleaning of the flue passages is a must. Most manufacturers provide a means of cleaning by removing the top jacket panel and the collector hood which exposes all flue-ways for easy, straight through cleaning with a steel brush, usually furnished with the boiler.

Figure 8-12. Flue gas travel. (*Courtesy Weil-McLain Co., Inc., Hydronic Division*)

Fire Tube Boilers

The immersion (fire tube) boiler is also built to conform to the ASME code. Standard design pressures are 100 and 150 psi.

The simple design of the boiler shell lends itself to modern, precision fabrication procedures—full round cylinder rolling, production cutting, and bevelling of heads, automatic welding of the head and shell seams and automatic controlled tube rolling (Figure 8-13).

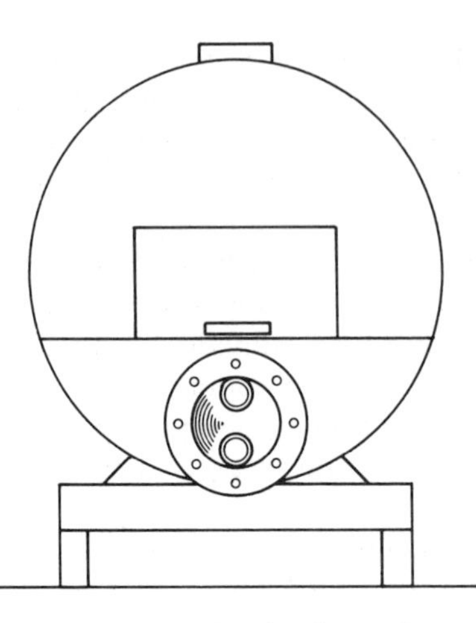

Figure 8-13. Immersion tube boiler. (*Courtesy Sellers Engineering Co.*)

The FiringTube: The firing tubes are copper clad on the water side. The 2 inch OD by number 13 gauge tubes (Figure 8-14) are formed by telescoping a .030 inch hard phosphorized copper tube over a .065 inch seamless steel tube and drawing over a mandrel and through a die for a tight bond. Long tube life is positively assured under the most aggressive conditions.

Straight Through Single Pass Firing: No traps or pockets for gas accumulations (Figure 8-15). No brick work or refractories anywhere. The identical firing tubes constitute individual combustion chambers—

Figure 8-14. Firing tubes. (*Courtesy Sellers Engineering Co.*)

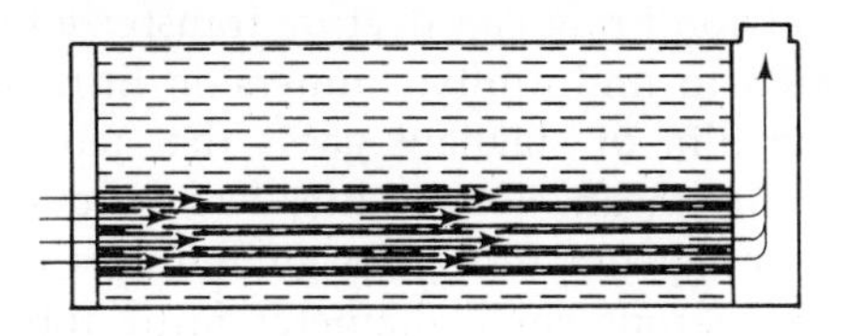

Figure 8–15. Flue gas passage. (*Courtesy Sellers Engineering Co.*)

water backed all the way. There are no strains or stresses. The uncrowded shell permits a generous ligament, never less than 1 inch, between tubes (Figure 8–16). This removes the hazards of cracked heads and loosened tubes.

Only in immersion fired boilers can the total gas input be widely dispersed among numerous, small firing tubes (Figure 8–17). This eliminates hot spots and excessive scale deposits in hard water usage. It provides uniform expansion stresses with no likelihood of tube loosening. More importantly, every square foot of heating surface is put to work in a more uniform manner.

There is no simpler combustion system, in fact, no simpler construction, than the straight through, single pass design of immersion firing.

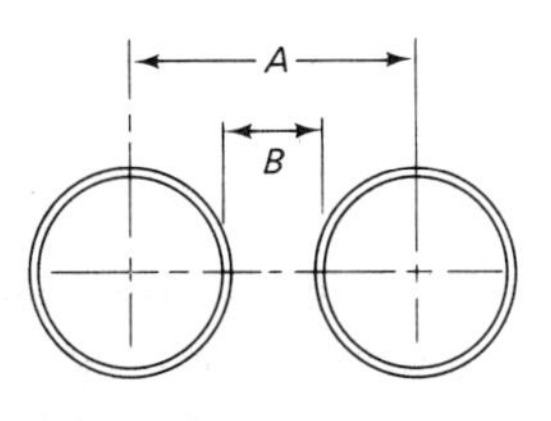

Figure 8–16. Uncrowded firing tubes. (*Courtesy Sellers Engineering Co.*)

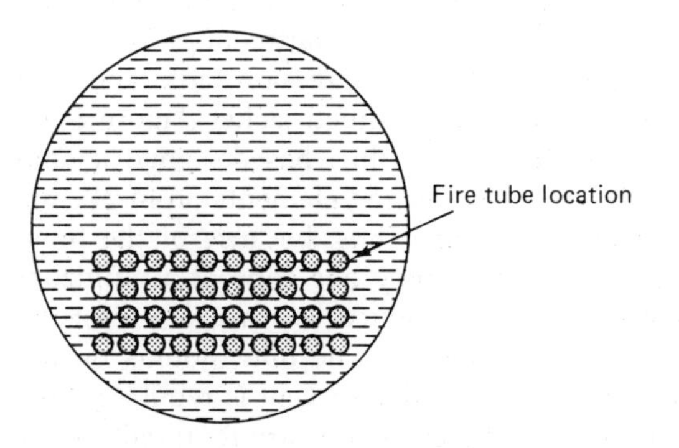

Figure 8–17. Fire tube nest location. (*Courtesy Sellers Engineering Co.*)

Only in immersion firing can heat be transferred to the water with so close an approach to uniformity. Immersion firing effectively scours away the insulating film of stagnant gases that tends to cling to heat transfer surfaces. This means longer life, less hard water scaling, and greater efficiency. Immersion firing design requires distribution of the gas input among numerous small diameter firing tubes. The tubes are alike in all respects. They carry gases with equal temperature gradients. There are no strains set up. There is no loosening of tubes. There is no ignition of a large volume of gas at a single point, but instead, smooth and safe ignition of a series of small burners (Figure 8–18). There is no combustion chamber to be insulated against heat loss. Each burner fires into a separate, water-backed combustion chamber.

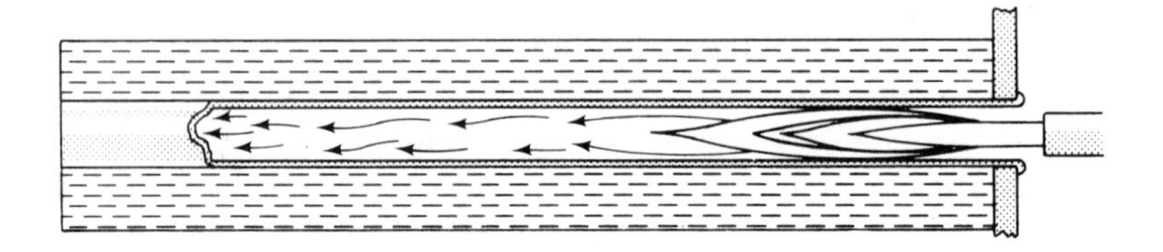

Figure 8–18. Fire tube and its burner. (*Courtesy Sellers Engineering Co.*)

Electric Boilers

Electric boilers have been in use for many years. The first ones were crude in appearance, but they accomplished what they were designed to do. The lack of electrical power, especially in rural areas, has been the most outstanding hindrance in electric heating. In modern times, however, electric boilers are gaining popularity.

Electric hydronic boilers are designed specifically for hot water heating systems in homes and apartments. They are factory-assembled and wired with circulator, compression tank, and controls; all parts are enclosed in a compact, clean-lined jacket (Figure 8–19).

The Weil-McLain Co., Inc. has designed a unique boiler control system which stages the heating elements on in 2½ minute intervals, and off in 10 second intervals. Thus, because of the time delay, full boiler capacity is not used to satisfy one zone of a multizone system.

These boilers are designed for fast, low cost installation, for new buildings or for replacement. The unit mounts on the wall, saving valuable living space, and no flue or vent is required. Standard components are used and electrical connections are required only for the power supply and thermostat. Piping connections are the same as for any hot water boiler.

Figure 8-19. Electric hydronic boiler. (*Courtesy Weil-McLain Co., Inc., Hydronic Division*)

Construction: The one-piece cast iron boiler sections are built to meet the requirements of ASME boiler and vessel code. Large water content (5.2 gallon) eliminates rapid internal temperature changes to assure better control response and a nearly constant supply water temperature.

The sections are insulated to reduce heat loss. A built-in air eliminator diverts air bubbles to an automatic air vent. No separate air eliminating device is necessary. Internal separators in the section assure full flow of water over the heating elements.

Heating Elements: The incoloy-sheathed, low-density elements (approximately 55 W/in^2) resist the corrosive effects of all chemicals found in domestic water systems. If it is ever necessary to replace an element, a standard water heater element may be used.

Compression Tank and Circulator: Most boilers are furnished with a large-capacity compression tank that has a flexible diaphragm to prevent water from contacting the tank charge for positive system protection. The circulator is an industry standard, noted for its long life and dependability (Figure 8-19).

Internal Fusing: Electric boilers are normally supplied with a separate fuse for each element leg plus a fuse for the circulator and control circuit.

Zoning: Because of the unique control systems employed in electric boilers, zoning may be easily accomplished with either zone valves or additional circulators. Indoor-outdoor controls and timing devices, necessary for many zoning applications, are not required with most electric boilers.

Contactors: Double line-break contactors close both legs of the power source on a call for heat, and open both legs of each element when the thermostat is satisfied.

STEAM HEATING

The widespread use of steam for space heating today points up a long recognized fact that steam, as a heating medium, has numerous basic characteristics which can be advantageously employed. Some of the most important advantages are as follows:

Steam's Ability to Give Off Heat

Properties of saturated steam may be found in steam tables (see Table 8-1) which give much information regarding the temperature and heat contained in 1 pound of steam for any pressure. For example, to change 1 pound of water to steam at 212° F at an atmospheric pressure of 14.7 psia requires a heat content of 1150.4 Btu which is made up of 180.1 Btu of sensible heat (the heat required to raise one pound of water from 32° F to 212° F) and 970.3 Btu of latent heat. Latent heat is the heat added to change the one pound of water at 212° F into steam at 212° F. This stored up heat is required to transform the water into steam and it reappears as heat when the process is reversed to condense the steam into water.

Because of this fact, the high latent heat of vaporization of 1 pound of steam permits a large quantity of heat to be transmitted efficiently from the boiler to the heating unit with little change in temperature.

Steam Promotes Its Own Circulation

For example, steam will flow naturally from a higher pressure (as generated in the boiler) to a lower pressure (existing in the steam lines). Circulation or flow is caused by the lowering of the steam pressure along the steam supply mains and the heating units due to the pipe friction

Table 8-1

Properties of Saturated Steam (Approximate)

Absolute pressure	Gage reading at sea level	Temp. ° F	Heat in water Btu per lb.	Latent heat in steam (Vaporization) Btu per lb.	Volume of 1 lb steam ft³	Wgt. of water lb. per ft³
0.18	29.7	32	0.0	1076	3306	62.4
0.50	28.4	59	27.0	1061	1248	62.3
1.0	28.9	79	47.0	1049	653	62.2
2.0	28	101	69	1037	341	62.0
4.0	26	125	93	1023	179	61.7
6.0	24	141	109	1014	120	61.4
8.0	22	152	120	1007	93	61.1
10.0	20	161	129	1002	75	60.9
12.0	18	169	137	997	63	60.8
14.0	16	176	144	993	55	60.6
16.0	14	182	150	969	48	60.5
18.0	12	187	155	986	43	60.4
20.0	10	192	160	983	39	60.3
22.0	8	197	165	980	36	60.2
24.0	6	201	169	977	33	60.1
26.0	4	205	173	975	31	60.0
28.0	2	209	177	972	29	59.9
29.0	1	210	178	971	28	59.9
30.0	0	212	180	970	27	59.8
14.7	0	212	180	970	27	59.8
15.7	1	216	184	968	25	59.8
16.7	2	219	187	966	24	59.7
17.7	3	222	190	964	22	59.6
18.7	4	225	193	962	21	59.5
19.7	5	227	195	960	20	59.4
20.7	6	230	196	958	19	59.4
21.7	7	232	200	957	19	59.3
22.7	8	235	203	955	18	59.2
23.7	9	237	205	954	17	59.2
25	10	240	208	952	16	59.2
30	15	250	219	945	14	58.8
35	20	259	228	939	12	58.5
40	25	267	236	934	10	58.3
45	30	274	243	929	9	58.1
50	35	281	250	924	8	57.9
55	40	287	256	920	8	57.7
60	45	293	262	915	7	57.5
65	50	298	268	912	7	57.4
70	55	303	273	908	6	57.2
75	60	308	277	905	6	57.0
85	70	316	286	898	5	56.8
95	80	324	294	892	5	56.5
105	90	332	302	886	4	56.3
115	100	338	309	881	4	56.0
140	125	353	325	868	3	55.5

(Courtesy ITT Hoffman Specialty)

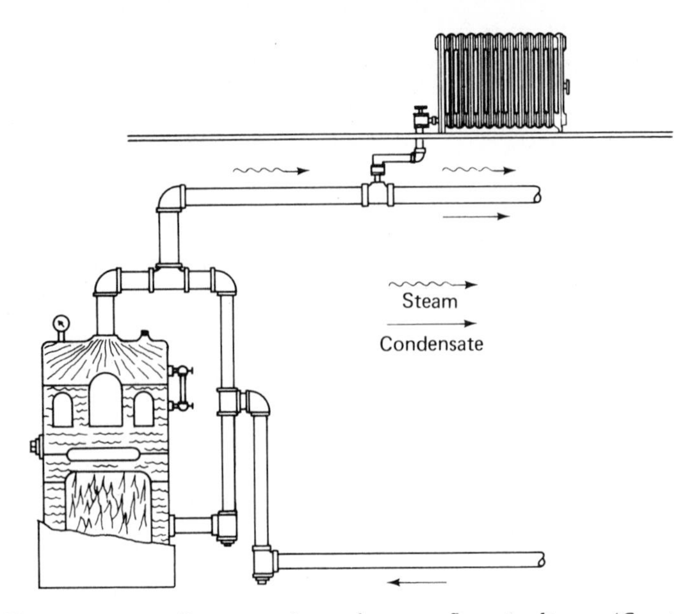

Figure 8–20. Steam and condensate flow in lines. (*Courtesy ITT Hoffman Specialty*)

and to the condensing process of the steam as it gives up heat to the space being heated (Figure 8–20).

Because of this fact, the natural flow of steam does not require a pump, such as that needed for hot water heating, or a fan, as employed in warm air heating.

Steam Heats More Readily

Steam circulates through a heating system faster than other fluid mediums. This can be important where fast pick-up of the space temperature is desired. It will also cool down more readily when circulation is stopped. This is an important consideration in spring and fall when comfort conditions can be adversely affected by long heating-up or slow cooling-down periods (see Figure 8–21).

Steam Heating Is More Flexible

Other advantages in using steam as a heating medium can be found in its easy adaptability to meet unusual conditions of heat requirements with a minimum of attention and maintenance. Here are some examples:

1. Temporary heat during construction is easily provided without undue risks and danger of freeze-ups.

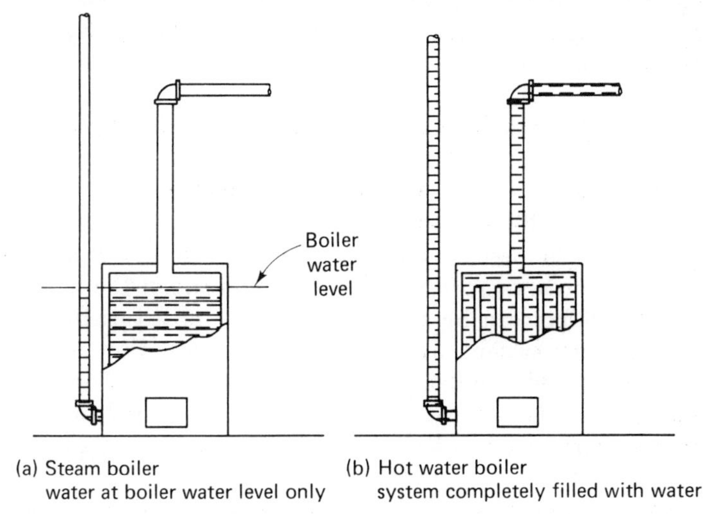

Figure 8–21. Boiler water comparison. (*Courtesy ITT Hoffman Specialty*)

2. Additional heating units can be added to the existing system without making basic changes to the system design.

3. Increased heat output from heating units can be easily accomplished by increasing the steam pressure the proper amount.

4. Steam heating systems are not prone to leak; however, leaks that may occur in the system piping, pipe fittings, or equipment cause less damage than leaks in systems using hot water. One cubic foot of steam condenses into a relatively small quantity of water. In many cases, a small leak does not cause any accumulation of water at the location of the leak; instead it evaporates into the air and causes no damage.

5. Repair or replacement of system components such as valves, traps, heating units, and similar equipment, can be made by simply closing off the steam supply. It is not necessary to drain the system and to spend additional time to reestablish circulation. There is less need to worry about freezing since the water in a steam heating system is mainly in the boiler (Figure 8–21). The boiler water can easily be protected from freezing during shut-downs, during new construction, and during repairs or replacement of parts by installing an aquastat below the boiler water level to control water temperature.

6. Steam is a flexible medium when used in combination processes and heating applications. These often require different pressures that are easily obtained. In addition, exhaust steam, when available, can be used to the fullest advantage.

7. Steam heating systems are considered to be lifetime investments. Many highly efficient systems are in operation today after more than 50 years of service.

Steam Is Easy to Distribute and Control in a Heating System

The distribution of steam to heating units is easy to accomplish with distribution orifices located at the steam inlet to the unit. Metering orifices can also be used along with proper controls to maintain steam pressure in accordance with the flow characteristic of the metering orifices. These orifices can be either a fixed-type or a variable-type. The controls for steam systems are simple and effective. They include those used to control space temperature by the application of "on-off" valves. Modulating controls can also be applied which respond to indoor-outdoor temperature conditions to control the quantity of steam flowing to orificed radiators.

Steam heating systems fall into two basic classifications—one-pipe systems and two-pipe systems. These names are descriptive of the piping arrangement used to supply steam to the heating unit and to return condensate from the unit. It is a one-pipe system when the heating unit has a single pipe connection through which steam flows to it and condensate returns from it at the same time. In a like manner, it is a two-pipe system when the heating unit has two separate pipe connections—one for the steam supply and the other for the condensate return.

One-Pipe System

Modern automatically-fired boilers promote rapid steaming and assure quick pick-up of the space temperature from a cold start. The natural circulation of steam in the system, in combination with the simplicity of piping and air venting, makes this type of system the most economical as well as a desirable method of heating.

A one-pipe system (Figure 8–22), properly designed for gravity return of the condensate to the boiler with open type air vents on the ends of mains on each heating unit, requires a minimum of mechanical equipment. The result is a low initial cost for a very dependable system.

The modern one-pipe system is a simple up-feed system. The basic equipment used is described as follows:

Steam Boiler: It is automatically fired and equipped with suitable controls to maintain system pressure. It also has the required safety devices for proper burning of fuel and should be equipped with an automatic water feeder and low water cut-off (Figure 8–23).

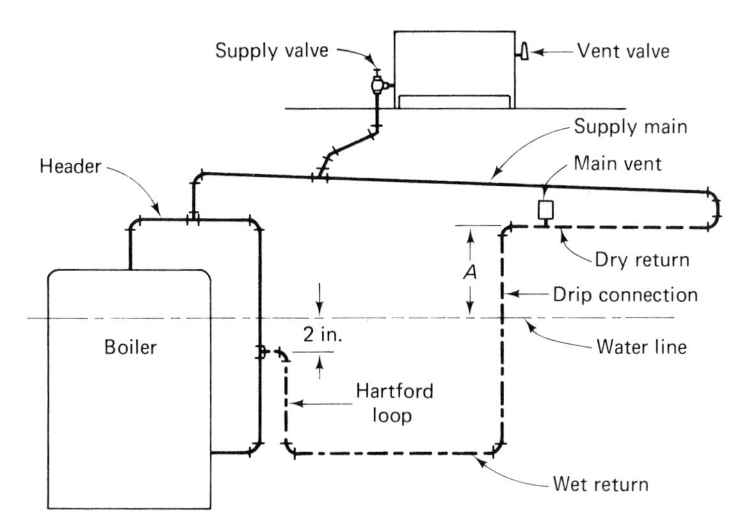

Figure 8-22. Basic one-pipe up-feed system. (*Courtesy ITT Hoffman Specialty*)

Heating Units: One-pipe systems can be designed to use convectors, wall fin-tube, and similar heat out-put units (Figure 8-24).

Air Vents: Steam cannot circulate or radiate heat until all the air has been vented from the system. Each heating unit and the end of each steam main must be equipped with an air vent valve (Figure 8-25).

Radiator Valves: Each radiator must be equipped with an angle pattern, radiator supply valve installed at the bottom inlet tapping (Figure 8-26).

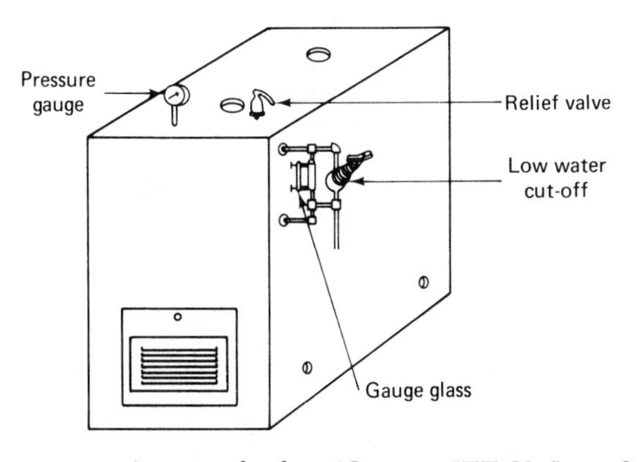

Figure 8-23. A steam boiler. (*Courtesy ITT Hoffman Specialty*)

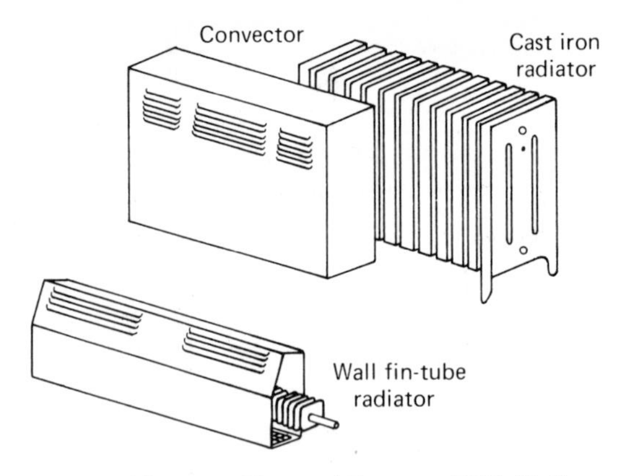

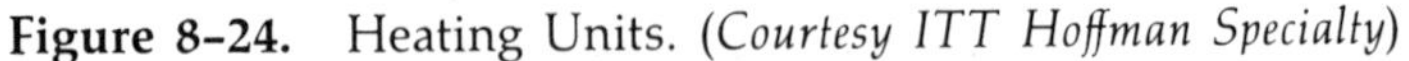

Figure 8–24. Heating Units. (*Courtesy ITT Hoffman Specialty*)

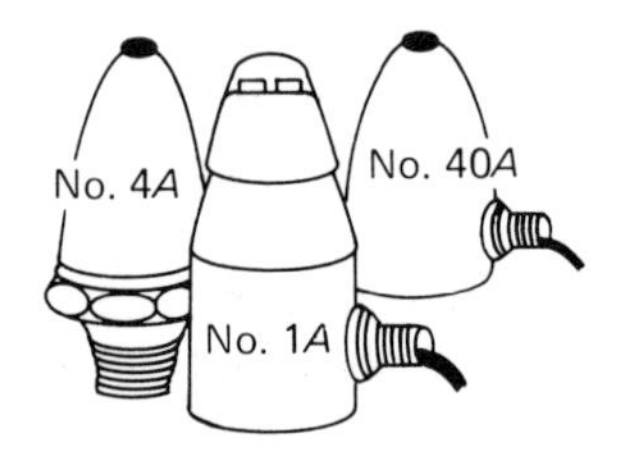

Figure 8–25. Air vents. (*Courtesy ITT Hoffman Specialty*)

Figure 8–26. Radiator supply valve. (*Courtesy ITT Hoffman Specialty*)

The typical system illustrated in Figure 8–22 also shows the piping components of a one-pipe system. The names and functions of these components are as follows:

Header: Boilers, depending on their size, have one or more outlet tappings. The vertical steam piping from the tapped outlet joins a horizontal pipe called a *header.* The steam supply mains are connected to this header.

Steam Supply Main: The supply main carries steam from the boiler to the radiators connected along its length. It also carries condensate accumulation from these units back to the drip connection (Figure 8-22). When the condensate flow in the supply main is the same direction as the steam flow, the system is called a *parallel flow system.*

Drip Connection: The drip connection is the vertical length of pipe connecting the remote end of the steam supply main to the wet return (Figure 8-22).

Wet Return: The return piping that carries the condensate accumulation back to the boiler and is installed below the level of the boiler water line is called a *wet return* (Figure 8-22). It is completely filled with water and does not carry air or steam. When the system is first filled with water or is cold, the pressure throughout the system is the same, or balances. Therefore, the water is at the same level in the boiler and drip connection as indicated by the boiler water line.

Dry Return: The dry return is that portion of the return line located above the boiler water line. In addition to carrying condensate, it also carries steam and air. The end of the dry return must be located at the proper height to maintain the minimum required distance above the boiler water line (Figure 8-22).

The illustration shown in Fig. 8-27 will be used to describe important operating principles of a one-pipe system. Steam is generated in the boiler when fuel is burned and heat added to the boiler water. This causes an increase in steam to the heating units. As steam flows through the steam supply main, there is a loss in steam pressure due to the re-

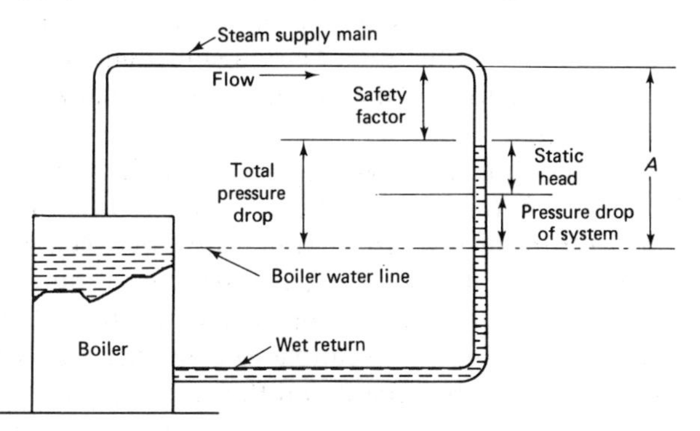

Figure 8-27. Pressure drop in a one-pipe steam system. (*Courtesy ITT Hoffman Specialty*)

sistance to flow caused by pipe friction. An additional pressure loss is caused by the condensing process in the steam main, other piping, and radiators. The sum of these pressure losses is the pressure drop which results in a lower pressure on the surface of the water in the drip connection at the end of the steam supply main as compared to the higher pressure acting on the surface of the water in the boiler. This difference in pressure causes the water to rise in the drip connection. The measured difference between these two levels is the pressure drop of the system. The end of the steam supply main must be a minimum distance above the boiler water line for any gravity return one-pipe system. Two other factors must be considered and the distance they represent must be added to the system pressure drop to obtain the proper distance. The static head represents the height of water required to return the condensate to the boiler. The safety factor represents an additional height to allow for unusual heating up conditions.

The Hartford Loop: The Hartford loop is a special arrangement of return piping at the boiler. Its purpose is to reduce the likelihood of an insignificant quantity of water creating a low water condition that can cause damage to a steam heating boiler. It came into general use in 1919 and was primarily designed for use with heating systems having gravity wet returns.

In effect, it consists of two loops of pipe forming two U-tubes (Figure 8–28). The first loop is around the boiler and the second loop is composed of the *drip connection,* the *wet return,* and a short riser called the *loop riser.* In the first loop, an equalizer line runs from the boiler header down the side of the boiler to the boiler return connection. A pressure balance is maintained in this loop because the steam pressure on the top of the water in the equalizer pipe is the same as the pressure on top of the water in the boiler.

In the second loop, the short riser is connected to the equalizer pipe with a close nipple, the center of which must be not less than 2 inches below the boiler water line. Here again, there is a balance between the pressure on top of the water in the equalizer pipe, in the boiler, and in the drip.

Should a leak occur in the wet return or should bad operating conditions be experienced, the boiler could drain down only until the water line fell to the bottom of the nipple connection between the loop riser and the equalizer pipe. Sufficient water will remain in the boiler to cover the crown sheet or section and prevent damage to the boiler.

It is important to use a close nipple to construct the Hartford loop. If a long nipple is used at this point and the water line of the boiler becomes low, water hammer noise will occur. Some designers prefer using a *Y* fitting that, if used, also must be 2 inches below the boiler water line.

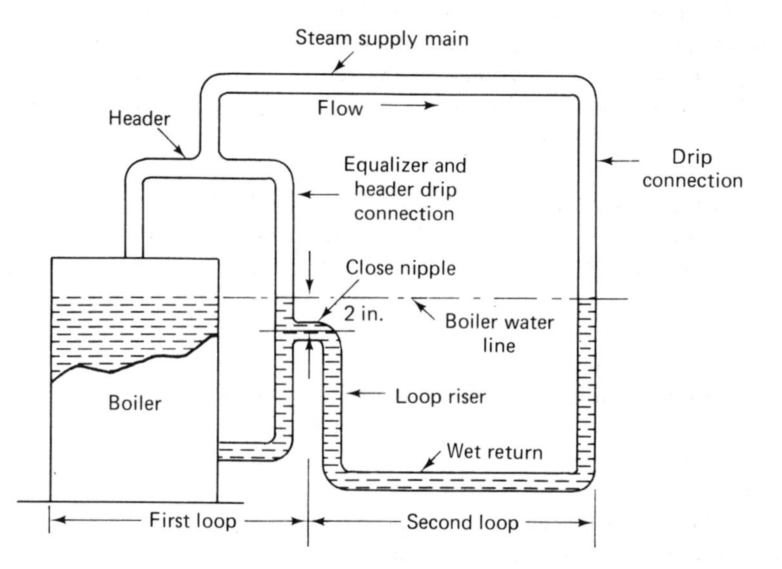

Figure 8-28. The Hartford loop. *(Courtesy ITT Hoffman Specialty)*

The use of a Hartford loop is not recommended where the condensate is returned to the boiler by a condensate pump. It can be a source of noise resulting from the introduction of relatively cold water to the boiler at the hottest boiler water temperature.

Two-Pipe Steam Heating Systems

Although a two-pipe system has fundamental differences from those of a one-pipe system, many components and piping installation practices are common to both systems. Also, the two-pipe system employs many advantages of using steam as a heating medium. They are applicable to a variety of structures from small residences to large commercial buildings, office buildings, apartment buildings, and industrial complexes.

Two-pipe systems are designed to operate at pressures ranging from subatmospheric (vacuum) to high pressure. Although they use many practical piping arrangements to provide up-flow or down-flow systems, they are conveniently classified by the method of condensate return to the boiler. Condensate can be returned to the boiler by gravity or by use of any one of several mechanical means.

By definition, it is a two-pipe system when the heating unit has two separate pipe connections—one used for the steam supply and the other used for the condensate return.

Basic equipment for two-pipe systems are used in various combinations depending on the type system or its design. Some of these same components are used for one-pipe systems and were described for these systems. However, all equipment used for two-pipe systems will be described and discussed as follows.

A Steam Boiler: Boilers are manufactured as either a sectional cast iron boiler or as a steel boiler. The modern boiler is automatically fired using electricity, oil, or gas as the fuel. They are provided with suitable pressure controls, safety firing devices, and protection against damage caused by low water conditions (Figure 8–29).

Heating Units: Two-pipe systems use a variety of heat output units such as cast iron radiators, convectors, unit ventilators, and heaters (Figure 8–30).

Thermostatic Traps: Thermostatic traps are the most common of all types used in two-pipe steam heating systems. They are used because they (1) are simple in construction; (2) are small in design and weight; and (3) have adequate capacity for usual heating system pressure. They are designed to open in response to pressure and temperature to discharge air and condensate and to close against the passage of steam. The temperature at which a thermostatic trap will open is variable but is at the required number of degrees below the saturated temperature for the existing steam temperature; the trap opening temperature is called *temperature drop.*

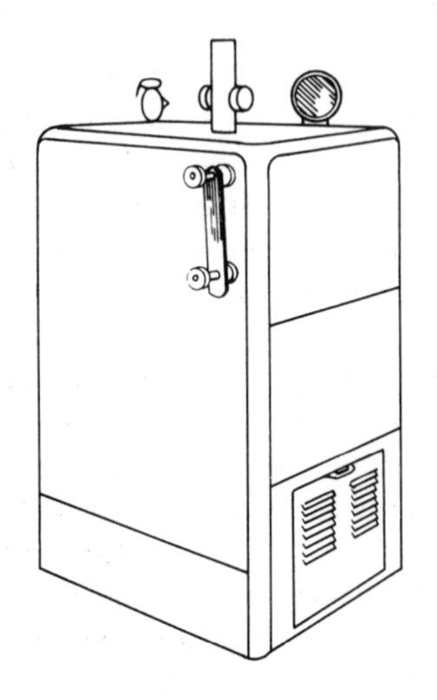

Figure 8–29. Steam boiler. (*Courtesy ITT Hoffman Specialty*)

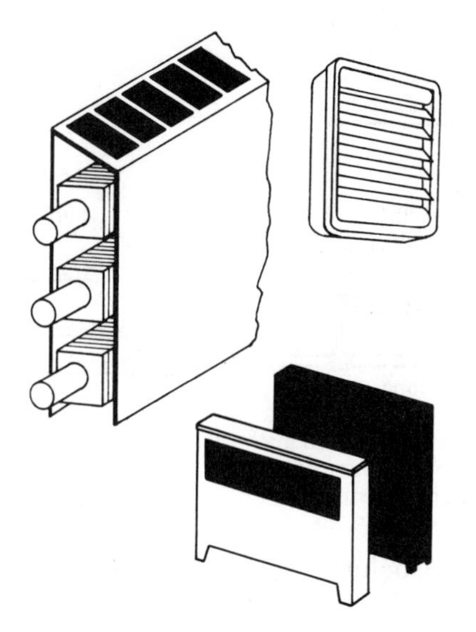

Figure 8–30. Heating units. (*Courtesy ITT Hoffman Specialty*)

Balanced pressure-type thermostatic traps are those which have been described. They employ thermal elements made from metal bellows, a series of diaphragms, or special cells made from diaphragms (Figure 8–31). There are some types that employ special shapes of bimetal for the thermal elements.

The radiator return connection is equipped with a thermostatic trap. They are also used as drip traps and to handle condensate from fan-coil units such as a unit heater. For these and similar applications, a cooling leg (an adequate length of pipe) must be installed between the equipment or drip and the thermostatic trap (Figure 8–32). The cooling leg permits the condensate accumulation to cool sufficiently to open the trap in order to discharge the condensate, thus preventing flooding of the equipment.

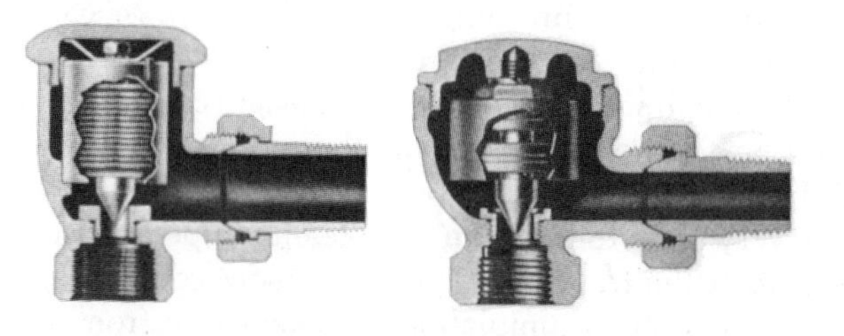

Figure 8–31. Radiator thermostatic traps. (*Courtesy ITT Hoffman Specialty*)

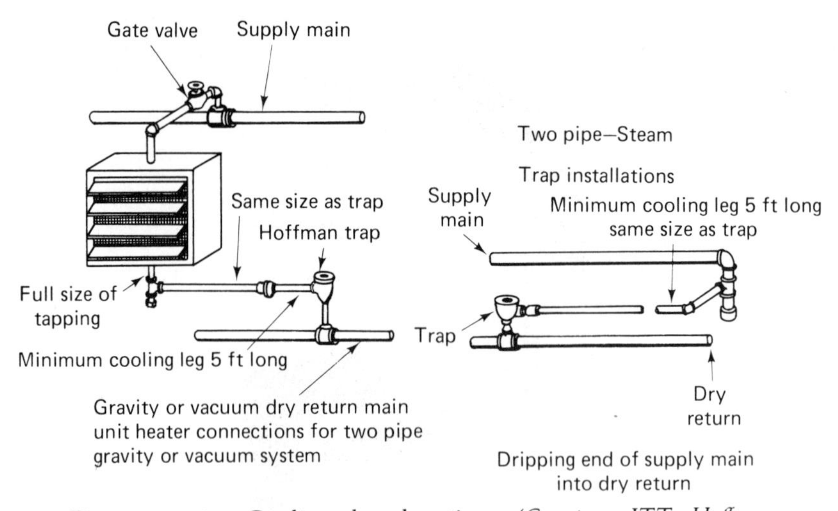

Figure 8–32. Cooling leg location. (*Courtesy ITT Hoffman Specialty*)

Thermostatic traps are made in angle, straightway, swivel, and vertical patterns and can be used over a wide range of pressures from subatmospheric (vacuum) to high pressure steam.

Mechanical Steam Traps: There are several different mechanical traps used in two-pipe heating systems. They include float and thermostatic traps, float traps, inverted bucket traps, and open or upright bucket traps. All of these have different operating characteristics that make them applicable to a variety of heat output units, steam mainlines, or riser drips.

1. *Float and thermostatic traps.* A float and thermostatic trap is often called an F and T trap (Figure 8–33). This trap opens and closes in response to the rising and lowering of the float caused by changes in the level of condensate entering the trap body. The discharge of condensate is continuous due to the throttling action of the pin or valve in the seat port. The thermostatic air bypass remains open when air or condensate is present at a temperature below its designed closing temperature. When steam enters the trap body, the thermostatic air bypass is closed. An F and T trap will discharge condensate at any temperature up to a temperature very close to the saturated steam temperature corresponding to the pressure at the trap inlet. For this reason, cooling legs are not required and condensate accumulations are kept free from steam lines or from equipment being served. An F and T trap is a first choice selection for dripping the end of a steam main, the heels of upfeed steam risers and the bottom of downfeed steam risers (Figure 8–34).

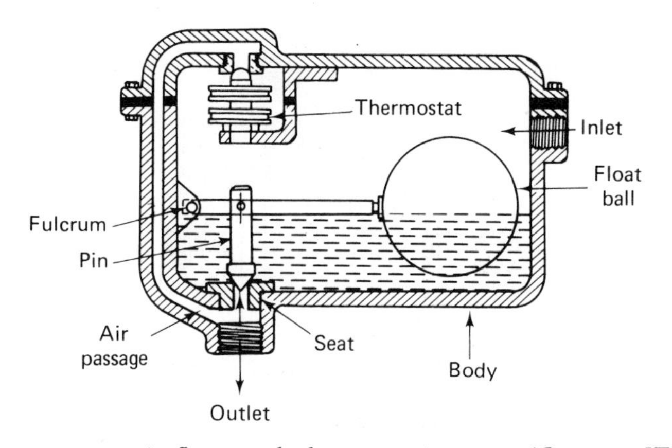

Figure 8–33. A float and thermostatic trap. (*Courtesy ITT Hoffman Specialty*)

They are also excellent choices for handling the condensate from fan-coil units such as unit heaters, unit ventilators, and ventilating coils.

2. *Float trap.* A float trap does not have an internal thermostatic air bypass. There are applications where air is not a problem and a float trap can therefore be used to handle condensate only. There are specifications where float traps are called for, using an external thermostatic air bypass. The external bypass is a thermostatic trap piped around the inlet and outlet of the trap body.

3. *Inverted bucket trap.* An inverted bucket trap is so named because the float is an inverted bucket that operates the leverage mechanism to open and close the trap (Figure 8–35). This trap will operate to discharge condensate at any temperature up to the saturated temperature corresponding to the steam pressure at the trap inlet. It operates in cycles at a

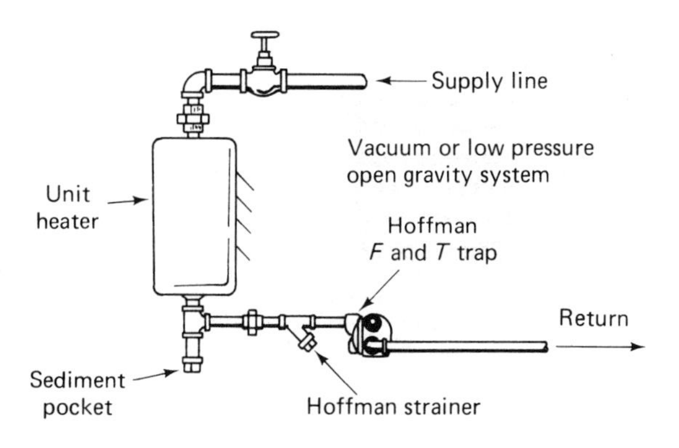

Figure 8–34. F and T trap application. (*Courtesy ITT Hoffman Specialty*)

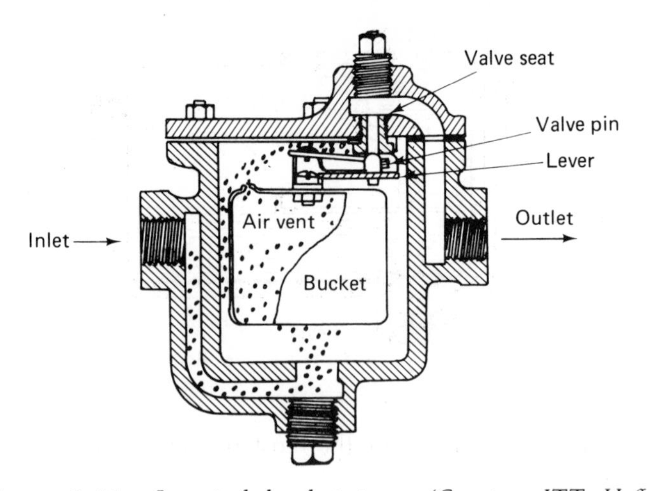

Figure 8-35. Inverted bucket trap. (*Courtesy ITT Hoffman Specialty*)

frequency that depends on the condensate load being handled. This inverted bucket trap is simply constructed and can handle condensate for many industrial types of requirements. It will handle the condensate from heating fan-coil units that must be lifted to discharge to the return mains located above the equipment (Figure 8-36). Before an inverted bucket trap is put into operation, it must be *primed* or filled with water. It operates best at or near full load conditions and where loads do not vary over a wide range.

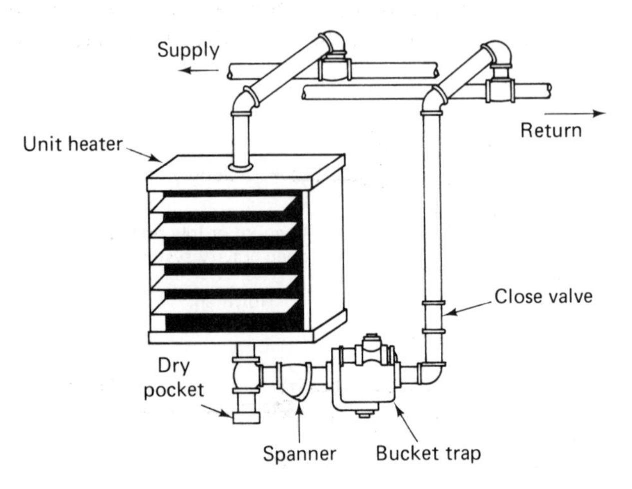

Figure 8-36. Installation of a bucket trap. (*Courtesy ITT Hoffman Specialty*)

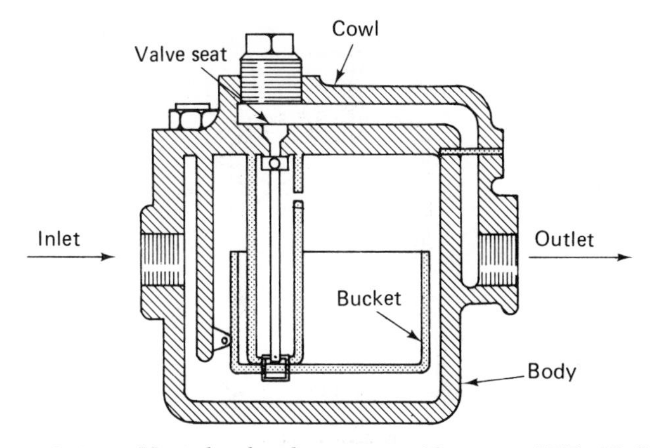

Figure 8-37. Upright bucket trap. (*Courtesy ITT Hoffman Specialty*)

4. *Upright bucket trap.* This trap operates on the principle of condensate entering the trap body to "float" the upright bucket (Figure 8-37). This action closes the discharge port. As condensate continues to enter the trap body, it rises to fill the space surrounding the bucket until it overflows into the bucket. When this occurs, the bucket sinks and the discharge port is opened. Steam pressure then discharges the condensate and the operating cycle is repeated.

Although the construction of this trap is not so simple as that of an inverted bucket trap, the upright bucket trap will handle applications having wide variations of load or pressure.

Main Vent Air Valves: For small two-pipe installations designed to return condensate directly to the boiler by gravity, air must be vented from the system. A main vent air valve must be used at the proper location. These vents are the same as those used for one-pipe systems and are installed at the end of the return main ahead of the point where it drops below the boiler water line to become a wet return.

The main vent valve is designed to permit air accumulations to be discharged (vented) to the atmosphere and to close against the passage of steam or water (Figure 8-38). The proper installation of a main vent is shown in Figure 8-39. The incorrect method is also shown.

Condensate Pump: When systems increase in size and higher steam pressures are required to circulate the steam, the condensate cannot be returned to the boiler by gravity. Some type of mechanical means must be used to perform this return function. The accepted method for modern steam heating systems is to use a condensate pump (Figure 8-40). The condensate pump consists of a receiver on which is mounted single or multiple electric motor-driven pump assemblies, float switches,

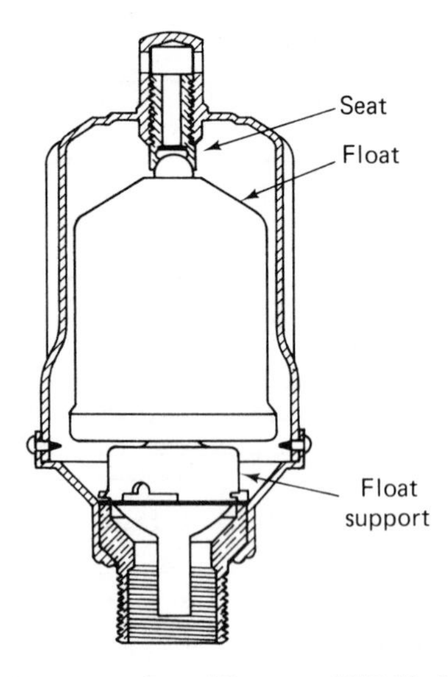

Figure 8-38. Main vent valve. (*Courtesy ITT Hoffman Specialty*)

and other electrical controls as required by the heating system. A condensate pump must be located at the low point of return and as close to the boiler as possible. The return main must be uniformly pitched ¼ inches to 10 feet to the receiver inlet so condensate can flow by gravity into the receiver. This is important so that condensate and air can be separated as it flows in the piping and the air can be vented to the atmosphere through the vent connection.

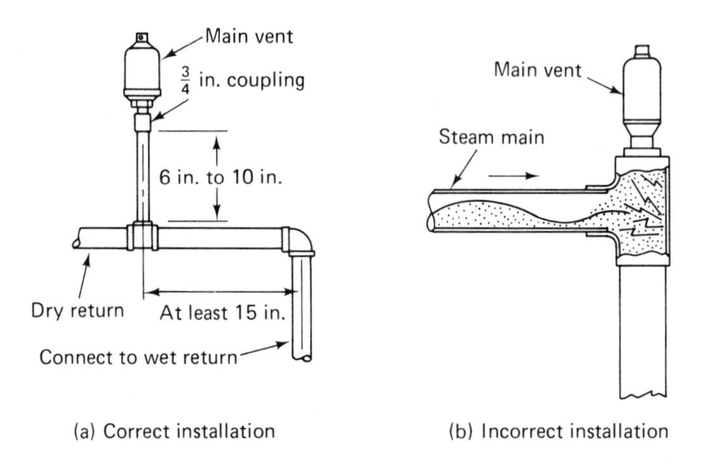

Figure 8-39. Correct and incorrect method of installing a main vent. (*Courtesy ITT Hoffman Specialty*)

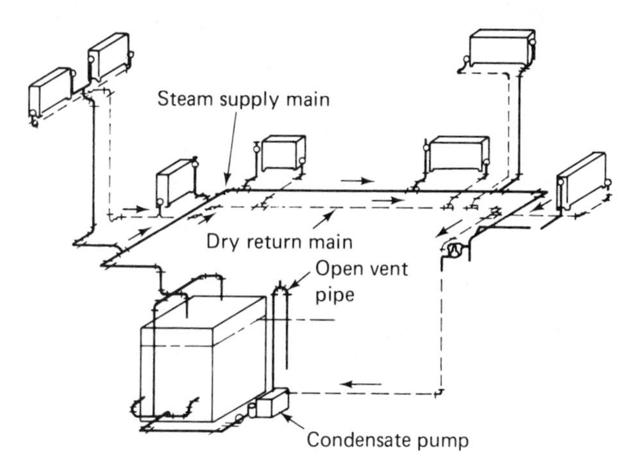

Figure 8-40. Mechanical condensate return. (*Courtesy ITT Hoffman Specialty*)

When it is necessary to locate the condensate pump below the floor level to obtain uniform pitch to the receiver inlet, the condensate pump can be installed in a pit or an underground type can be used (Figure 8-41).

The condensate discharge piping between the pump and the boiler must be equipped with a discharge valve located in the discharge piping adjacent to the pump discharge connection. A shutoff valve must be installed adjacent to the check valve and between it and the boiler to permit easy servicing of the pump unit.

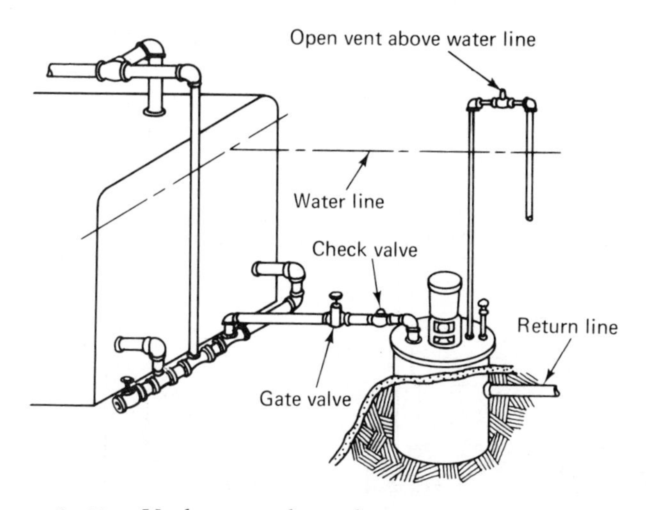

Figure 8-41. Underground condensate pump. (*Courtesy ITT Hoffman Specialty*)

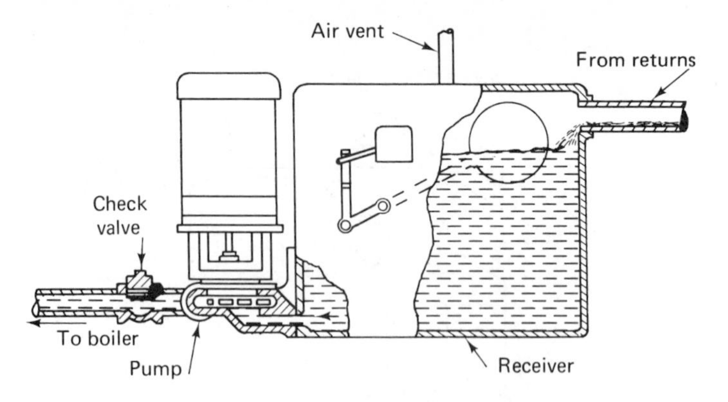

Figure 8–42. Condensate tank full. (*Courtesy ITT Hoffman Specialty*)

As condensate flows into the receiver (Figure 8–42), the water level rises until the float reaches its top position, at which time the electrical contacts are closed to start the pump motor. As condensate is pumped from the receiver, the water level is lowered sufficiently to cause the electrical contacts to open and stop the pump motor when the float has reached its lowest position (Figure 8–43). This cycle is repeated as often as necessary to handle the condensate from the system and maintain boiler water level at the proper height.

The float switch for a condensate pump is adjustable so that the quantity of water being discharged for each operating cycle can be set to satisfy the water level condition required for the boiler. Duplex condensate pumps are often equipped with mechanical alternators that will cause the lead pump to operate every other time. The alternator is designed to cause both pumps to operate together to provide double capacity when the system condensate load demands such operation.

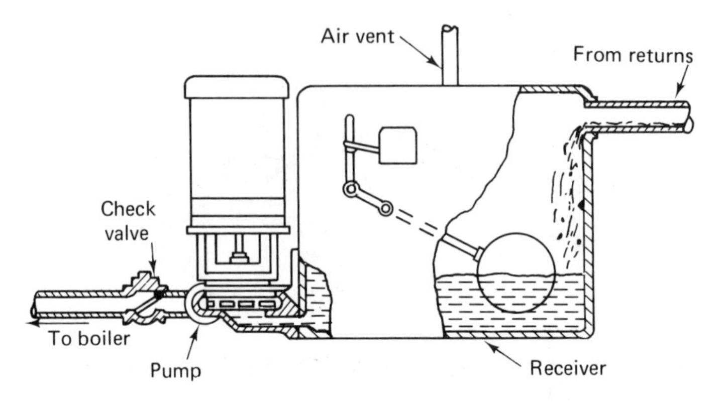

Figure 8–43. Condensate tank empty. (*Courtesy ITT Hoffman Specialty*)

REVIEW QUESTIONS

1. What is the main problem involved in heat transfer?
2. How is the stagnant gas film removed in a boiler?
3. Name four basic cast iron boiler designs.
4. What does the variety of nipple arrangements indicate in a boiler?
5. Where is the proper place to install a pump on a hot water boiler?
6. What can be installed to provide a means of adjustment for the desired division of positive and negative pressure gradient?
7. Does reversing the flow increase or decrease the heat transfer in a cast iron boiler?
8. What will be some indications of air in a hot water system?
9. What type of fire-box does not require a refractory?
10. What makes the sections and fire-box of a boiler air-tight?
11. Why are several short draw rods desired over long ones?
12. What is the purpose of the uptakes in a boiler?
13. What makes the fire-tube boiler simple in design?
14. What type of metal are the tubes made from?
15. What makes the fire tube boiler economical?
16. What eliminates hot spots in a fire tube boiler?
17. How is the gas ignited in a fire tube boiler?
18. How are rapid internal temperatures avoided in electric boilers?
19. How is a full flow of water assured in an electric boiler?
20. What type of heating element is used in electric boilers?
21. What makes electric boilers adaptable to zoning applications?
22. Why does steam release so much heat?
23. For what reason does steam promote its own circulation?
24. What does the pressure have to do with steam temperature?
25. What is used to make steam easy to control in a heating system?
26. Name two classifications of steam heating systems.
27. What is the function of a header?
28. What is the function of the Hartford loop?

29. What is the function of a steam trap?
30. What are two ways to install a condensate pump?
31. What is the purpose of the main vent air valve?
32. Where is a cooling leg used?

9

Comfort Heating Furnaces

The objectives of this chapter are:

- To bring to your attention the air requirements for gas furnaces.
- To introduce you to the furnace components and their sequence of operation.
- To acquaint you with the combustion gas flow through a gas furnace.
- To acquaint you with the air flow path through a furnace.
- To describe to you the temperature rise through a furnace and its use in heating systems.
- To instruct you in the insulation requirements of furnaces.

Heating furnaces and equipment are probably the most abused and misused machinery in the modern home. Normally, they are purchased and installed in too small a space and that is the last thought given to them until they stop performing.

VENTILATION AIR REQUIREMENTS

Gas furnaces, like people, are designed to function in a sea of air. Take away the air supply and problems occur—heating units that operate erratically or not at all.

In the "good old days," warm air furnaces were installed in the basement, surrounded by plenty of air to breathe. Modern day furnaces, however, are crammed into a space in the modern home the size of a broom closet. The furnace would still be all right, though, if someone would only remember that it takes lots of air.

For every 1,000 Btu of rated input, a gas unit requires a total of 45 ft^3 of replacement air, i.e., 15 ft^3 for complete combustion; 15 ft^3 dilution air at the draft hood; 15 ft^3 ventilation air for the control of the temperature within the furnace room.

If we take a warm air furnace, for example, with a rated input of 100,000 Btu, we would see that it needs 4,500 ft^3; this means that to satisfy proper operating conditions for the furnace requires 45 complete roomsful of air each hour. A good task for a crack under the door!

Modern architectural design created this problem by putting equipment in rooms much too small. Because replacement air and venting are interdependent, the responsibility for correcting this situation rests with the installer or, later, with the service technician. Fortunately, this situation can easily be corrected.

Following are some easy-to-use illustrations for replacement air supply.

Divide the combined rated appliance inputs by 1,000 to get the number of in.^{2} of free area required for each of two grills (Figure 9-1). Install both grills, one grill low and the other high above the draft

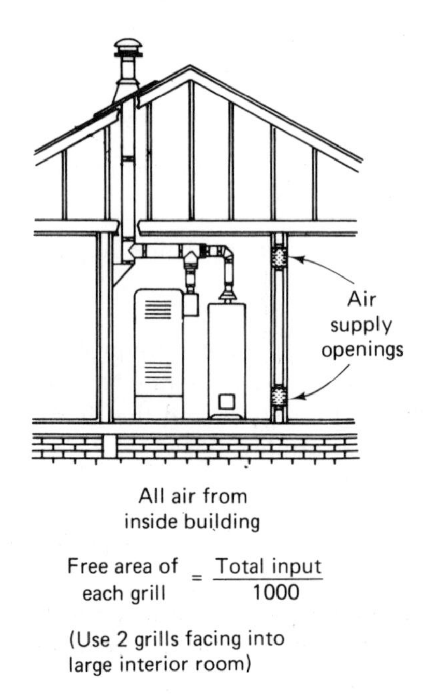

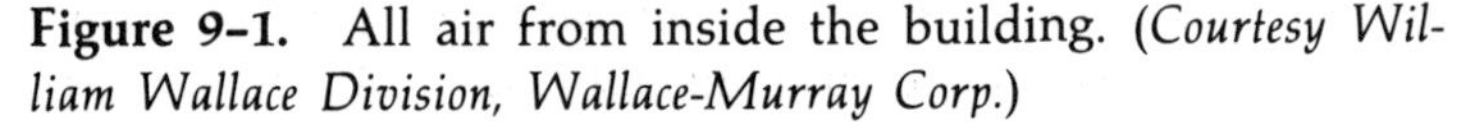
Figure 9-1. All air from inside the building. (*Courtesy William Wallace Division, Wallace-Murray Corp.*)

hood opening, to connect the appliance room to the large interior room as shown.

Divide the combined rated appliance by 2,000 (if openings are located in an outside wall, divide by 4,000) to get the number of in.^{2} of free area required for each of two grills (Figure 9-2). Install both grills, one high and one low, to connect the appliance room to the outdoors as shown.

Divide the combined rate appliance input by 4,000 to get the number of in.^{2} of free area required for each of two grills (Figure 9-3). Install both grills, one high and one low, to connect the appliance room to ventilated attic as shown.

Divide the combined rated appliance input by 4,000 to get the number of in.^{2} of free area required for each of two grills (Figure 9-4). Install both grills, one high and one low, to connect the appliance room to ventilated spaces as shown.

Sequence of Operation (Gas Furnace)

It should be evident that when the furnace is not in operation there are no air requirements. However, when the thermostat demands heating, demands are made of several components of the furnace and related equipment.

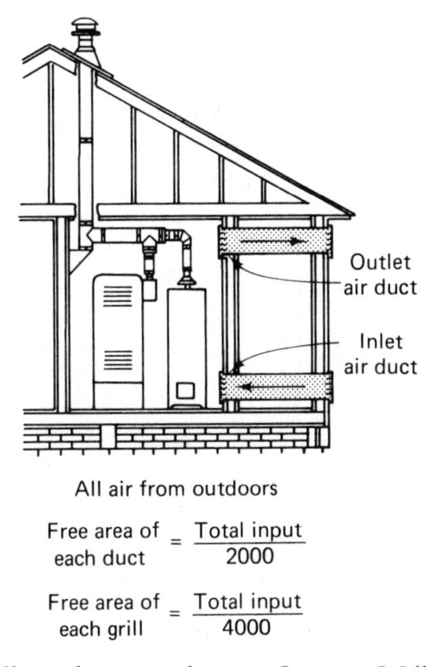

Figure 9-2. All air from outdoors. (*Courtesy William Wallace Division, Wallace-Murray Corp.*)

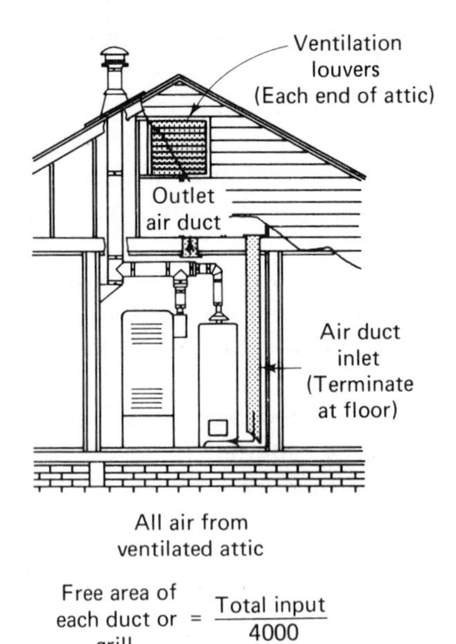

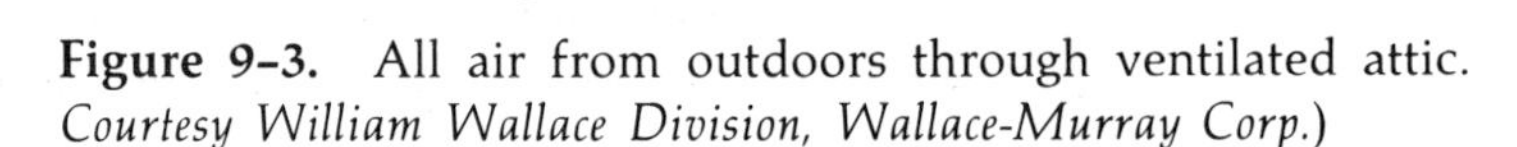
Figure 9-3. All air from outdoors through ventilated attic. *Courtesy William Wallace Division, Wallace-Murray Corp.*)

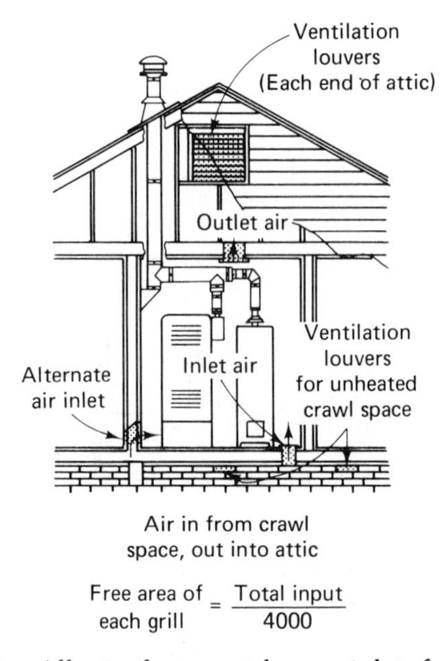

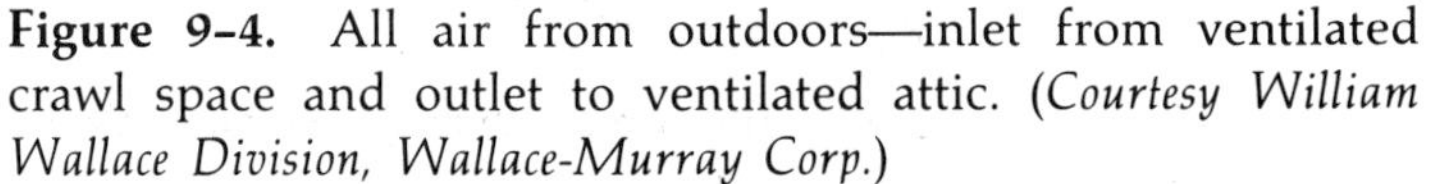
Figure 9-4. All air from outdoors—inlet from ventilated crawl space and outlet to ventilated attic. (*Courtesy William Wallace Division, Wallace-Murray Corp.*)

In operation, while the furnace is at rest the only functioning part is the pilot burner. When the thermostat calls for heat, gas is admitted from the gas distribution system to the furnace. Before the gas can enter the combustion area, it must pass through a gas pressure regulator where the pressure is reduced to approximately 3½ inches of water column. From the pressure regulator, the gas must go through a main gas valve, which functions on demand from the thermostat. These two components are usually combined into one housing and termed a combination gas control (Figure 9–5).

After the main gas valve has opened, the gas is admitted to the gas manifold pipe and through the main burner orifices. From the main burner orifices, the gas goes into the main burner, where it is mixed with primary air for combustion. As the gas flows from the main burner head, it is ignited by the pilot burner flame and a tremendous amount of heat is released.

At this point, secondary air is mixed with the flame to insure complete combustion. The gas has now been changed to vent gases that leave the combustion area (fire box) and release additional heat while passing through the heat exchanger (Figure 9–6). On leaving the heat exchanger the vent gases enter the venting system and flow to the outside atmosphere.

As the temperature within the circulating air passages of the heat exchanger is raised, the fan control completes an electrical circuit to the indoor fan motor, thus causing the air to be forced through the air filter, the furnace heat exchanger, and into the duct system where it is distributed to the individual rooms (Figure 9–7).

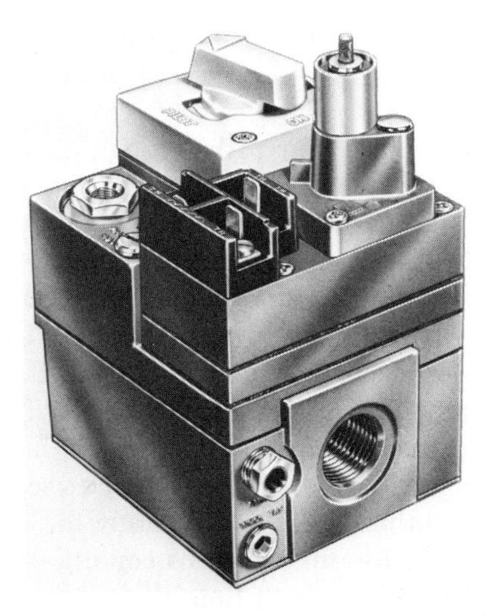

Figure 9–5. Combination gas control. (*Courtesy Honeywell Inc.*)

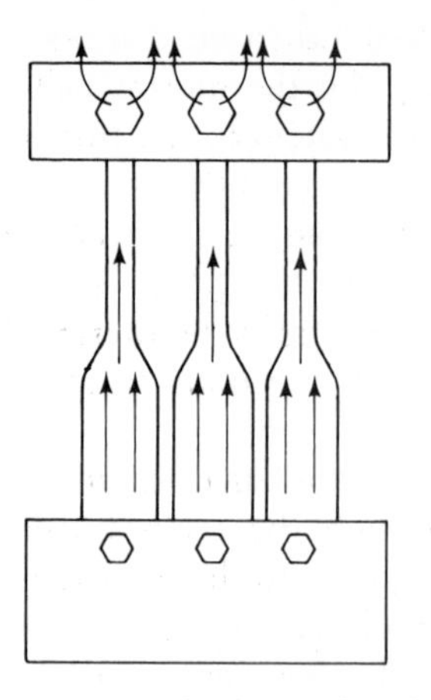

Figure 9–6. The vent gas path through a heat exchanger.

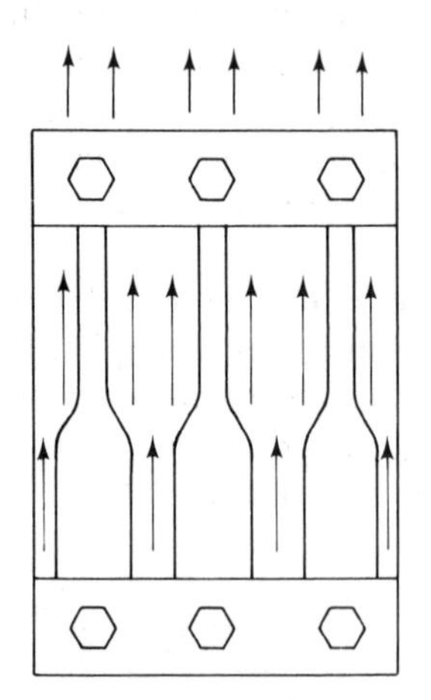

Figure 9–7. Circulating air path through a furnace.

The heating system is now in normal operation. Should the temperature within the furnace become unsafe, the limit control will interrupt the control circuit causing the supply of gas to stop.

As the thermostat becomes satisfied, the sequence of operation is reversed. The thermostat contacts open, allowing the main gas valve to close, thus stopping the flow of gas to the burners. The fan will continue to circulate the air until the furnace temperature has fallen sufficiently. When the furnace temperature reaches the lower set point, the fan control contacts open and the fan quits circulating the air. The heating system is now at rest and will remain in this condition until the thermostat demands additional heat, at which time the sequence of operation will be repeated.

Basically, there are three types of gas furnaces: the upflow; the counterflow; and the horizontal furnace. The direction of air flow through the furnace is the determining factor for its name. It should be noted, however, that these are single purpose, or single mounting, furnaces. That is, a horizontal furnace cannot be used in any other position. To do so would upset the burner and venting action.

ELECTRIC FURNACES

Even though electric heating is not new, forced air electric furnaces (Figure 9-8) have been available in large quantities for only a few years. Electric furnaces, although similar in appearance to gas furnaces, are quite different. They do not require gas piping or venting. They do, however, require that high capacity electric circuits be installed to the furnace (see Table 9-1). Electric furnaces are more flexible than gas furnaces because there is no specific position in which to install them. The only requirement is that all components be accessible for installation and service.

Sequence of Operation (Electric Furnace)

Although there are few differences in the operation of an electric furnace and a gas furnace, we will go through the sequence of operation for clarity. As the temperature in the conditioned area falls, the thermostat contacts close the electrical control circuit. The control circuit energizes some relays which completes the line voltage electrical circuit to

Figure 9-8. Forced air electric furnace. (*Courtesy Electric Products Manufacturing Corp.*)

Table 9-1
Ratings and Specifications

MODEL NUMBER	BTU	KW	ELECTRICAL RATING	AMPS	MIN. WIRE SIZE	NUMBER CIRCUITS
EFS-04	13660	4	240/1/60	16.6	#8	1
EFS-05	17065	5	240/1/60	20.8	#8	1
EFS-08	27300	8	240/1/60	33.3	#6	1
EFS-10	34130	10	240/1/60	41.7	#6	2
EFS-12	40960	12	240/1/60	50.0	#6	2
EFS-15	51200	15	240/1/60	62.5	#6	2
ERS-16	54610	16	240/1/60	66.7	#6	2
EFS-20	68260	20	240/1/60	83.3	#4	2
EFS-24	81920	24	240/1/60	100.0	#4	3
EFS-25	85325	25	240/1/60	104.1	#4	3
EFS-30	102390	30	240/1/60	12.0	#4	3

(Courtesy Dearborn Stove Co.)

the heating elements. When sufficient time has passed or the temperature has risen, the fan control turns on the blower to circulate the air as with the gas furnace. The limit controls on electric furnaces are on each individual heating element and respond to that temperature.

When the thermostat is satisfied, the control circuit is interrupted and the heating relays remove the elements from the line voltage circuit. After the required time has lapsed, or the temperature has fallen enough, the fan motor will stop. Thus, the heating unit is at rest.

The component parts of an electric furnace are shown in Figure 9–9. Note that relays are used in place of a main gas valve and that electric heating elements replace the main burners, pilot, and heat exchanger. Also ventilation air is minimal and no combustion air is required.

BTU COMPARISON

Gas Furnaces

The Btu rating of gas furnaces is certified by the American Gas Association. The input rating per hour is placed on the name plate, a permanent part of the furnace. The input rating is always higher than the output or bonnet capacity, which is usually about 80% of the input rating of the furnace. This 20%, known as waste gas, is used to operate the venting system. Vent gas is not actually wasted; however, it is not used to heat the conditioned area and is, therefore, termed *waste heat*.

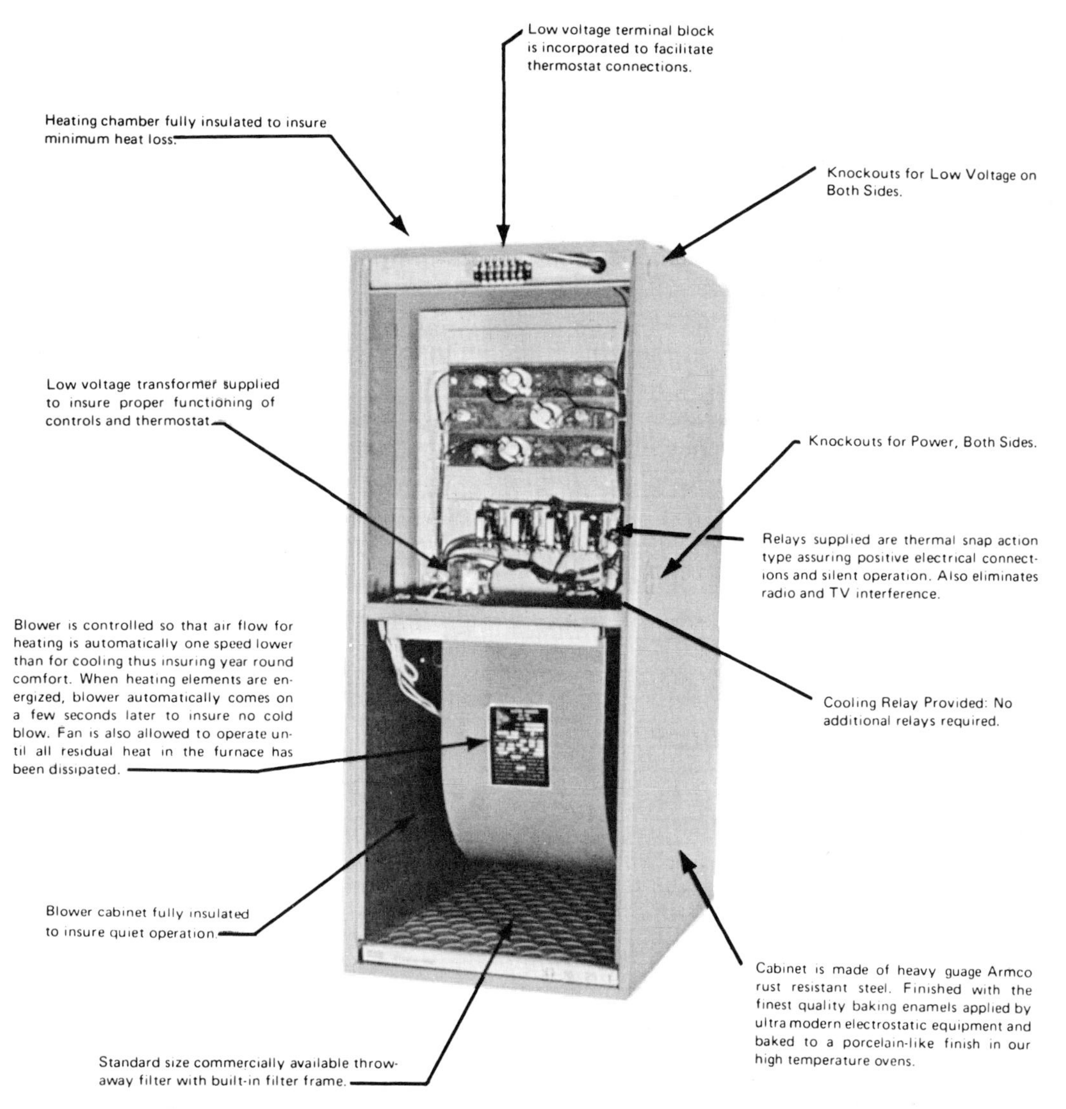

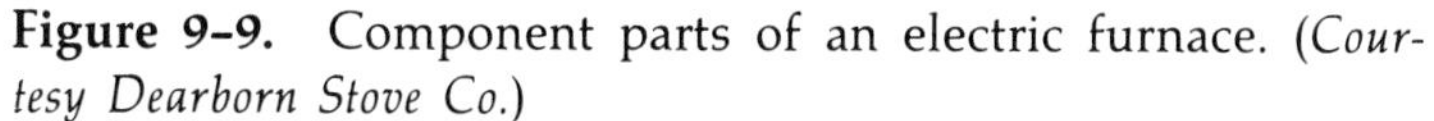
Figure 9–9. Component parts of an electric furnace. (*Courtesy Dearborn Stove Co.*)

The input rating of a gas furnace is the maximum amount of Btu input that can safely be allowed to that furnace. Most manufacturers design their equipment to operate close to these ratings. However, these furnaces can be operated safely on about 80% of the input rating. Any input less than this is dangerous because there will not be enough heat to properly operate the vent system. If the input Btu is increased above the name plate rating, two hazards exist: (1) the vent system cannot remove the extra gas properly; and (2) the extra heat at the burner head will tend to overheat it along with the heat exchanger, thus resulting in possible burnout of both the burners and the heat exchanger. Both conditions are costly and hazardous.

Electric Furnaces

The Btu rating of electric furnaces is certified by the National Electrical Manufacturers Association (NEMA). Electric furnaces are rated by the kilowatt (kW) input per hour. The kilowatt are then converted to obtain the Btu rating (see Table 9–2). Electric furnaces are virtually 100% efficient. That is, for each Btu input to the furnace, almost 1 Btu of heat is delivered to the conditioned area. No heat is used for vent operation. The Btu rating of an electric furnace can be changed by adding or removing heating elements to meet the requirements. These elements may be added as long as there is a space available to install them.

Table 9–2
Btu/kW Rating

Model Number	kW @ 240 V Nom.	Btu @ 240 V Nom.	Total 1-Ph Amp.	Max 3-Ph Amp.
PT-2-1-03	3	10,681	16	NA
PT-2-1-04	4	14,244	22	NA
PT-2-1-05	5	17,805	26	NA
PT-2-1-06	6	21,365	29	NA
PT-2-1-08	8	28,487	39	NA
PT-2-1-10	10	35,610	48	40
PT-2-1-12	12	42,730	54	44
PT-2-1-13	13	46,292	60	NA
PT-2-1-15	15	53,415	70	40

(Courtesy Electric Products Mfg. Co.)

TEMPERATURE RISE

The temperature rise through a furnace is the rise in temperature of the air as it passes through the heating elements. The Btu input and the air delivery of the blower are the major factors determining the temperature rise.

The temperature rise in gas furnaces will normally be rated from 60° to 100° F, depending on the unit design conditions. It is usually stated on the furnace name plate. Thus, the discharge air temperature of a heating furnace warming an 80° F area will be 80° F + 60° F = 140° F minimum.

An electric furnace will be rated from 40° to 80° F temperature rise. This lower temperature rise has several advantages. First, the conditioned area temperature will be more uniform due to the longer running time. Second, the relative humidity of the conditioned area will be higher, resulting in lower conditioned area requirements. The temperature rise on an electric furnace will also depend on the number of elements operating at a given time.

REVIEW QUESTIONS

1. Under what ambient conditions are furnaces designed to operate?
2. How many ft^3 of air per hour does a gas furnace require?
3. Where should the combustion and ventilation air openings be located?
4. What is the purpose of ventilation air?
5. What is circulating air?
6. Should the ventilation air and the circulating air be allowed to combine?
7. What causes the indoor fan motor to operate?
8. Can a gas furnace be installed in any position?
9. What is the purpose of the main burners?
10. Can electric furnaces be installed in any position?
11. The heating elements take the place of what parts of the gas furnaces?
12. Why can electric furnaces be installed in a tighter space than gas furnaces?
13. What organization certifies gas furnaces?
14. What organization certifies electric furnaces?
15. What is the maximum allowable percentage above the rated input Btu at which a gas furnace can be safely operated?
16. What is the minimum allowable percentage Btu at which a gas furnace can safely be fired?

17. How are electric furnaces rated?
18. How may the input rating of an electric furnace be changed?
19. Define the temperature rise through a furnace.
20. Will the relative humidity be higher with a gas furnace or an electric furnace?

10

Infrared Heating

The objectives of this chapter are:

- To instruct you in the theory of infrared heating.
- To acquaint you with different methods used in infrared heating.
- To bring to your attention the desired location of infrared heating units.
- To introduce you to how space heating is accomplished with low air temperatures.

Large open buildings and areas exposed to strong drafts, such as shipping docks, can create many problems for the heating contractors responsible for installing the equipment used for human comfort and for the protection of materials and equipment.

When conditions such as these exist, infrared heating will often provide the desired solution. Infrared heating is a form of radiation that resembles and behaves like light rays. Its invisible rays race through space at the speed of light to warm solid objects such as floors, equipment, and people, before the air is heated.

Like the sun, heat generated by an infrared energy source is absorbed rapidly by solid objects that in turn release heat to the air around them.

INFRARED THEORY

Heat may be transmitted from one body to another without altering the temperature of the intervening medium. That this is true may be proven by a simple, everyday experience. If you stand before a fire or a hot radiator, you experience a sensation of warmth that is not due to the temperature of the air. This is evident if a screen is placed between you and the fire: the sensation immediately disappears. Such would not be the case if the air had a higher temperature.

This phenomenon, which is called thermal radiation, is but one of the many forms of radiant heat energy that is continuously being emitted and absorbed in various degrees by all bodies. All radiant energy may be regarded as a form of wave motion known as *electromagnetic phenomena*. This type of wave motion should not be confused with sound waves, spring vibrations, and other elastic mechanical waves that may occur in solids, liquids, and gases only. Radiant energy waves may be transmitted even through a vacuum.

Infrared radiation is a form of energy that is propagated as an electromagnetic wave, like radio waves, light waves, and X-rays. The energy is transmitted at a velocity of 186,272 mps; the common source of infrared radiation is a hot radiating body. While being transmitted this way, energy is called *radiant energy* and is not heat. The energy will be instantaneously converted to heat on being absorbed by an absorbing medium.

A given body under a fixed set of conditions will emit radiation of various wave lengths. The intensity of the radiation in the various wave lengths is different. The type of radiation emitted is characterized by the band of wave lengths having the greatest intensity. The curves in Figure 10–1 show the distribution of intensity of radiation with wave lengths for a *black body*. (A black body is one that emits the maximum possible radiation at a given temperature. The adjective has nothing to do with the color of the body.) Table 10–1 provides the same information in table form. An examination of this table or of the curves shown in Figure 10-1 would result in the following conclusions:

1. An increase in temperature causes a decrease in the wave length at which maximum energy emission occurs.

2. An increase in temperature causes a rapid increase in energy emission at any given wave length and the total energy emission.

3. The total rate of energy emission at any given temperature and for any range of wave lengths is given by the area under the curve for that temperature taken over the wave length range being considered.

The differentiation among various types of radiation, such as light

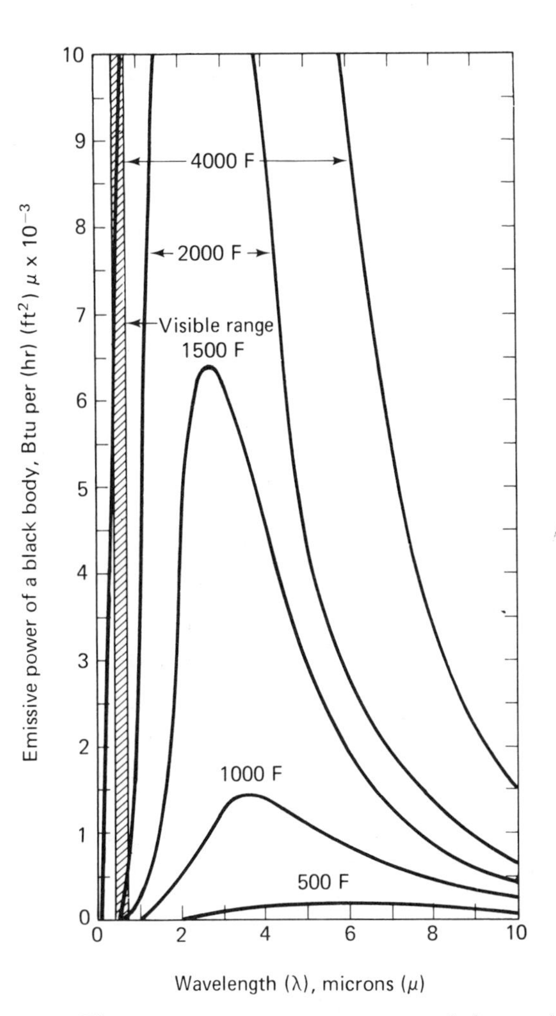

Figure 10–1. The emissive power generated by a black body. (*Courtesy Thermal Engineering Corp.*)

radiation, thermal radiation, etc., is rather indefinite since radiation of all wave lengths will heat bodies to some extent.

In the use of high intensity gas fired infrared generators, we will be primarily concerned with the infrared portion of the spectrum (Figure 10–2). A continuous interchange of energy among bodies results from the reciprocal process of radiation and absorption. Thus, if two bodies are at different temperatures within an enclosure, the hotter body receives from the colder body less energy than it radiates. Consequently, its temperature decreases. The colder body receives more energy than it radiates, and its temperature increases.

Table 10–1
Black Body Thermal Radiation—Btu/hr-ft^2
(1) In Various Wave Length Bands (Quality)
(2) Total Thermal Radiation (Quantity)
(3) Wave Length at Maximum Radiation

Source Temp °F	(1) Wave Length Band—Microns 0–0.7	0.7–1	1–2	2–3	3–4	4–5	5–6	6–8	8–10	10+	(2) Total Thermal Rad.	(3) Wave Length At Max. Rad. Microns
500			1	27	99	155	174	308	215	482	1460	5.43
600			3	70	201	278	285	442	300	590	2170	4.92
700			12	162	376	445	414	620	376	705	3110	4.50
800			32	325	620	672	575	809	468	840	4340	4.14
900			76	576	963	952	752	1035	565	960	5880	3.84
1000			174	960	1400	1305	975	1250	660	1095	7820	3.57
1100			316	1525	1982	1628	1245	1510	765	1220	10,190	3.34
1200		1	556	2250	2673	2053	1518	1763	876	1370	13,060	3.14
1300		3	973	3200	3424	2550	1830	2010	1025	1483	16,500	2.96
1400		8	1564	4404	4380	3090	2059	2340	1095	1650	20,590	2.80
1500		18	2388	5960	5335	3600	2415	2640	1245	1780	25,380	2.66
1600		34	3557	7720	6410	4340	2715	2910	1365	1920	30,970	2.53
1700		71	5059	9840	7570	5050	3035	3215	1540	2060	37,440	2.41
1800		126	7040	21,310		5525	3450	3590	1620	2200	44,870	2.31
1900	5	225	9550	25,440		6300	3840	3890	1760	2350	53,360	2.21
2000	12	13,090		29,935		11,210		4220	1965	2455	62,990	2.12
2500	157	43,125		58,650		17,300		6015	2500	3155	131,200	1.76
3000	1281	109,660		96,850		23,900		7640	3205	3940	246,400	1.51
4000	20,415	405,560		194,620		38,110			21,780		680,500	1.17
5000	129,900	1,005,400		311,750		51,950			29,935		1,528,000	0.96

(Courtesy Thermal Engineering Corp.)

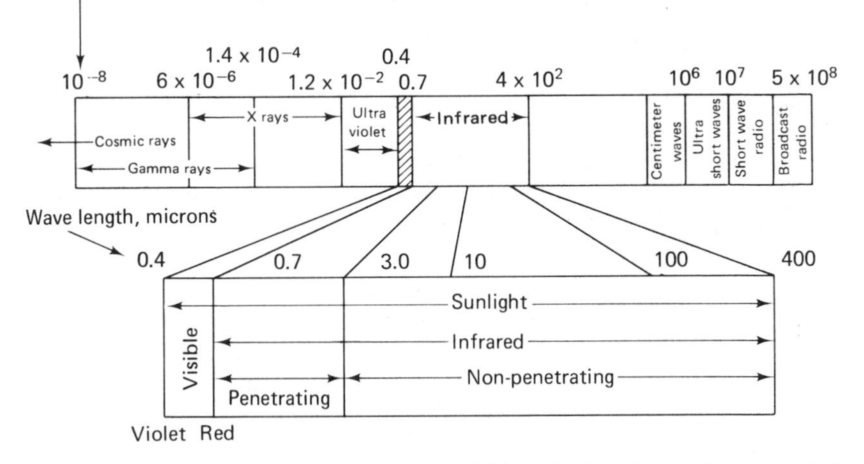

THERMAL RADIATION, which includes both visible and infrared waves, is but a part of the spectrum of the family of electromagnetic waves. The quantity and quality of thermal radiation depends solely upon the temperature of the surface emitting the radiation. All electromagnetic waves are transmitted at the speed of light. All heated solids and liquids, and some gases emit thermal radiation. Conversely, these materials absorb a part of the visible and infrared waves and convert them to heat. At surface temperatures greater than 1000 F, a part of the radiation is visible

Figure 10-2. Thermal radiation. (*Courtesy Thermal Engineering Corp.*)

This interchange of energy continues even after thermal equilibrium is reached, except that both bodies then receive as much energy as they radiate. According to this theory, which agrees well with observations, body would cease to emit thermal radiation only when its temperature had been reduced to absolute zero.

The higher the temperature of an emitting body, the faster it radiates infrared energy. Obviously, the amount of energy radiated per unit of time will also be proportional to the amount of exposed surface.

Relationships involving thermal radiation are expressed by a number of radiation formulas. Some have been deduced theoretically and have been proven true by experiment, whereas others are purely empirical. Only the equation that relates to the emission of energy from infrared heaters will be discussed here. However, to understand these formulas, their uses and their limitations, it is necessary to understand some of the definitions and phenomena of radiation theory.

Absorptivity and Emissivity

Absorptivity is defined as the ratio of radiant energy by an actual surface at a given temperature to that absorbed by a black body at the same temperature. A black body can also be defined as one that will absorb all incidental radiation. In nature there is no truly black body for thermal radiation; however, the black body concept is useful in comparing the absorption and emission of radiant energy by different materials.

The *emissivity* of a surface is the ratio of the energy it emits to that emitted by a black body at the same temperature. The emissivity of most materials increases with the temperature of the material.

In nature, absorptivity and emissivity for most materials are approximately equal and have a value of less than 1.0. This is true for most materials at temperatures normally encountered. However, absorptivity depends on the quality of the incident radiation as well as the surface temperature. If the incident radiation is from a high source such as the sun, the emissivity and absorptivity may differ markedly. In space heating applications, the quality of the radiation is an important factor. For this reason high intensity gas-fired heaters operate at a surface temperature that allows the energy to be emitted at wave lengths that will be readily absorbed by most materials encountered in space, spot, or process heating applications.

Total Energy Emission

Infrared energy emitted by a black body depends on its absolute temperature only and can be simply computed by use of the Stefan-Boltzman law which states:

$$E_{bb} = aT^4$$

where

E_{bb} = total thermal radiation emitted by a black body
a = Stefan-Boltzman Constant
= 0.1713×10^{-8}.
T = Absolute temperature °R

Column 2 of Table 10-1 provides E_{bb} for a range of temperatures from 500°-5,000° F (960°-5460° R).

The foregoing has all been related to a perfect radiator. Since all infrared devices are less than perfect, some modification of the black body laws is needed to provide good engineering estimates of the thermal radiation emitted by the device.

By including the emissivity of the actual emitter and its effective area in the Stefan-Boltzman law, a good approximation of the thermal radiation emitted can be made. The formula then becomes:

$$E_{act} = AeaT^4.$$

where

E_{act} = total thermal radiation emitted by the actual emitter surface

A = effective or solid area of the emitter surface

e = emissivity of the emitter surface

a = Stefan-Boltzman constant of 0.1713×10^{-8}.

T = absolute temperature of the emitter surface °R.

The Thermal Engineering Corp. has developed a radiation calculator (see Table 10–2), which can be used instead of the above calculation. The dashed heavy line shows the total thermal radiation that can be expected from the TEC B45 or PB45 ceramic surface (Figure 10–3).

The burner used in these or any other models is designed to produce a ceramic surface temperature of 1,739° F.

The overall area of a TEC B45 burner is 156 in.^{2}. However, the slots occupy 29% of the total area; therefore, the effective area or solid area is 0.71×156, or 111 in.^{2} A scale for the solid area percentage is included on the Thermal Engineering Corp.'s *Thermal Radiation Calculator* (see Table 10–2). Emissivity of the Thermal Engineering Corp. ceramic is approximately 0.90.

If the surface temperature of the actual emitter has been taken as the direct reading of an optical pyrometer, no correction need be made for emissivity.

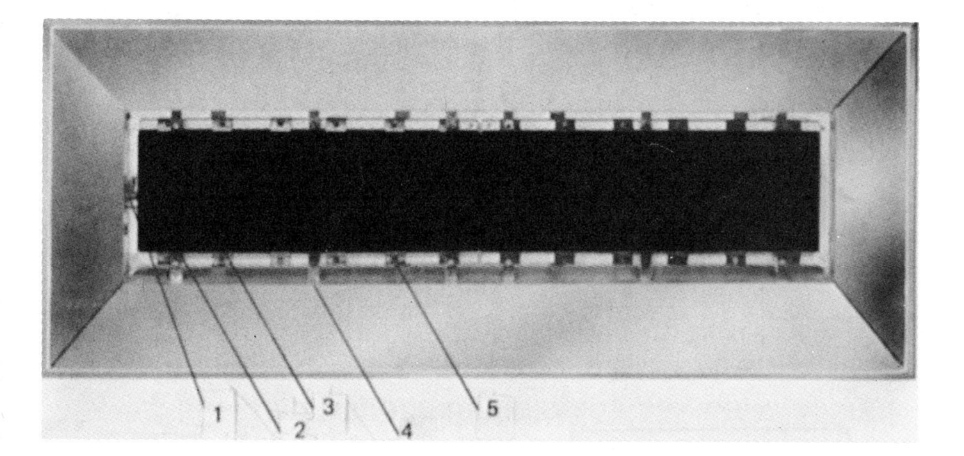

Figure 10–3. Infrared burner unit. 1. Pilot burner assembly. 2. Ceramic burner surface. 3. Full nichrome screen. 4. Screen mounting bracket. 5. Ceramic assembly mounting clip. (*Courtesy Thermal Engineering Corp.*)

Table 10-2
Thermal Radiation Calculator

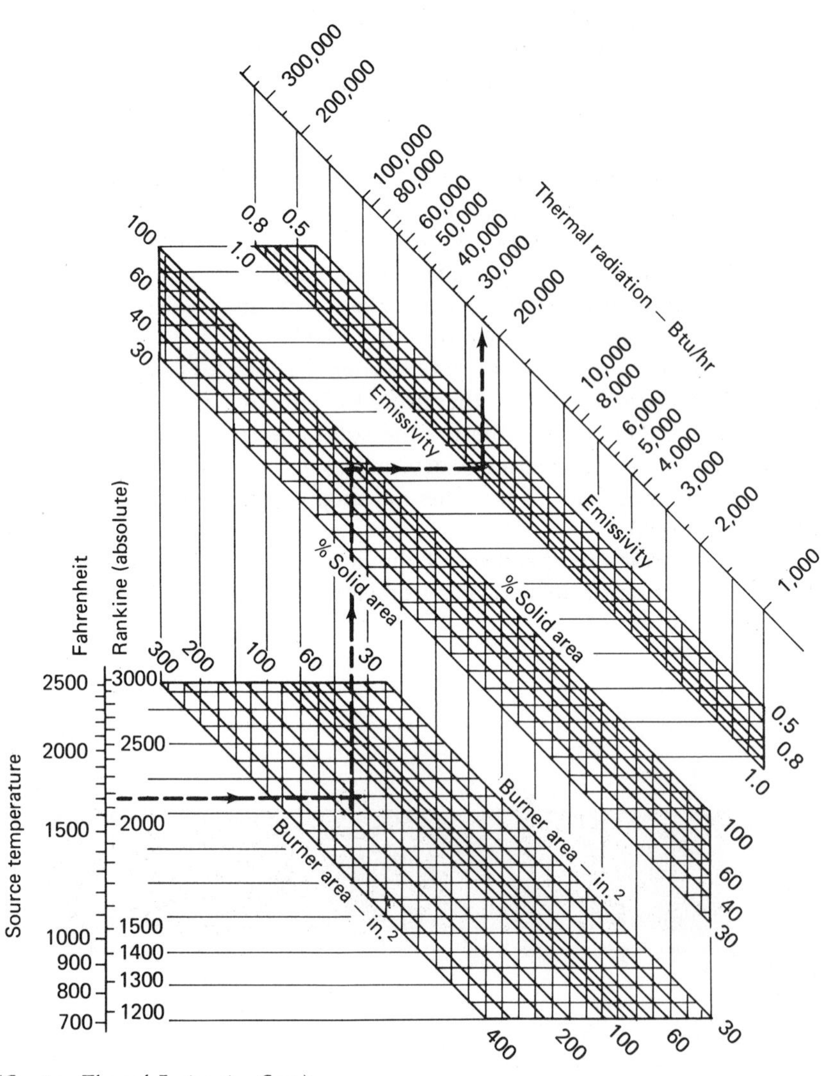

(Courtesy Thermal Engineering Corp.)

(Note: An optical pyrometer reading must be corrected for the emissivity of the surface being measured. Therefore, if the direct reading is taken, the emissivity can be taken as 1.0 and the actual emissivity need not be known.)

Infrared Energy Exchange

To determine the actual energy transferred between two objects, additional factors must be considered. One is the emissivities of both objects and the other is the portion of energy leaving one object that strikes the other. Because of the difficulty in obtaining these factors, no attempt will be made here to describe their determination and use.

The Thermal Engineering Corp. has used these principles in determining energy deliveries of each heater, and this data appears in Tables 10-3 and 10-4.

Table 10-3
Energy Transfer Between Objects
Btu hr/ft^2 delivery of B30—30° angle—to a vertical surface 1 ft × 1 ft—midpoint 3 ft above floor

	d FT									
h FT	2	4	6	8	10	12	14	16	18	20
7	125	119	81	52	35	25	18	14	12	9
9	44	61	54	42	31	23	18	14	12	9
11	19	28	33	30	25	20	16	13	10	9
13	10	18	21	21	19	17	14	12	10	9
15	5	11	14	15	15	13	12	10	9	8
17	4	7	10	11	11	10	10	9	8	7
19	3	4	6	8	9	9	8	7	6	6
21	2	3	4	6	6	6	6	6	6	5
23	1	3	3	4	5	5	5	5	5	4

(Courtesy Thermal Engineering Corp.)

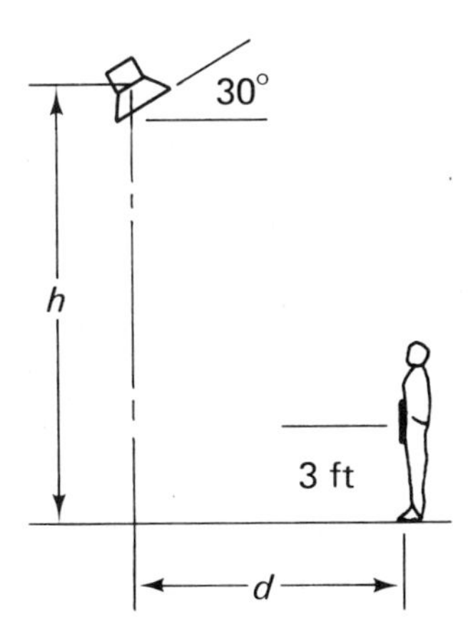

Table 10-4
Energy Transfer Between Objects
Btu hr/ft² delivery of PB 30—30° angle—to a vertical surface 1 ft × 1 ft—midpoint 3 ft above floor

h FT	*d* FT 2	4	6	8	10	12	14	16	18	20
7	405	242	110	57	27	13	7	3		
9	104	194	109	64	38	25	15	10	6	4
11	37	91	92	61	40	27	19	15	9	7
13	18	48	74	55	39	28	20	15	12	10
15	8	26	46	49	37	27	21	16	12	11
17	6	15	26	39	32	25	21	16	13	10
19	4	9	16	27	30	25	20	14	13	9
21	3	6	10	16	22	20	17	13	12	9
23	2	4	7	12	17	18	15	13	12	9

(Courtesy Thermal Engineering Corp.)

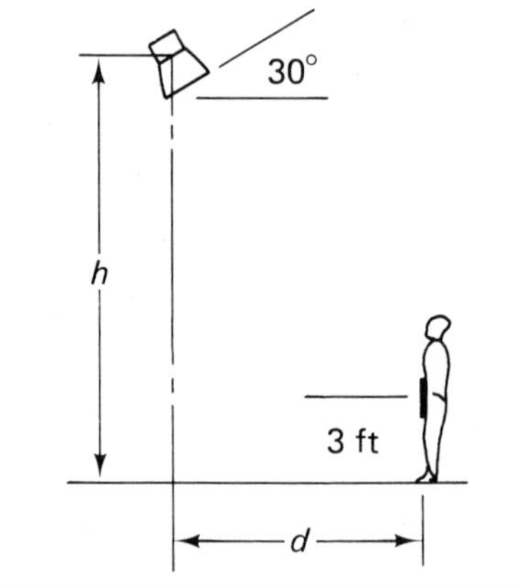

MODEL NO.	1 UNIT	4 OR MORE UNITS IN LINE** 10 FT APART	14 FT APART
PB30	1.0	2.16	1.52
PB45	1.5*	3.24	2.28
PC55	1.8*	3.89	2.74
PB60	2.0*	4.32	3.04
PB75	2.5*	5.40	3.80
PB90	3.0*	6.48	4.56
PC110	3.6*	7.78	5.47

*For section through short axis of unit. Units deliver elliptical pattern. Delivery is greater to sides.
** For average delivery

SPACE HEATING WITH GAS-FIRED INFRARED HEATERS

An intervening heat transfer medium is not required in the transfer of energy by radiation. This phenomenon allows high intensity radiant heaters to be employed for space heating very effectively. Three methods exist by which heat may be transferred: conduction, convection and radiation.

In the conventional methods of space heating, air is used as the intervening medium to transfer the energy from its source to the people or objects to be heated. In such a system, convection and, to some lesser degree, radiation are employed. The energy is transmitted from the flame by conduction, radiation, and convection through a heat exchanger, generally to air or water. The energy is then distributed to the area to be heated by the use of convection.

Air currents passing over an individual subject at temperatures less than body temperatures will cause a cooling sensation to be experienced by the individual subject. Consequently, when convection means are employed to heat a space, the air leaving the heat exchanger must be at a relatively high temperature in comparison to the body temperature. This is essential to prevent a chilling effect on the occupants of the building.

The temperature required to create this condition makes the air buoyant or lighter than the surrounding air and causes it to rise toward the ceiling. In most applications where convection heat is used, the temperatures at the top of the building are generally higher than those at floor level because of the buoyant effect of the warm air.

In gas-fired infrared heaters, actually, two heat exchangers are employed in the distribution of energy. The emitting surface of the burner is a radiant heat exchanger that transmits the infrared energy to the various absorbing mediums in the surrounding area below the heater. The energy is instantaneously converted to heat on absorption of the infrared energy by these absorbing mediums.

The air circulating in close proximity to the various materials and objects that have absorbed the infrared energy is heated. Thus, in effect, the floor and the other equipment and structures in the building become a very large heat exchanger, reflector and radiator for the infrared energy.

One important difference in the operation of an infrared system is that the floor and other objects that become the heat exchanger are heated to a much lower temperature than is the heat exchanger in a convection heater. The air circulating in close proximity to the floor and other objects that have absorbed the energy cannot be heated any higher than the temperature of the objects themselves. Consequently, the air heated in an infrared system does not have the buoyant effect found in a conventional convection system. Because the air is not buoyant and because it cannot be heated any warmer than the heat exchanger from which it receives its heat, the heat must be concentrated in the working area of the building instead of rising to the roof.

Since the air is warmest closest to the floor and other absorbing mediums, it must of necessity decrease in temperature as it attempts to rise. Consequently, the heat loss through the walls and ceilings of the building is decreased because of the lower ambient temperatures prevailing on the upper portion of the walls and ceiling.

Since the energy has been distributed by a radiant method of heat transfer, no high velocity air circulation is required that would otherwise create uncomfortable conditions because of the relatively low air temperature.

Naturally, the exhaust gases escaping from the infrared heaters are buoyant. They will rise toward the ceiling and will help offset the ceiling or roof losses. The mounting height of the infrared heaters will determine the temperature of the ceiling. Table 10-5 accounts for variations in installation height as related to ceiling height. The use of this table will be explained as follows:

Obviously, lower ceiling temperatures will result when the heaters can be installed at low levels in relationship to the ceiling height. When

Table 10-5
Installation Height as Related to Ceiling Height

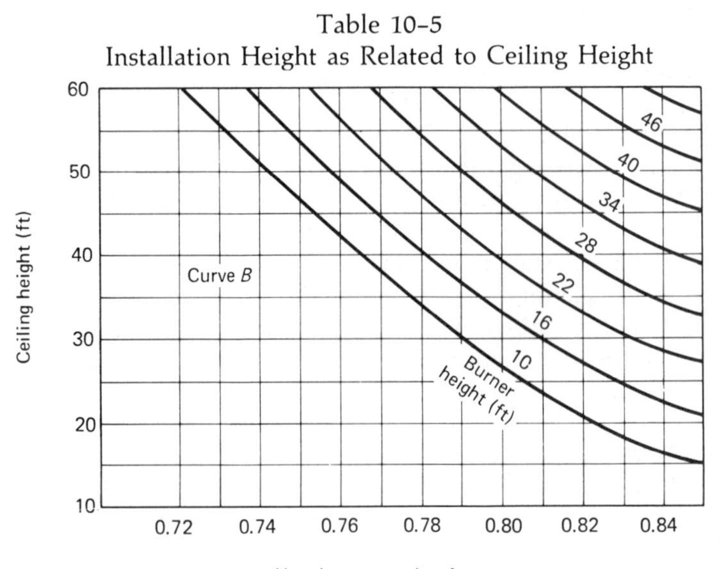

(Courtesy Thermal Engineering Corp.)

the heaters are installed at a low level, the surrounding air will have an opportunity to dilute the exhaust gases and lower their temperature before the gases progress to the level of the ceiling. Under such conditions, the energy of the exhaust gases becomes contained in a greater volume of air resulting in relatively low ceiling temperatures.

The radiant energy accounting for more than one-half of the total energy input to the heater is available to offset heat losses of the floor, lower walls, and infiltration. Stated in another manner, infrared energy emitted from high bay heaters does not heat the air first. This results in heat conservation. To a lesser degree, additional heat is conserved since the wall and roof internal surface coefficients are decreased because of reduced air movement resulting from use of infrared heaters

This decrease lowers the *U* values or, in other words, provides a measure of insulation without changing construction.

Table 10–5 gives consideration to the above factors and allows the conventional heat loss to be corrected where a conservation of energy can be realized with the installation of an infrared heating system.

It has been shown that where the occupants of an area are exposed to direct radiation from the heaters and/or the reduction of radiation losses to warmer floors and surrounding materials, lower ambient air temperatures can be used. In fact, in most cases it is insisted on. This further reduces energy consumption and, if anticipated, can result in lower initial system costs.

SPACE HEATING AT LOW AIR TEMPERATURES

The previous section on space heating dealt only with the heating of a building where air temperatures are maintained at or slightly below traditional levels.

Another approach to heating an entire space is to provide a thermally comfortable environment and yet allow the inside air temperature to fall well below traditional levels.

When heating an entire space at lower than normal air temperature, the comfort of the space is governed by two factors: (1) the direction of the radiation received by the subject from every heater and surface that his body "sees" and (2) the warming of the air in the space that results from convection heat transfer from the floor and other surfaces that are heated. A balance must be achieved between the two heating effects within which the body can adjust to attain comfort. Where infiltration rates are high, direct radiation can be depended on to minimize the cost of heating the infiltrating air. At very high air change rates caused by industrial ventilation, consideration must be given to the use of make-up air systems to temper the incoming air required to maintain a suitable pressure balance.

Where infiltration is minimal, the building is fairly well insulated and has relatively low ceilings and the air temperature closely approximates normal room temperatures, direct radiation is of minor consequence. However, when these conditions are absent, the direct radiation becomes a major factor in the total comfort.

As heat loss increases per ft^2 of floor space, the benefits of direct radiation absorptivity by the occupants becomes of greater importance in establishing comfortable conditions. Table 10-6 is intended to correct normal heat losses where the absorption of direct radiation is beneficial to the extent that ambient temperatures on the inside of the building can be lowered, resulting in lower heat losses. Also, less energy would be lost due to air changes when lower ambient temperatures exist inside the building. Table 10-6 also corrects for energy conserved in the air changes due to the lower inside temperatures.

Table 10-7 corrects for the energy conserved due to lower ceiling temperatures when the heaters are installed at a lower level in relation to the ceiling.

Determination of Energy Input

1. Calculate the conventional heat loss of the building using the normal inside air temperatures.

Table 10-6
Normal Heat Loss Correction

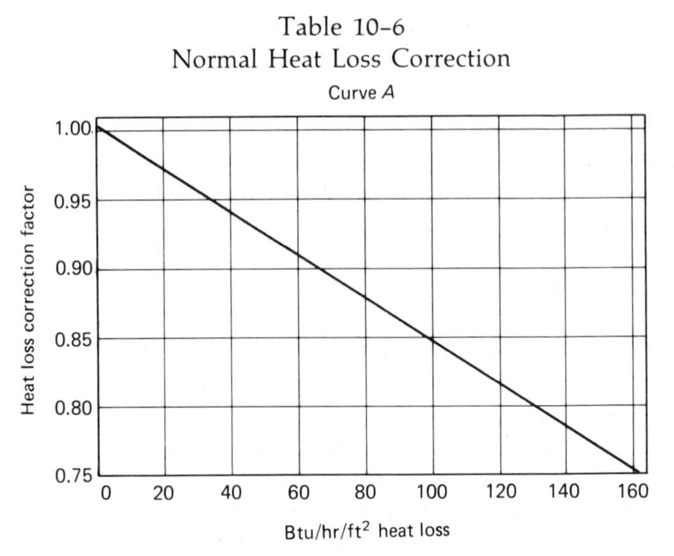

(Courtesy Thermal Engineering Corp.)

Table 10-7
Lower Ceiling Temperature Correction

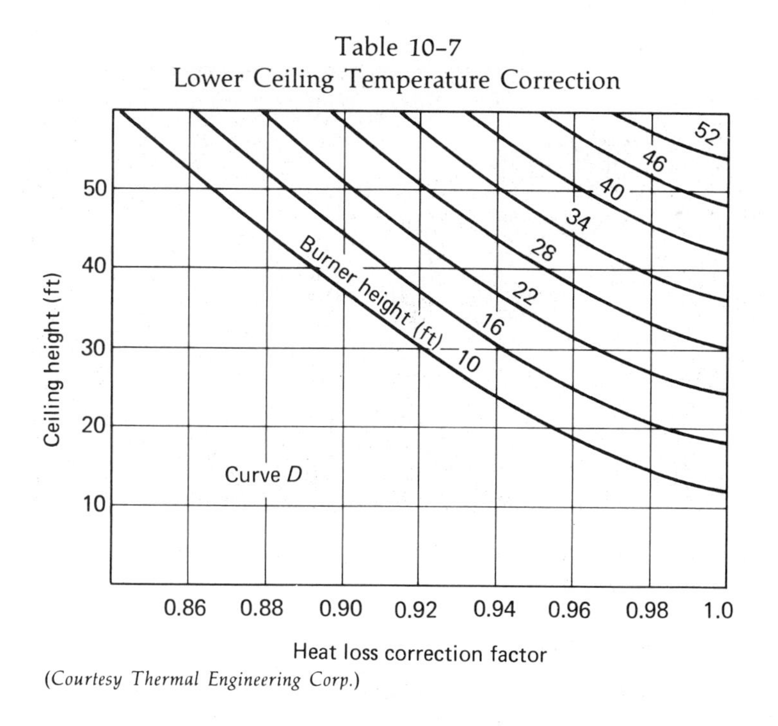

(Courtesy Thermal Engineering Corp.)

2. Determine the heat loss per ft² by dividing the total heat loss by the total floor space of the building.

3. Use Table 10-6 to determine the correction factor *A*.

4. Use Table 10-7 to determine the correction factor *D*.

5. Multiply the correction factor *A* by *D* and by the total heat loss to obtain the proper energy input for an infrared heating system.

6. Be sure that the infiltration rates for ventilation are sufficient to dilute the CO_2 to 5,000 parts per mil (ppm) or less.

7. Check for condensation possibilities.

LOCATION OF INFRARED HEATING UNITS

Because comfort depends on the radiation being received, the heaters must be placed so that all areas are covered. This differs from the perimeter system where the heat loss of the building is the prime consideration. Areas adjoining exposed walls should have greater energy to overcome body losses to cold surfaces (Figure 10-4).

Figure 10-4. Perimeter heating system. (*Courtesy Modine Manufacturing Co.*)

Spot and Area Heating

When the occupancy of a building is low or work stations are static and the entire area does not require general heating, spot heating with infrared heaters will provide the most economical solution, from the standpoint both of first cost and of operational cost (Figure 10-5).

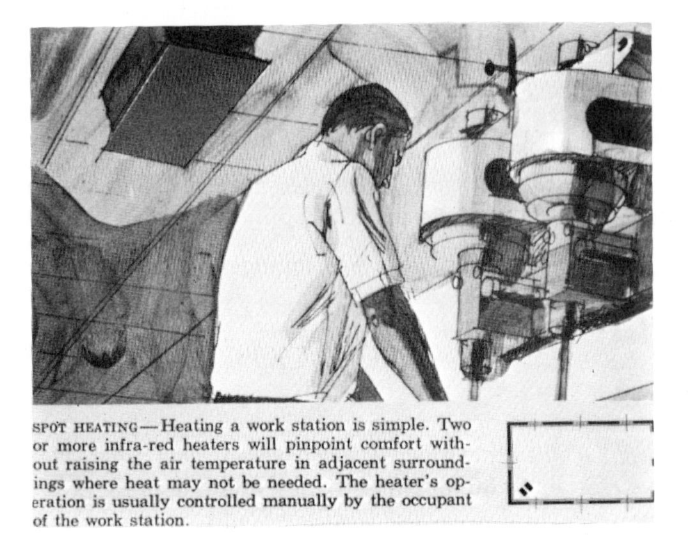

Figure 10-5. Spot heating. (*Courtesy Modine Manufacturing Co.*)

The major purpose is to heat the user. A person loses heat through radiation losses to surrounding surfaces that are lower in temperature than his clothing. His clothing also loses heat by convection to the surrounding air temperature. The prime factor here is air velocity.

Other factors that influence a person's heat losses are the amount of clothing he wears and the heat he loses as a result of his degree of activity (internal heat gains).

Design Procedure

The Thermal Engineering Corp. has developed a surface heat loss chart (see Table 10-8) that provides a reasonably accurate estimate of a person's surface losses. The procedure is summarized below:

1. Determine the design air, the surrounding temperature, and the clothing to be worn.
2. Determine the design air velocity.
3. Enter at the top or bottom horizontal scales both the air temperature and the clothing to be worn.
4. Follow the vertical line up or down until the air velocity line is crossed.
5. At this point extend a horizontal line to the left and read the surface loss from the vertical scale.
6. Adjust the reading (if necessary) according to the adjustment table on the chart.

The amount in step 6 gives the surface heat loss of the person in Btu/hr/ft^2. This amount of heat must be delivered to his body or clothing surface in order that he remain in thermal comfort.

Tables 10-9 and 10-10 give the delivery of heaters at varying heights and differences from the person. Several solutions can be found, but the most economical should be used.

For larger areas where spot heating is desired, the same design procedure is followed but the multipliers under Table 10-10 are used. Higher design temperatures for the air and surroundings may be possible in these applications because the greater number of units may increase these temperatures. These factors must be considered and the requirements adjusted.

For optimum design, the person or persons being heated should be heated by infrared energy from at least two sides.

Table 10-8
Surface Heat Losses—Btu/hr/ft^2,
Radiation and Convection Losses for Man Doing
Light Bench Work

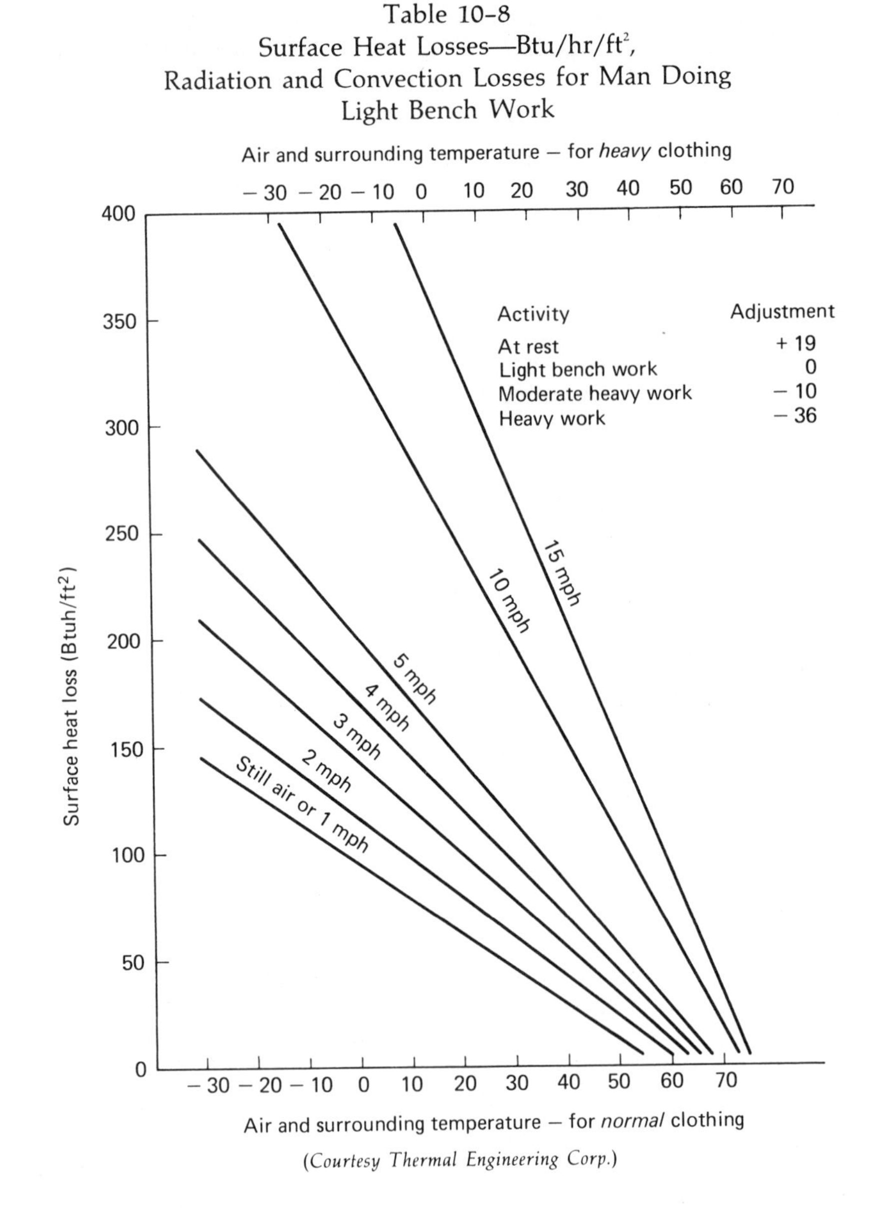

(Courtesy Thermal Engineering Corp.)

ROBERTS-GORDON CO-RAY-VAC

The Roberts-Gordon CO-RAY-VAC is a compact, highly efficient, self-vented, infrared gas heating system working on a patented vacuum-firing principle that actually simulates the healthful warm rays of the

Table 10–9
U Factors for Various Fuels and Types of Equipment

Fuel	Equipment	Unit	*U* Factor
Natural Gas	CO-RAY-VAC	ft^3	0.295
Natural Gas	Unit Heater	ft^3	0.430
Natural Gas	Boiler	ft^3	0.482
LP Gas	Unit Heater	lb	0.020
LP Gas	Boiler	lb	0.023
LP Gas	CO-RAY-VAC	lb	0.014
Fuel Oil	Boiler	gal	0.00437
Coal	Boiler	lb	0.0592

(Courtesy Roberts-Gordon Appliance Corp.)

Table 10–10
Correction Factor for Outdoor Design Temperatures

Outdoor design temperature F	−20	−10	0	+10	+20
Correction factor	0.778	0.875	1.00	1.167	1.400

(Courtesy Roberts-Gordon Appliance Corp.)

sun. The CO-RAY-VAC basically consists of small overhead gas burners (each having a gas input of 40,000 Btu/hr) connected from one combustion chamber to the next by a standard 2½ inch steel pipe (Figure 10–6), usually suspended from 10 to 15 feet above the area to be heated and directly under a bright-surface metal reflector. Generally, the combustion chambers are spaced 15 to 21 feet apart, while the reflectors direct

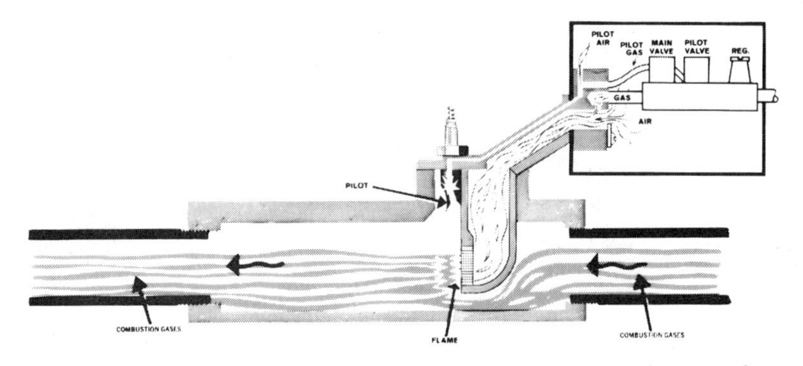

Figure 10–6. CO-RAY-VAC combustion chamber with burner and controls assembly. *(Courtesy Roberts-Gordon Appliance Corp.)*

Figure 10-7. Typical CO-RAY-VAC installations. (*Courtesy Roberts-Gordon Appliance Corp.*)

the "draftless" infrared rays emitted by the steel pipes downward to blanket the entire work area. The system uniformly heats a span, not a spot. No heat is wasted to the ceiling! And there are no air blasts to stir up dust, dirt and germs, as with conventional heating systems. Greatly improved employee productivity is also reported by users (Figure 10-7).

Vacuum Combustion Principle

The products of combustion from the connected burners are totally exhausted to the outdoors by means of a vacuum pump or exhauster. Each burner depends on vacuum to introduce a filtered air-gas mixture to complete the combustion process. The system fires and operates with

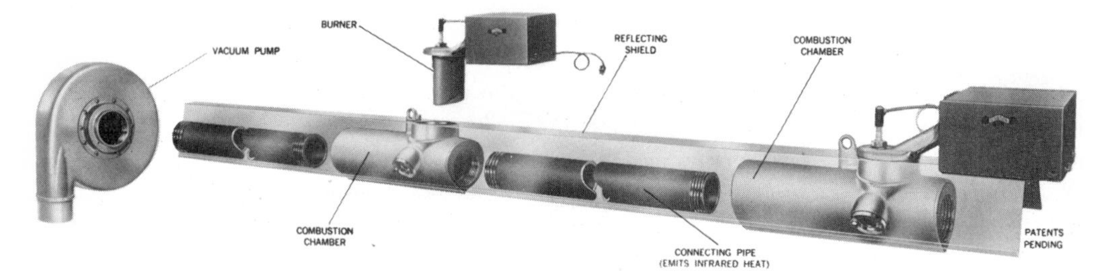

Figure 10-8. CO-RAY-VAC vacuum system. (*Courtesy Roberts-Gordon Appliance Corp.*)

extreme fuel economy under a unique, highly-efficient vacuum principle (Figure 10-8).

Combustion efficiency is approximately 93%. Almost every Btu is extracted from the combustion products. Exhaust temperatures are extremely low, proving CO-RAY-VAC's unmatched heat transfer efficiency.

No Venting or Condensation Problems

The vacuum pump in a CO-RAY-VAC system discharges all flue products directly to the outdoors through a single vent. No make-up air for flue product dilution is required. Moisture from the flue products, which may cause condensation damage, is eliminated. The problem of negative air pressures in the building, which may create toxic fumes or other hazardous conditions with gravity-vented heating units, is not encountered with CO-RAY-VAC's method of venting. Normally, air for combustion is taken from the enclosed space, but in special cases the equipment is designed to provide for outside air to be ducted to the burners.

High Operating Safety

The blue flame in each burner is visible through a small glass-sealed port. The flame is completely enclosed and protected from sudden drafts. Moreover, the vacuum principle guards against combustion products or gas leaking into the building. A vacuum failure would produce an automatic positive mechanical interlock to provide a 100% gas shut-off.

The maximum operation temperature of the CO-RAY-VAC is about 900° F. The infrared rays invisibly emitted are below glowing temperature and are readily and safely absorbed by people and most common materials.

Rugged Durability

Each combustion chamber (containing the firing device) is a heavy-duty alloy casting which, together with the standard 2½ inch steel pipe, constitutes an exceptionally rugged heat exchanger of extremely long life. The metal reflectors will last indefinitely.

Infrared heating has wide use. It can be found in all kinds of buildings—from sprawling factories to large indoor sports arenas.

The heat can be directed with pinpoint accuracy into a concentrated work area to solve spot heating problems. Cold doorways or hallways, for example, are made more comfortable by an adequate number of infrared heaters properly located. Heaters located along the outside walls of a building where heat loss is usually greatest will provide comfortable working conditions.

Versatile infrared heaters may be installed in groups for zone control to maintain the temperature demands of certain areas. They may be controlled automatically or manually.

REVIEW QUESTIONS

1. When infrared heating is used, is the air between objects heated?
2. What is the form of wave motion known as in infrared heating?
3. Is this form of motion the same as sound waves?
4. Is the energy transmitted by an infrared heater known as heat?
5. Are wave form lengths the same for all bodies and conditions?
6. What is a *black body?*
7. How is the emissivity of a body affected by a change in temperature?
8. What is the average value of the absorptivity and emissivity of a body?
9. Does the distance between bodies affect their emissivity and absorptivity?
10. For what purpose is the Stefan-Boltzman law used?
11. What is the ceramic surface temperature of the TEC infrared heater?
12. With what is a direct reading for emissivity taken?
13. What two factors must be considered when determining the infrared energy exchange?
14. What are the three methods of heat transfer?
15. How many heat exchangers are used in infrared heating?
16. Why is the heat loss less with infrared heating than with conventional furnace heating methods?
17. Does the installation of an infrared heater affect the ceiling temperature?
18. What two factors govern the comfort of a space when heating with lower than normal air temperatures?

19. When is the direct absorption of radiation by occupants most important?
20. When would the perimeter system best be used?
21. When would spot heating best be used?
22. What principle does the Roberts-Gordon CO-RAY-VAC unit use?
23. What is used for a connection between the CO-RAY-VAC units?
24. Why are there no condensation problems with the CO-RAY-VAC system?
25. What is the maximum operating temperature of the CO-RAY-VAC units?
26. What percentage of the building heat loss is used in determining the number of CO-RAY-VAC units to install?

11

Amana Heat Transfer Module

The objectives of this chapter are:

- To acquaint you with the operating cycle of the heat transfer module (HTM).
- To acquaint you with the HTM heat exchanger.
- To acquaint you with the heating fluid cycle.
- To acquaint you with the burner used in the HTM.
- To acquaint you with the method of gas ignition in the HTM.

Amana Refrigeration, Inc., a subsidiary of Raytheon Co., has introduced a combination electric-gas central heating-cooling system that combines its revolutionary, compact gas Heat Transfer Module (HTM) with electric cooling to provide what is probably the most efficient forced air system in existence today.

The heat transfer module is a major breakthrough in heating technology, and the first real change in gas furnace design in many years.

The heat transfer module is actually a miniature heat exchanger with enough capacity to heat an average size home, but small enough to hold in your hand. It consists of a newly-designed compact burner mounted in the core of a matrix of thousands of small steel balls fused together with oxygen-free copper. Steel tubing is embedded in the matrix, through which a liquid is pumped to carry away the heat.

BASIC HTM OPERATION

When the indoor room thermostat demands heat, the HTM heat exchanger 9000 burner is ignited by a spark plug.

The two-stage heating system begins operation at two-thirds capacity of 80,000 Btuh input. When needed, it will automatically go to full capacity, 120,000 Btuh input with the blower at the relative speed.

The blower, controlled by a water stat, begins operation as soon as the circulating solution warms up. The blower then forces air through the coil, extracting heat, and into the home or building.

A retard relay permits the combustion blower to continue running for 5 seconds after shut down to purge the HTM heat exchanger of any residual combustion products. It also permits the pump to run an extra 5 seconds so that the solution in the HTM heat exchanger can cool down slightly, but still retain some heat for the next operating cycle. The fluid pump (A) moves the liquid into the heat exchanger (Figure 11-1).

The extremely high temperature combustion products in the HTM heat exchanger (B) pass through a porous matrix to create a high turbulence and transfer heat quickly to the tubes embedded in the matrix that carry the circulating liquid. The liquid makes six passes through the matrix and out of the HTM heat exchanger.

The hot liquid from the HTM heat exchanger is pumped into a heating coil (C), an air conditioning-type coil located above the cooling coil on this model. The blower forces air through the fins, extracting the

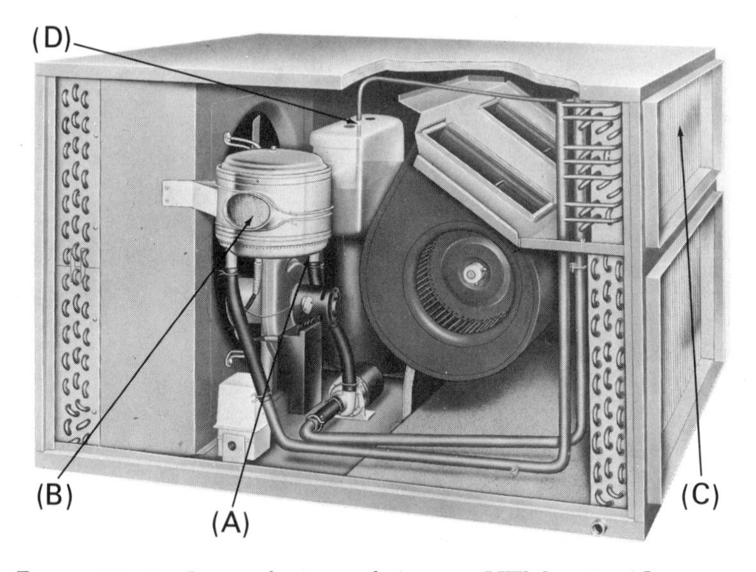

Figure 11-1. Internal view of Amana HTM unit. (*Courtesy Amana Refrigeration, Inc.*)

heat, and into the house or building. The cooled liquid is then returned to the pump and begins another cycle.

The increased volume of the heated liquid moves into the expansion tank (D), which provides reserve storage.

HTM COMPONENTS

The components of this exclusive heating cycle were developed for or adapted for use with the HTM.

Gas Valve

The gas valve is a combination valve-regulator (servo valve-regulator) that takes gas at line pressures and regulates it to a negative pressure (Figure 11-2).

Pressures at the valve are nonadjustable, as the valve is capable of handling line pressures from 3 to 16 inches of water column. The diaphragm of the valve is made of a special material that allows for proper operation down to −40° F. Because the gas is reduced to a negative pressure, it must be drawn out of the valve. This is done by the combustion blower motor assembly.

This combustion blower motor assembly is designed to draw gas out of the valve, mix the proper amount of air with the gas, and deliver

Figure 11-2. Amana Servo Valve-Regulator. (*Courtesy Amana Refrigeration, Inc.*)

this combustible mixture under pressure to the burner. It is here, at the combustion blower, where the 80,000 Btuh or the 120,000 Btuh firing rate is determined.

Orifice Plate

Gas entering the combustion blower is metered by an orifice plate located at the end of the gas manifold pipe. Entering air is metered by two air orifices. One, the low-fire orifice, is located at the end of the mixing chamber; the other, the high-fire orifice, is located in the orifice plate with the gas orifice. The low-fire air orifice is used when firing at the 80,000 Btuh rating, and both air orifices are used when firing on high-fire, or 120,000 Btuh (Figure 11-3).

A solenoid-operated damper plate rests on the orifice plate completely covering the high-fire orifice and restricting the gas orifice to a smaller opening. The unit always starts with the damper plate in the low-fire or down position.

If the solenoid is energized (this is done by using a two-stage thermostat) the damper plate is lifted, opening the high-fire orifice and increasing the size of the gas orifice. Changes in the combustion blower speed because of voltage or any other variation in blower capacity, have little effect since both gas and air enter the mixing chamber at essentially zero pressures.

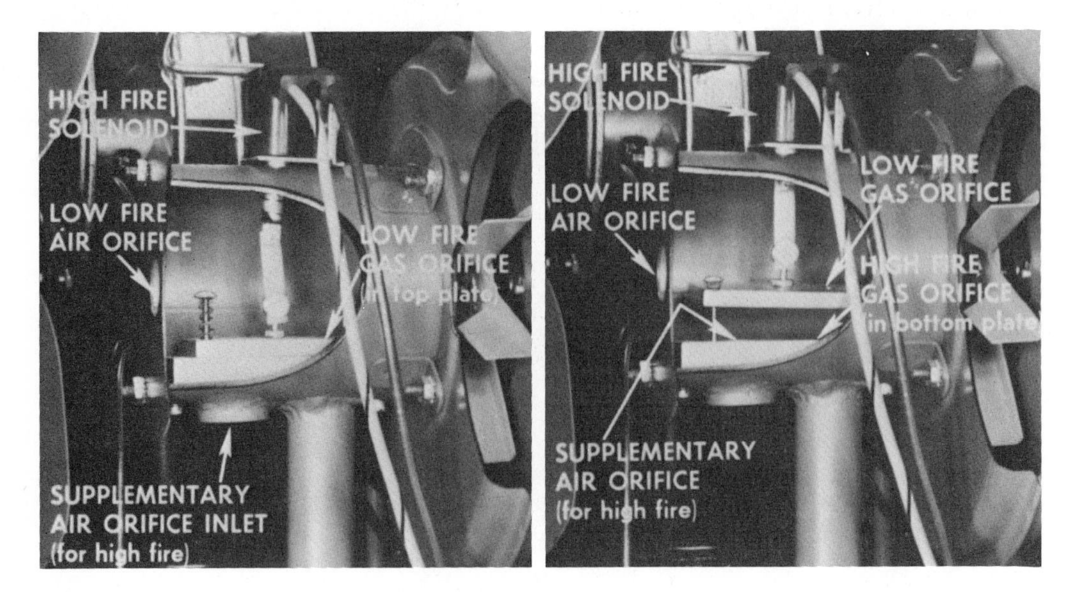

Figure 11-3. High and low fire damper plates. (*Courtesy Amana Refrigeration, Inc.*)

Heat Transfer Module

The heart of the system is the heat transfer module (HTM). The HTM is a miniature heat exchanger which, along with the burner, makes this compact heating system possible.

The HTM is made up of (1) the mixing tube, (2) a spark plug, (3) a flame probe, (4) a stainless steel burner, (5) a cylindrical heat exchanger, (6) the solution carrying tubes, (7) a limit control that monitors the temperature of the solution, and (8) the exhaust flue connection (Figure 11–4).

Burner

The burner, 4¼ inches high and 2¼ inches in diameter, is designed for high heat release with quiet operation (Figure 11–5). It is made of stainless steel and perforated with 9,980 openings or burner ports. This burner represents a level of power density never before achieved in equipment of this type. Because of the number of ports, there are thousands of small, short flames, which are quieter than larger flames. Also, the module holds only 0.11 ft^3 of combustible mixture (gas and air). This is equal to approximately 8 Btuh. Small flames, plus the burning of gas in small volumes, give very quiet combustion.

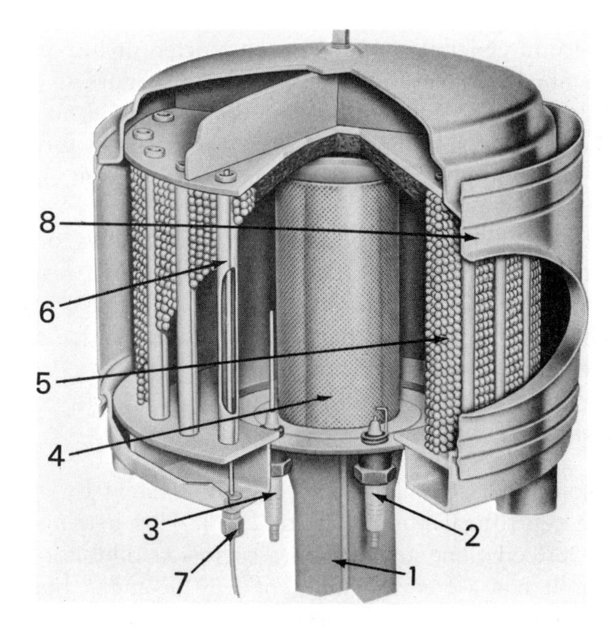

Figure 11–4. Amana HTM. (*Courtesy Amana Refrigeration, Inc.*)

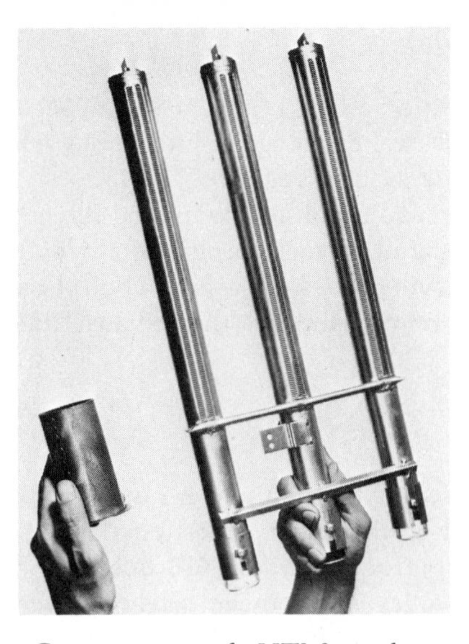

Figure 11-5. Comparison of HTM and regular burner. (*Courtesy Amana Refrigeration, Inc.*)

Matrix

The second generation HTM is constructed of heavy gauge fins in which are embedded steel liquid passage tubes (Figure 11-4, number 5). There are 24 passage tubes arranged so that the liquid makes six passes through the module. The fins heated by the gases in turn transfer the heat to an ethylene glycol solution circulating through the steel liquid passage tubes.

Because of the ingenious arrangement of the tubes to the fins, there is a great deal of turbulence of the flue gases, which results in a heat transfer coefficient typically ten times greater than those in conventional designs. The flue products are then exhausted out the flue outlet with the power provided by the combustion blower.

Heating Solution

The solution is circulated through the system by a small motor driven by a centrifugal pump (Figure 11-6). This heating solution is a mixture of 50% ethylene glycol with a corrosion inhibitor and 50% distilled water. It has a freezing point of −39° F and a boiling point of 224° F. Ethylene glycol is an automotive-type antifreeze and summer coolant liquid and does not contain an antileak material. The brand name of the ethylene glycol Amana uses is Dupont Telar. The entire heating system holds 1.2 gallons of solution. This liquid, circulated at

Figure 11-6. Centrifugal pump location. (*Courtesy Amana Refrigeration, Inc.*)

approximately 4.7 fps, gives a turbulent flow in the circulating tubes that provides for a high rate of heat transfer.

Expansion Tank

Since the solution, when heated, expands as much as 7%, an expansion tank is provided. It is located on the bulkhead (Figure 11-7) and is open to the atmosphere. It provides storage for 1 gallon of reserve liquid to replace any evaporation. Because it is open to the atmosphere, it is also at a point where any air in the system can escape during the heating up period. Between cycles, as the solution cools, it is siphoned out of the expansion tank and back into the system.

Heating Solution Circuit

The liquid goes from the HTM (1) to a heating coil (3) (Fig. 11-8), located in the discharge area of the evaporator section and the cooling coil in the return air area of the evaporator section, through the tubing (2). The expansion tank (4) holds any overflow of the ethylene glycol.

Indoor Blower

The indoor air circulating blower is direct drive with a PSC (permanent split capacitor) motor. In the 2 and 2.5 ton units, the blower operates on high speed in heating and low speed in cooling. In the 3 ton

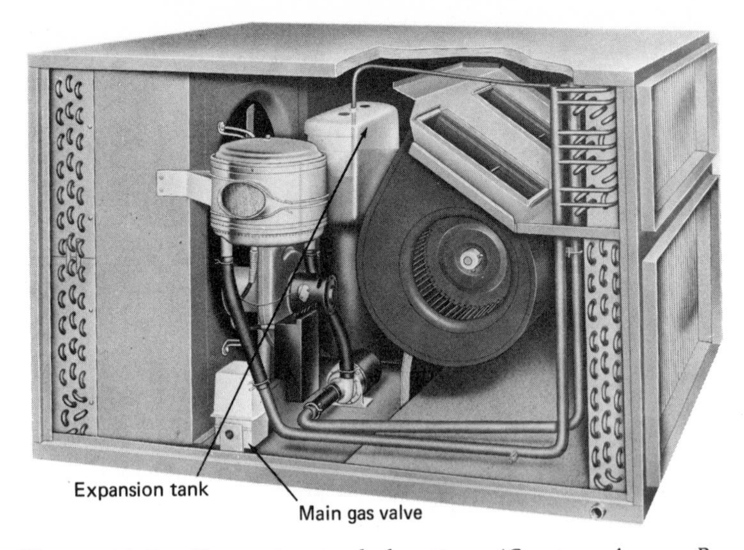

Figure 11-7. Expansion tank location. (*Courtesy Amana Refrigeration, Inc.*)

unit, the blower is a single speed. The blower operates on low speed in heating and high speed in cooling in the 3.5, 4, and 5 ton units. Heating is on high speed in the 2 ton unit and on low speed in the 5 ton unit because the heating capacities remain the same while the cooling capacities increase and require a change in the cfm of air.

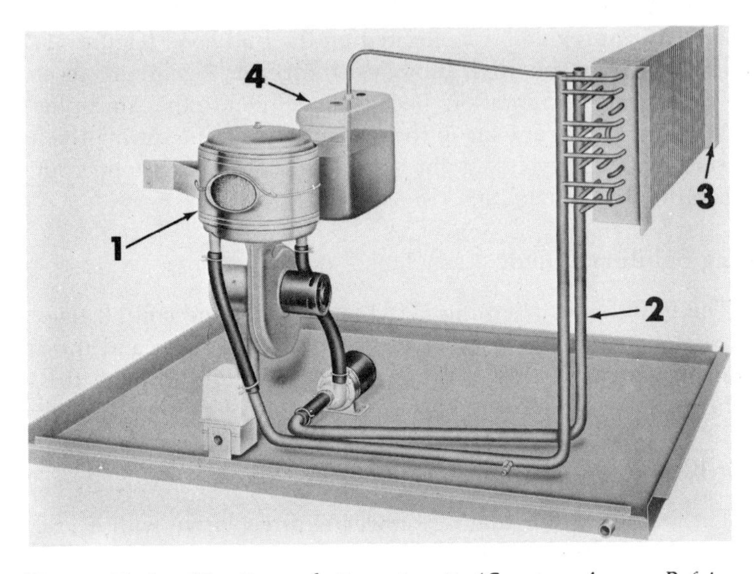

Figure 11-8. Heating solution circuit. (*Courtesy Amana Refrigeration, Inc.*)

HEATING CYCLE

Operation of the unit is controlled by a two-stage heat, one-stage cool thermostat. As the thermostat calls for heat, it energizes the heating relay which starts the combustion blower, the liquid circulating pump, and the air circulating blower. It also energizes the direct spark ignition system (DSI). The DSI energizes the spark plug ignitor and opens the gas valve. The DSI supplies approximately 19,000 V to the spark plug. It also sends a signal to the flame probe asking for proof of flame.

On burner ignition, the flame probe terminates the trial for ignition and discontinues the ignition spark. At this point the unit fires with the damper plate down, or on low-fire.

If the temperature continues to fall, the second stage of the thermostat closes and a time-delay is energized. Forty seconds later the orifice plate solenoid energizes, lifting the plate and allowing the unit to fire at the high, or 120,000 Btuh firing rate. As the room conditions are satisfied, the second-stage heat contacts of the thermostat open and the firing rate of the unit is reduced to 80,000 Btuh. As the room temperature continues to be satisfied, the first stage of the thermostat opens. This deenergizes the heating relay, the DSI control, the gas valve, and shuts off the combustion blower.

Whenever the heating requirement is equal to or below 80,000 Btuh, the use of a single-stage heat and single-stage cool thermostat is permissible. In applications of this type, the solenoid is never energized, the high fire damper is never raised, and the unit never fired at 120,000 Btuh.

The solution circulating pump and air circulating blower are kept in operation by a solution stat until the temperature of the solution has dropped to about 105° F.

If the water limit control, located in one of the HTM solution tubes, senses an excessive solution temperature, or if the stack fuse senses a temperature above 600° F, the thermostat is over-ridden and the system goes on safety lockout. Also, if the flame is not proven by the flame probe within 15 seconds after the thermostat calls for ignition, the system will go on safety lockout. The system will remain on lockout until it is reset. To reset the system if the water limit control or flame probe has put it on lockout, turn down the room thermostat until it no longer calls for heat. Then reset it to the desired room temperature. Of course, if the stack fuse is broken, a service technician must be called.

PLACING THE HTM IN OPERATION FOR HEATING

The unit is placed in operation by closing the main electrical disconnect switch and opening the main gas valve (Figure 11-9). The gas valve control knob is located under the plastic weather cover. This cover

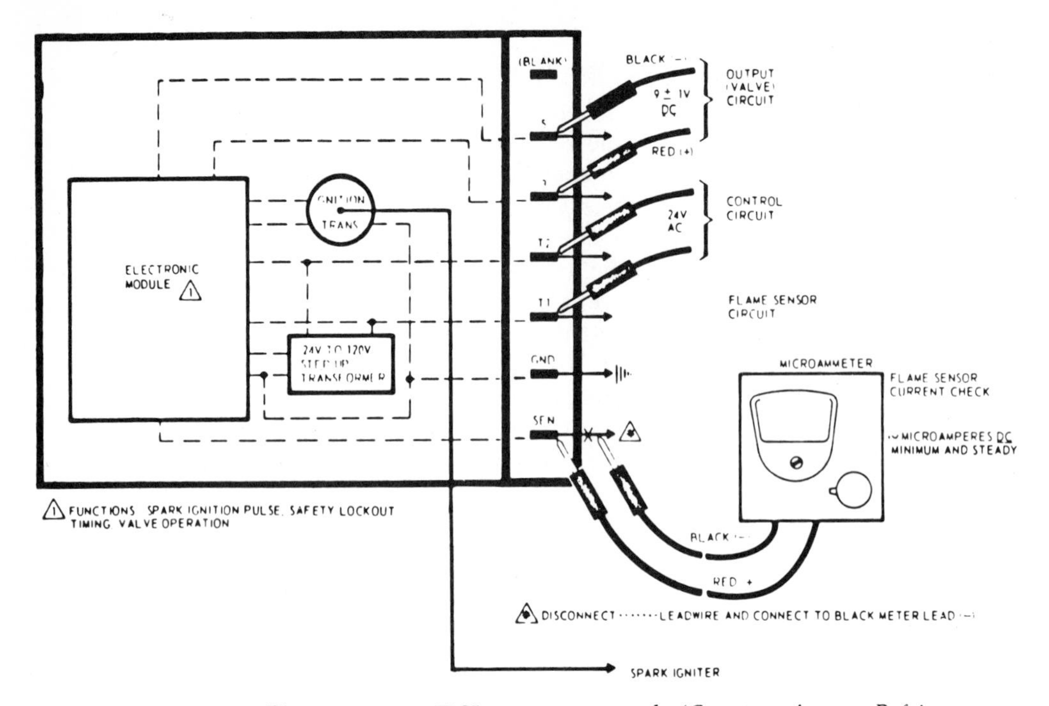

Figure 11–9. DSI system control. (*Courtesy Amana Refrigeration, Inc.*)

must be replaced after the unit has been placed in operation. On a new installation, there will probably be air in the gas supply line. The air can be bled off by cracking open the ground joint union until gas is coming through. Tighten the union and wait for at least 5 minutes until all gas has been dissipated. Be sure there is no flame in the vicinity.

Turn the thermostat setting to a point above the room temperature, set the selector switch to HEAT and the fan switch to AUTO. The solution pump and combustion blower motors should all start. There probably is still some air in the gas supply line if ignition does not take place. The unit will shut off in about 15 seconds. It will be necessary to lower the room thermostat setting below the room temperature or disengage the electrical disconnect switch for at least 1 minute to let the safety switch reset. It may take several cycles before the flame is established.

The unit must stay in operation for several minutes before lowering the thermostat setting for shut off after the initial start-up. The air circulating blower should keep running after the heat transfer module is off, that is, until the blower timer contacts open. Put the unit through several cycles of about 3 minutes on and 5 minutes off. The resultant expansion and contraction of the solution will expel any air that might be in the system. The system is now ready for operation.

COMPONENT CHECKS

IMPORTANT

1. If any condition exists which will cause the system to go into safety lockout, meter readings (except control circuit T1-T2 24 V) must be taken quickly after restart; that is, within the trial ignition period.

2. Always de-energize the system for at least 1 minute before recycling for further tests.

DSI System Control

Check as follows:

1. *Control Circuit—Terminal T1:* The DSI system is powered by the unit transformer. Using a voltmeter, restart the system in the heating cycle and check for 24 volts AC between terminal T1 and ground. This reading should be 24 volts AC. If not, test the thermostat wiring, high limit control, and secondary limit control.

2. *Output Circuit:* Start the unit in the heating cycle. Check the output voltage between terminals T3 and T5 with a DC voltmeter. The voltmeter should indicate 9± 1 volts DC. This reading must be made within 21 seconds, otherwise if the system should go into lockout, no voltage will be present.

If the reading is approximately 33 volts DC and then gradually decreases, the trouble is in the gas valve circuitry. Check the wiring to the gas valve, then proceed to check the combination control.

If no voltage is present except for 24 volts AC between terminal T1 and ground, then replace the Direct Spark Ignition Control.

CAUTION

Do not short across terminals T3 and T5 because to do so will open an internal fuse in the T1 circuit of the DSI control.

3. *Flame Rod Circuit:* It is necessary to have combustion to check the flame rod circuit because the burner flame completes the circuit on the flame rectification principle.

With the power off, complete the following checks (Figure 11-9):

(a) Remove the flame rod and ignitor from the HTM and inspect for loose connections or physical damage.
(b) Reinstall and connect a volt-ohm meter which will read microamperes or a microammeter (with a microampere DC

scale) in series with the DSI control terminal S and the flame rod.

(c) The red positive (+) lead on the meter must be connected to terminal S of the control and the black negative (−) lead to the flame rod.

(d) Start the system in the heating cycle and observe the reading on the microammeter.

(e) The reading should be a minimum of 9 microamperes.

(f) If unable to obtain the proper reading, or a fairly steady reading, recheck the flame rod, wiring, and connections visually.

(g) A low microampere reading could indicate underfiring or too lean a gas-air mixture. A high reading would indicate overfiring or too rich a gas-air mixture. Recheck the gas input and/or air leaks.

(h) Try a new flame-rod-ignitor assembly before replacing the DSI control.

4. *Spark Ignition Circuit:* The electronic module and step-up transformer in the DSI control provides spark ignition at 19,000 volts. This circuit can be checked at the recessed stud terminal on the front of the DSI control.

With the power off, check as follows:

(a) Close the manual gas cock knob on the gas valve to prevent a flow of gas.

(b) Remove the ignition cable from the spark ignitor and slide the rubber boot back, exposing the metal surface.

(c) While holding the rubber boot and cable, start the system in the heating cycle and remove the free end toward the spark ignitor.

CAUTION

Do not come in contact with the pigtail of the high limit control or the flame sensor wire.

(d) Observe the length of the spark gap when arcing first occurs between the ignitor and the cable. The spark length should be at least 11/64 inch.

(e) If no arc is found, check the complete cable for continuity with an ohmmeter.

(f) If the cable checks good, hold one end firmly to the ground terminal *G* on the DSI control and move the free end toward the ignition transformer stud on the control. Note the length of the arc. It should be at least 11/64 inch.

If there is no arc, replace the control. In all instances, remember the system must be in the heating cycle, and each test conducted in less than 15 seconds, before the system goes into lockout. If system lockout occurs, the power must be turned off to the DSI control for at least one minute before restarting.

If nuisance shutdowns have been reported, be sure to check the ground connection from the DSI control terminal *G* to its ground connection at the unit.

Spark Ignitor

With the power off, make the following checks:

1. Remove the spark ignitor from the HTM and visually check the following:
   (a) The spark plug gap must be 11/64 inch. Check and adjust by carefully bending the angle tip of the outer (ground) electrode (Figure 11–10).
   (b) Check the porcelain for cracks or dirt that can cause a path to the ground.

2. Check the ignition cable for poor insulation. See that the connections to the DSI control and on the ignitor are tight and clean. Also, be sure that the cable does not touch any metal surface.

3. If no spark or a weak spark is found after the above steps are

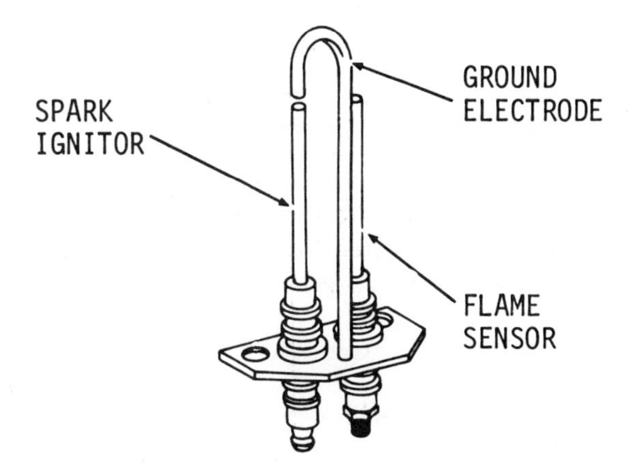

Figure 11–10. Spark plug igniter. (*Courtesy Amana Refrigeration, Inc.*)

completed, recheck for satisfactory power at the DSI control before replacing.

Combination Gas Control

This unit is equipped with a negative regulating gas valve of a specific type. With the power off, make the following checks:

1. Disconnect the valve operator leads from the valve.

2. Using an ohmmeter, check for continuity at the valve terminals.

3. If the operator coil is open, replace the entire valve. Do not attempt to replace components or repair the valve. Always replace with the same model valve as listed in the parts manual for that unit.

CAUTION

This is a negative regulated valve with a 9± 1 volt DC operator. Never connect 24 volts or higher voltage and try to operate.

If the gas input should exceed or be less than 1 cubic foot of natural gas in 46 seconds for 80,000 Btu or 22 seconds for 170,000 Btu input, proceed as follows:

1. Check the attachment of the combustion blower assembly and placement of the asbestos gasket to the Heat Transfer Module. Also, check the seal between the motor and housing, gas-air chamber to housing for no air leaks.

2. If the input is still wrong, remove the ⅛ inch pipe plug on the side of the gas valve while the unit is in the OFF cycle.

3. Install a fitting, so a slope gauge (inclined manometer) may be installed. Attach the hose connection from the fitting to the negative side of the gauge. Also, install a water manometer to the drip leg fitting to read the inlet pressure (Figure 11–11).

CAUTION

Pinch the hose connection to the slope gauge until the unit is operating and also before stopping, or the solution may be drawn out of the gauge at unit start-up and shut down.

4. Start the system in the heating cycle in order to take a reading. The negative pressure should read −.20±.15 inches of water column.

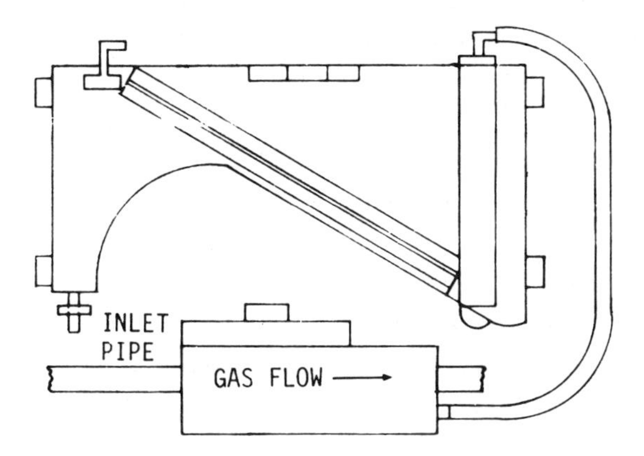

Figure 11-11. Water manometer used for reading inlet pressure.

5. If the pressure reading is higher than −.05 inches (closer to zero) the input will be higher, or if lower than −.35 (further from zero) the input will be reduced. Pressure readings within these tolerances (±.15) will show approximately ±5% Btu input change from the normal −.20 inches of water column capacity.

6. In either case adjust the valve.

7. The inlet pressure during the test should be a nominal 7 inches water column on natural gas and 11 inches on propane while the unit is operating.

CAUTION

The gas valve must always be in the horizontal position with the gas cock knob up. This is mandatory on controls designed for negative outlet pressure.

Low-High Fire Damper

Use for check out with a two-stage heating thermostat. To check out the low- and high-fire damper operation proceed as follows with the thermostat setting below room temperature and the unit off:

1. Move the thermostat setting to a point which is 5° F above room temperature. The unit should start normally (80,000 Btuh firing rate). In approximately 40 seconds, the solenoid on the top of the inlet chamber should operate and lift the damper plate off the orifice plate approximately ½ inch.

2. If the solenoid does not function, check to see if the time delay is receiving 24 V of power. Furthermore, check to see if the relay contacts are closed to provide 230 V power to the solenoid. These checks will determine if these components are functional.

3. If the damper plate does not lift to its full height or the solenoid is energized and the damper does not lift, check the linkage assembly.

4. To check out the low-fire damper operation, proceed by removing the cover from the room thermostat.

5. Carefully lower the room temperature setting until the second-stage heat mercury tube contacts open; however, the first-stage heat mercury tube contacts must remain in the closed position.

6. With the mercury tubes positioned as explained in step 5, the damper plate should drop in approximately 40 seconds. The burner should continue to operate as long as the first stage heat mercury tube contacts are closed.

7. The burner will go off when the contacts of both heating mercury tubes open; however, it still will take about 40 seconds for the damper plate to drop when the thermostat is adjusted from high fire to a call for no heat.

Module Heater

The module heater is strapped to the bottom portion of the heat transfer module. It has a 40 W element and is energized from *L*1 and *L*3.

With the power on check as follows:

1. Using a voltmeter, measure for line voltage on terminals *L*1 and *L*3 to which the heater leads are connected.

2. No voltage indicates blown fuses or bad wiring to the heater. Repair or replace wiring.

3. If voltage is present, turn the power OFF and disconnect the heater lead wires.

4. Using an ohmmeter, check heater continuity. It should test continuous. If not, replace the heater.

REVIEW QUESTIONS

1. How is the burner ignited in the HTM unit?
2. What is the capacity of the HTM on starting?
3. How is the circulating blower motor controlled?

4. Why is the circulating blower motor and the fluid circulating pump required to operate for 5 seconds after the unit is off?
5. How many passes does the fluid make in the HTM?
6. What type of heat exchanger is in the circulating air stream?
7. What is the purpose of the expansion tank?
8. Is the gas pressure regulated at the main gas valve?
9. Are the gas pressures adjustable at the gas valve?
10. How is the mixture metered at the combustion blower?
11. How is the high-fire rate obtained with the HTM?
12. Does the speed of the combustion blower affect the combustion effficiency of the HTM?
13. Name the components in the HTM.
14. How many ports are there in the HTM burner?
15. How much combustible material will the HTM hold?
16. What causes the turbulence of the flue gases in the HTM?
17. What is the heat transfer coefficient of the HTM compared to other types of equipment?
18. What is the velocity of the fluid in the HTM?
19. Why is the air circulating blower operated on high speed in the 2 and 2.5 ton units?
20. What is the voltage in the DSI circuit?
21. What is the time required between the low-fire and the high-fire positions of the orifice plate?
22. At what temperature does the stack fuse "blow"?
23. How much time is allowed by the flame probe before shutdown after the thermostat calls for heat?
24. How is the Amana HTM reset after the limit control or flame probe has put it on lockout?
25. Why is power combustion safe?
26. Is the circulating fluid ever under pressure?
27. What is a safety factor concerning the circulating air in the HTM unit?
28. Does changing the HTM from natural gas to propane affect its operation or safety?

12

Humidification

The objectives of this chapter are:

- To familiarize you with the terms used in humidification.
- To introduce you to the advantages of proper humidification.
- To give you the desired humidity requirements for comfort conditioning.
- To provide you with the different types of humidifier operation.

Humidification is, unfortunately, one of the least understood, most important aspects of comfort conditioning. It is understandable that it is misunderstood because humidity is intangible: you cannot see it, you cannot touch it, it has no color, no odor, and no sound.

WHAT IS HUMIDITY?

To make sure we are starting out on common ground, let us define some of the words used in discussing humidity.

Humidity is the water vapor within a given space. *Absolute humidity* is the weight of water vapor per unit volume. *Percentage humidity* is the ratio of the weight of water vapor per pound of dry air to the weight of water vapor per pound of dry air saturated at the same temperature. *Relative humidity* is the ratio of the mol fraction of water vapor present in the air to the mol fraction of water vapor present in saturated air at the same temperature and barometric pressure. Approximately, it equals the ratio of the partial pressure or density of the water vapor in the air to the saturation pressure or density, respectively, of the water vapor at the same temperature.

Although there is a difference between percentage humidity and relative humidity, it is only very slight and is practically negligible at normal room temperatures. So, for our purposes, we can say that relative humidity indicates the amount of water vapor actually in the air expressed as a percentage of the maximum amount that the air could hold under the same conditions.

WHY WE NEED HUMIDIFICATION

The amount of moisture in the air has a direct bearing on personal comfort or discomfort. Just as extremely high relative humidity in the summer gives one a soggy feeling, low humidity in the winter in heated homes gives one a dried-out feeling.

When cold air enters the house during the winter, it is heated; thus its moisture holding capacity is increased. If moisture is not added to this air, indoor relative humidity drops below the minimum range for personal comfort. In most homes without humidification equipment, heated indoor air is much too dry for the well being of its occupants, their pets, plants, and furnishings. Proper humidity control is simple with a humidifier. Simply dial the desired humidity on the humidistat.

Properly humidified air enhances personal comfort and well being by helping to prevent throat irritations, nasal discomfort, bronchial aggravations, and itchy dry skin caused by hot, dry indoor air. Floors, doors, frames, and wood furniture will have a minimum of drying out, cracking, or warping. Draperies and upholstering stay fresh and wear longer. Annoying minor shocks caused by static electricity are greatly reduced. House plants stay fresher and prettier. A home can be kept more comfortable at lower temperatures with properly humidified air. This helps lower heating costs.

As stated before, the warmer the air, the more moisture it can hold. Air in a home heated to 70° F can hold about 8 grains of moisture per ft^3. That is 100% relative humidity. If there are only 2 grains/ft^3 in the home, this is one quarter of the capacity of the air to hold moisture. Therefore, the relative humidity is also one quarter or 25%. The air can hold four times as much water.

However, the important thing to remember is what happens to air when it is heated. The outdoor-indoor relative humidity conversion chart (Table 12–1) illustrates this.

Because of this capability of warm air to hold more water than cold air, a substantial reduction of relative humidity is taking place in every unhumidified or underhumidified home where winter heating is prevalent.

Table 12–1
Outdoor-Indoor Relative Humidity Conversion Chart

Outdoor relative humidity	-20°	-10°	-5°	0°	+5°	**+10°**	+15°	+20°	+25°	+30°	+35°	+40°	+45°	+50°
100%	2%	3%	4%	6%	7%	9%	11%	14%	17%	21%	26%	31%	38%	46%
95%	2%	3%	4%	5%	7%	8%	10%	13%	16%	20%	24%	30%	36%	44%
90%	2%	2%	4%	5%	6%	8%	10%	12%	15%	19%	23%	28%	34%	41%
85%	2%	2%	4%	5%	6%	8%	9%	12%	15%	18%	22%	27%	32%	39%
80%	2%	2%	4%	5%	6%	7%	9%	11%	14%	17%	20%	25%	30%	37%
75%	2%	2%	3%	4%	5%	7%	8%	10%	13%	16%	19%	23%	28%	36%
70%	1%	2%	3%	4%	5%	**6%**	8%	10%	12%	15%	18%	22%	26%	32%
65%	1%	2%	3%	4%	5%	6%	7%	8%	11%	14%	17%	20%	25%	30%
60%	1%	2%	3%	3%	4%	5%	7%	8%	10%	13%	15%	19%	23%	28%
55%	1%	1%	2%	3%	4%	5%	6%	8%	9%	12%	14%	17%	21%	25%
50%	1%	1%	2%	3%	4%	4%	6%	7%	9%	10%	13%	16%	19%	23%
45%	1%	1%	2%	3%	3%	4%	5%	6%	8%	9%	12%	14%	17%	21%
40%	1%	1%	2%	2%	3%	4%	4%	6%	7%	8%	10%	12%	15%	18%
35%	1%	1%	2%	2%	3%	3%	4%	5%	6%	7%	9%	11%	13%	16%
30%	1%	1%	1%	2%	2%	3%	3%	4%	5%	6%	8%	9%	11%	14%
25%	1%	1%	1%	1%	2%	2%	3%	3%	4%	5%	6%	8%	10%	12%
20%	+%	1%	1%	1%	1%	2%	2%	3%	3%	4%	5%	6%	8%	10%
15%	+%	+%	1%	1%	1%	1%	2%	2%	3%	3%	4%	5%	6%	7%
10%	+%	+%	+%	1%	1%	1%	1%	1%	2%	2%	3%	3%	4%	5%
5%	+%	+%	+%	+%	+%	+%	1%	1%	1%	1%	1%	1%	2%	2%
0%	0%	0%	0%	0%	0%	0%	0%	0%	0%	0%	0%	0%	0%	0%

Outdoor temperature

(Courtesy Research Products Corp.)

To solve this problem, we add moisture artificially so that more water is available for this thirsty air to hold. That is, we humidify because of benefits as important as heating to overall indoor comfort and well-being during the heating season. And these benefits are actually what must be maintained.

These benefits can be grouped into three general classifications:

1. Comfort
2. Preservation
3. Health

Benefit Number One: Comfort

Did you ever step out of the shower, start shaving, and notice how warm it is in the bathroom? It is actually muggy. It will probably be about 75° F in the bathroom and the relative humidity will probably be about 75–80%, because of the water vapor added to the air while showering. Now, the phone rings and you have to step out into the hall to answer it. What happens? You freeze. Although the temperature is probably about 70° F, just 5° F cooler than in the bathroom, you shiver. This is because you have just become an evaporative cooler. The air out in the hall is dry. The relative humidity is probably about 10–15%. You are wet and this thirsty air goes to work on your skin. As it evaporates the water, your skin is cooled. This same thing continues day after day, every winter, in millions of homes. People turn their thermostats up to 75° F and more in order to feel warm. Even then, it feels drafty and chilly because the evaporative cooling process is going on. Proper relative humidity levels make you feel more comfortable at lower thermostat settings.

But this cold feeling is not the only discomfort caused by too dry air. Static electricity, usually an indication of low relative humidity levels, is a condition that is constantly annoying. Proper relative humidity will alleviate this discomfort.

Benefit Number Two: Preservation

The addition or reduction of moisture drastically affects the qualities, the dimensions, and the weight of hygroscopic materials.

Wood, leather, paper, and cloth, although they feel dry to the touch, contain water—not a fixed amount of water, but an amount that will vary greatly with the relative humidity level of the surrounding air. Take, for example, 1 ft^3 of wood with a bone-dry weight of 30 pounds. At 60% relative humidity, the wood will hold over 3 pints of water. Now, if the relative humidity is lowered to 10%, the water held by the wood will not fill even a 1 pint bottle. Thus, in effect, we have withdrawn 2½ pints of water from the wood by lowering the relative humidity from 60% to 10% (1 pint of water = approximately 1 pound).

This type of action continues, not only with wood, but with every single material in the home that is capable of absorbing moisture. Paper, plaster, fibers, leather, glue, hair, skin, . . . , practically everything in the home. These materials shrink as they lose water and swell as they absorb water. When the water loss is rapid, warping and cracking take place. Also, as the relative humidity changes, the condition and dimensions of the materials change. This is why humidity must be added. This is why proper relative humidity is important.

What are the effects of this constantly changing or constantly low moisture content of the air? They are damaging. Furniture construction is affected. Glue dries out, joints separate, rungs fall out of chairs, and cracks appear. The plaster walls dry out and crack. Joints and wall studs shrink, causing the room to be out of square. The wood paneling separates and cracks. The boards in the floor separate. Musical instruments lose their tone. Pieces of art, books, and documents dry out and either break or crack. The rugs wear out quickly, simply because a dry fiber will break while a moist fiber will bend.

Now that the problems of too little humidity have been discussed, let us discuss the problem of too much humidity and the effect of vapor pressure.

Perhaps you have noticed windows that fog during the winter—maybe a small amount of fog on the lower corners or a whole window fogging or completely frosting over. These conditions are an indication of too high indoor relative humidity.

This condensation is due to the effect of vapor pressure. Dalton's law explains vapor pressure: *In a gaseous mixture, the molecules of each gas are evenly dispersed throughout the entire volume of the mixture.* Taking the house as the volume involved, water vapor molecules move throughout the entire home. Because of the tendency of these molecules to disperse evenly, or to mix, the moisture in the humidified air moves toward the drier air. In other words, in a house the moist indoor air attempts to reach the drier outside air. It moves toward the windows where there is a lower temperature and, therefore, causes an increase in relative humidity to a point at which the water vapor will condense out on the cold surface of the window. This is the dew point and it occurs at various temperatures, depending on the type of windows in the home.

With an indoor temperature of 70° F and an outdoor temperature of 20° F, for example (see Table 12-2) condensation begins on a single glass plane at about 24% relative humidity. At 38% for a single pane with a loose storm sash. At 50% for a thermopane with ½ inch space

Table 12-2
Condensation Temperatures

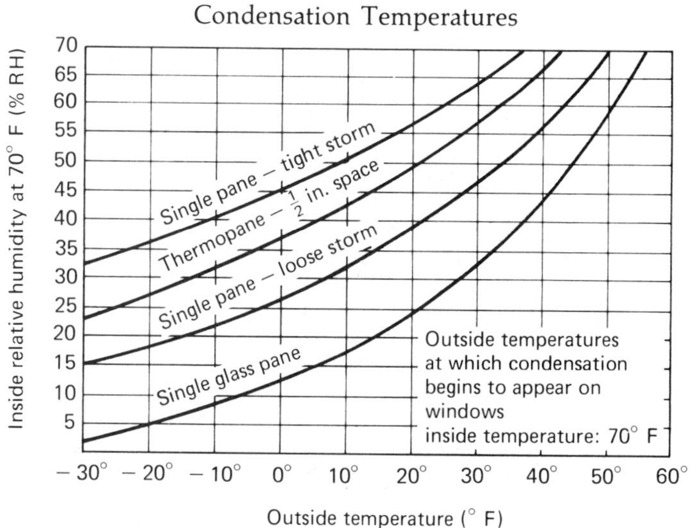

(Courtesy Research Products Corp.)

between the panes. And at 58% for a single pane with a tight storm sash.

Usually condensation on inside windows is a type of measurement of the allowable relative humidity inside a home. We can furthermore assume that if this condensation activity is taking place on windows, it may also be taking place within the walls if there is no vapor barrier.

A vapor barrier, as the name implies, is a material that restricts the movement of water vapor molecules (Figure 12–1). Examples of a typical vapor barrier are aluminum foil, polyethylene film, plastic wall coverings, plastic tiles, and some types of paint and varnish. Actually, practically every home has a vapor barrier of some type that at least retards the movement of the water molecules from a high vapor pressure area (inside) to a low vapor pressure area (outside).

The typical outside wall has a dry wall, or plaster, a vapor barrier (on the warm side of the insulation), the insulation, air space, sheathing, building paper, and siding. With an indoor temperature of 70° F, a relative humidity of 35%, and an outside temperature of 0° F, what happens to the temperature of the air passing through the wall? It drops to about 60° F at the vapor barrier, down to 17° F at the sheathing, and on down to 0° F outside. And, if we checked a psychrometric chart, we would find that with an indoor temperature of 70° F and an indoor relative humidity of 35%, the dew point would be 41° F. This temperature occurs right in the middle of the insulation in the wall. This is where we have condensation, and this is where we would have trouble without a vapor barrier and without controlled humidification.

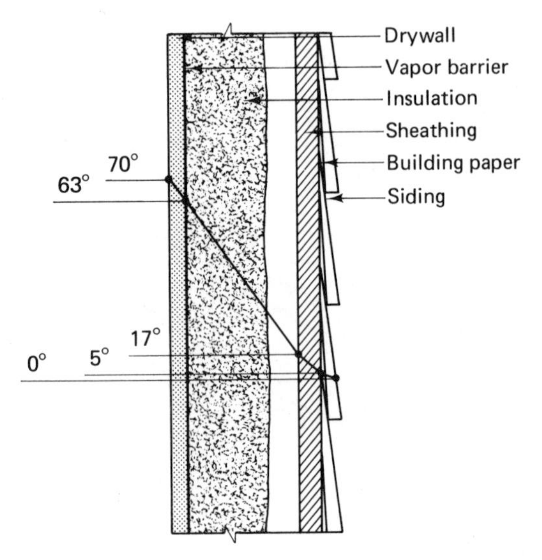

Figure 12–1. Typical outside frame wall construction. (*Courtesy Research Products Corp.*)

The important aspect is, then, properly-controlled relative humidity to avoid the damaging effects of too dry air, and equally as important, to avoid the damaging effects of too high relative humidity.

Benefit Number Three: Health

Dr. Arthur W. Proetz, eye, nose and throat specialist says, in the annals of *Otology, Rhinology and Laryngology:*

"In the struggle between the nose and the machinery in the basement, sometimes the heater wins and sometimes the cooler, but seldom the nose. The nasal mucus contains some 96% water. To begin with, it is more viscous than mucus elsewhere in the body and even slight drying increases the viscosity enough to interfere with the work of the cilia. Demands on the nasal glands are great even under usual conditions and they cannot cope with extreme dryness indoors in winter. Experience has shown that with approaching winter, the first wave of dry-nose patients appears in the office when the relative humidity indoors falls to 25%. It would seem, therefore, that 35% would be required as a passing grade but 40% something to shoot for. It boils down to this, a pint of water is a lot of water for a small nose to turn out. In disease or old age, it simply doesn't deliver and drainage stops and the germs take over."

The right-hand column in Table 12-3 is the number of cases of respiratory disease per 1,000 population, taken from the U.S. Public Health statistics. The left-hand column is a typical indoor relative humidity in Madison, Wisconsin during these same months. Note that there is an obvious and definite correlation between good health and high relative humidities and between poor health and low relative humidities. All the facts point toward a positive connection between humidity and health.

THE CORRECT INDOOR RELATIVE HUMIDITY

While some humidity conditions may be ideal for health and comfort, they are, in many cases, less ideal for other reasons. An indoor relative humidity of 60% may fulfill all the requirements for health and comfort, but it can result in damage to walls, furnishings, etc. The fogging of windows is usually an indication of too high relative humidity, and it must be remembered that this same condensation is taking place inside walls and other places vulnerable to damage by excessive moisture (Table 12-4).

Table 12-3
Respiratory Disease per 1,000 Population.

Month	Average Indoor Relative Humidity			Total Number of Cases of Respiratory Diseases per 1,000 Population
July	59%	High Humidity	Good Health	17
August	56%			20
September	51%			23
October	40%			29
November	20%	Low Humidity	Poor Health	40
December	12%			58
January	12%			92
February	14%			102
March	18%			89
April	25%			55
May	35%	High Humidity	Good Health	35
June	57%			22

(Courtesy Research Products Corp.)

Table 12-4
Condensation on Windows

Inside Temp. 70° F.	Outside Temperature at Which Condensation Will Probably Occur*	
Inside Relative Humidity	Single Glass	Double Glass Thermopane
50%	43°	18°
45%	38°	11°
40%	34°	2°
35%	28°	−8°
30%	22°	−20°
25%	15°	−30°
20%	8°	
15%	0°	
10%	−11°	

(Courtesy Research Products Corp.)

It is, therefore, necessary to set safe limits of indoor relative humidity levels to receive the maximum benefits from correct humidity, without making the structure itself susceptible to damage. It is recommended that Table 12-5 be followed to insure these benefits.

Table 12-5
Temperature-Humidity Table

OUTSIDE TEMPERATURE	RECOMMENDED R.H.
+20° and above	35%
+10°	30%
0°	25%
−10°	20%
−20°	15%

(Courtesy Research Products Corp.)

EFFECT OF WATER CHARACTERISTICS

Only distilled water or rain water caught before it reaches the ground is free from minerals. Water from wells, lakes, and rivers all contain varying amounts of minerals in solution. These minerals are picked up as the water moves through or across water soluble portions of the earth's surface. In many cases, the level of these minerals is sufficiently high to make water-conditioning equipment necessary to remove the objectionable minerals for normal domestic use.

It is common knowledge that water evaporated from a tea kettle leaves a residue known as lime. Since evaporation of water is the only way to create and distribute water vapor into the air present in homes, it is apparent that mineral residue resulting from evaporation presents a problem.

Water hardness varies in different localities. Drinking water contains some hardness, consisting primarily of calcium carbonate and/or magnesium carbonate. This hardness is expressed in grains per gallon (Table 12-6).

Table 12-6
Water Hardness Table

CLASS OF HARDNESS	GRAINS HARDNESS PER GALLON	% FIGURE IN U.S.
Low	3–10	30
Average	10–25	55
High	25–50	15

(Courtesy Research Products Corp.)

If 1 gallon of average water hardness is evaporated, a residue of 25 grains remains. If 100 gallons of water are evaporated to provide humidity, 2,500 grains or 5.7 ounces of solids will build up on the evaporating surface.

HOW RELATIVE HUMIDITY IS MEASURED

There are two types of instruments normally used in measuring relative humidity: hygrometers and psychrometers.

Hygrometers

Hygrometers are the instruments most commonly seen in homes and offices. They are manufactured in a variety of models. Most of them use a device that changes in dimensions as the relative humidity changes. This dimension change actuates a dial from which the relative humidity can be read. This type of instrument, when properly calibrated, will provide reasonably accurate relative humidity readings (Figure 12-2). The humidistat supplied with humidifiers is a type of hygrometer. The accuracy of these instruments can be checked by the second type of measuring device, the psychrometer. For convenience and maximum accuracy, the humidistat should be wall-mounted in the living area.

Sling-Type Psychrometer

This instrument is used by humidifier installers and others to determine the exact relative humidity readings necessary for calculations and recommendations (Figure 12-3).

For accurate results, correct sling psychrometer techniques should be utilized. The following hints are helpful to insure accurate readings. Use a clean wick and distilled water. Be sure the wet and dry bulb thermometers are matched, with no mercury separation. Rotate the psychrometer at approximately 600 fpm.

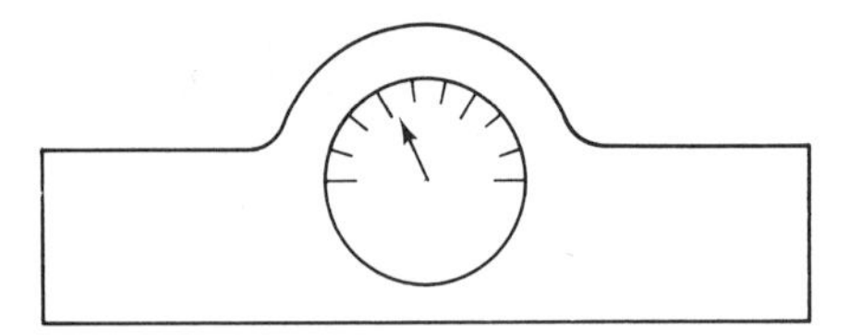

Figure 12-2. Desk hygrometer.

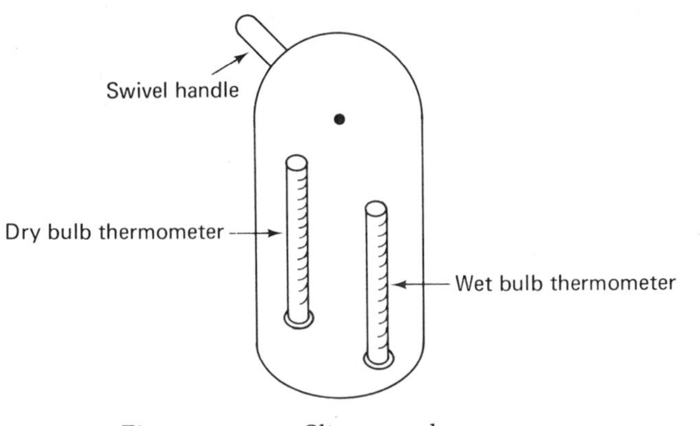

Figure 12-3. Sling psychrometer.

How to Take Readings

1. Dip the wick on the wet bulb thermometer in distilled water, preferably (only one dipping per determination of relative humidity, but never between readings). The progressive evaporation of the moisture in the wick, until it reaches equilibrium with the moisture in the air, is the determining factor of the wet bulb reading.

2. Whirl the sling psychrometer for 30 seconds. Take the readings quickly on the wet bulb thermometer first, then on the dry bulb, and jot them down. Whirl the psychrometer again, taking readings at 30 second intervals for five successive readings, and jot down the temperatures each time, or until the lowest reading shows a leveling off or a return curve (two or more nearly identical successive readings).

3. Use a psychrometric chart or table to obtain the relative humidity.

THE PSYCHROMETRIC CHART

The psychrometric chart simplifies the determination of air properties and eliminates many tedious calculations. We are concerned with five specific psychrometric terms that can be found on the chart. They are: (1) the dry bulb temperature; (2) the wet bulb temperature; (3) the relative humidity; (4) the absolute humidity; and (5) the dew point temperature. If we know any two of these air properties, we can, from the chart, determine the other three.

The first term, *dry bulb temperature,* is the temperature measure of an

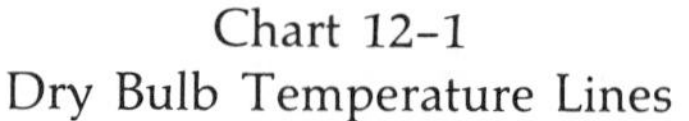

Chart 12–1
Dry Bulb Temperature Lines

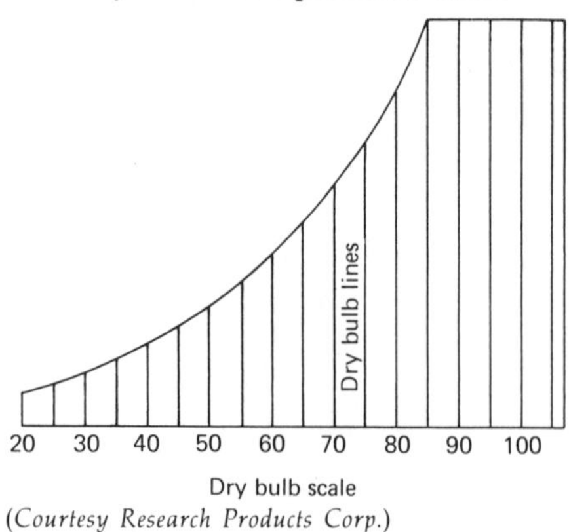

(*Courtesy Research Products Corp.*)

ordinary thermometer. This temperature scale runs horizontally across the bottom of the chart (see Chart 12–1). The dry bulb temperature lines are the straight vertical lines.

The second air property found on the psychrometric chart is the *wet bulb temperature,* or the temperature resulting when water is evaporated off a cloth covering an ordinary thermometer. As you can see, the wet bulb scale is measured along the curve portion of the psychrometric chart (see Chart 12–2), from the lower left to the upper right. The wet bulb lines run diagonally across the chart.

Next, we find the *relative humidity.* On a complete psychrometric chart, relative humidity lines are the only curved lines on the chart (see Chart 12–3). The various relative humidities are indicated on the lines. There is no coordinate scale as with the other air properties.

Chart 12–2
Wet Bulb Temperature Lines

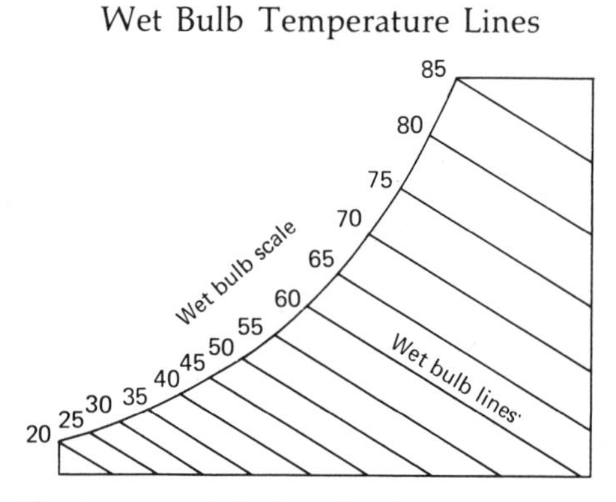

(*Courtesy Research Products Corp.*)

Chart 12–3
Relative Humidity Lines

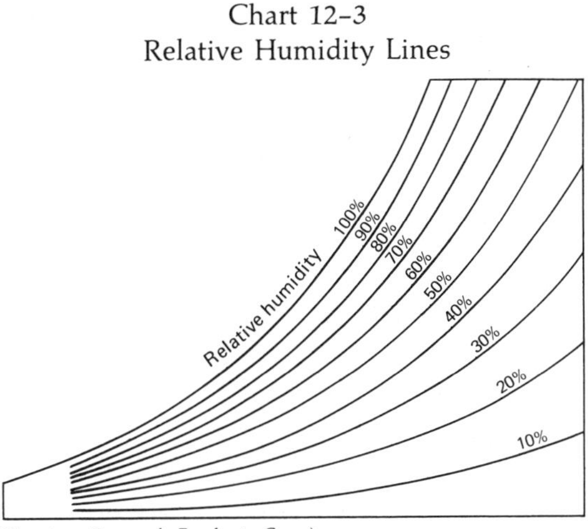

(*Courtesy Research Products Corp.*)

The fourth component of a psychrometric chart is the *absolute humidity,* or the actual weight of water vapor in the air. The scale for absolute humidity is a vertical scale on the right side of the psychrometric chart (see Chart 12–4). The absolute humidity lines run horizontally from this scale.

The *dew point temperature* is the fifth air property that is included on a psychrometric chart. This is the temperature at which moisture will condense on a surface. The scale for the dew point temperature is identical to the scale for the wet bulb temperature (see Chart 12–5). However, the dew point lines run horizontally across the chart, not diagonally as is the case with the wet bulb temperature lines.

When we put together the five charts we have just covered, we will have a complete psychrometric chart (see Chart 12–6).

Chart 12–4
Absolute Humidity Lines

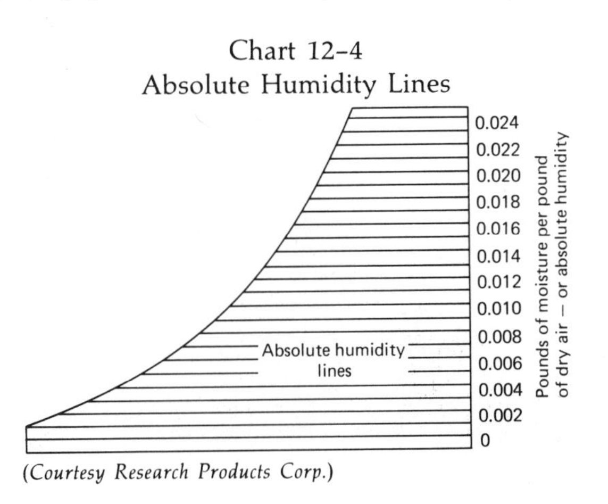

(*Courtesy Research Products Corp.*)

Chart 12-5
Dew Point Temperature Lines

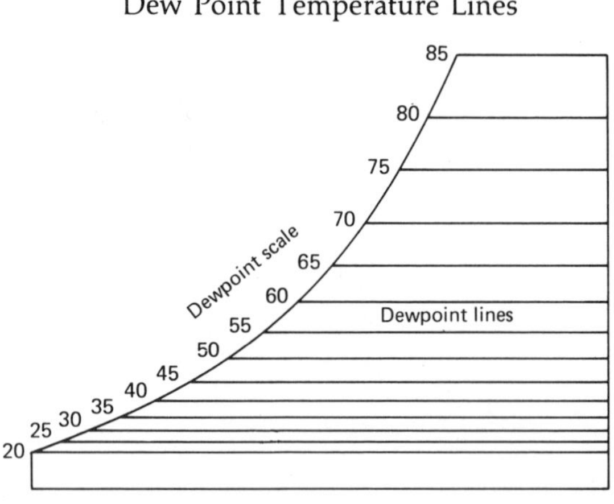

(Courtesy Research Products Corp.)

Let us take an example to explain the working of the chart. Let us assume that we have used a sling psychrometer and have taken readings of a dry bulb temperature of 72° F and a web bulb temperature of 56° F. We know two factors. The 72° F dry bulb temperature is found on the bottom scale. The wet bulb scale is on the curved outside line of the chart on the left. The wet bulb temperature line is 56° F. Extending the two lines, we find that they intersect. From this point we can determine any other information we need. The relative humidity is, for example,

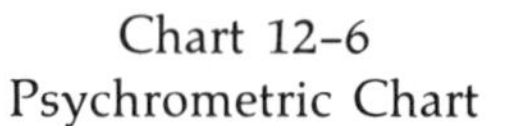
Chart 12-6
Psychrometric Chart

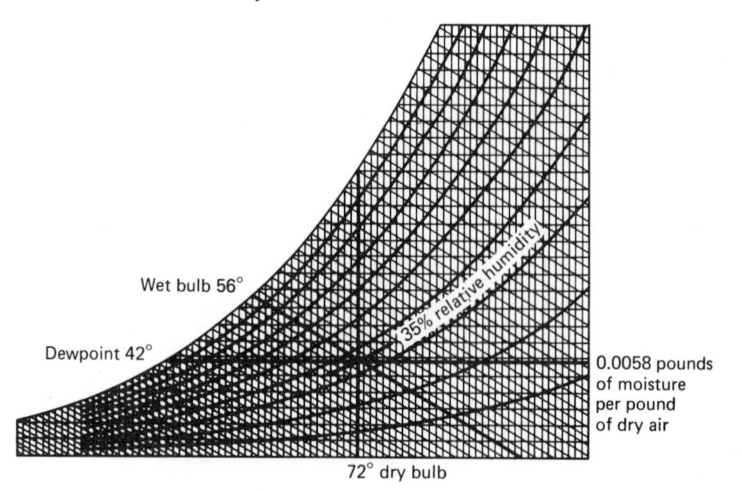

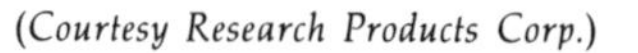
(Courtesy Research Products Corp.)

35%. The absolute humidity is .0058 pounds of water per pound of air. The dew point temperature is 42° F.

HUMIDIFIERS

There are many, many humidifiers available. They vary in price, they vary in capacity, and they vary in principle of operation.

For classification purposes, it is simpler, and more logical, to consider humidifiers in three general types:

1. Pan-type units
2. Atomizing-type units
3. Wetted element-type units.

The pan-type unit is the simplest type of humidifier (Figure 12-4). Its capacity is low. On a hot radiator, it might evaporate .0083 gallons of water per hour. In the warm air plenum of a furnace, it would evaporate approximately 0.18 gallons of water per hour. To increase the capacity, the air-to-water surfaces can be increased by placing water wicking plates in the pan. The capacity increases as the air temperature in the furnace plenum increases.

Greater capacity is also possible through the use of steam, hot water, or an electric heating element immersed in the water. A 1,200 W heating element, for example, in a container with water supplied by a float valve could produce .48 gallons of moisture per hour.

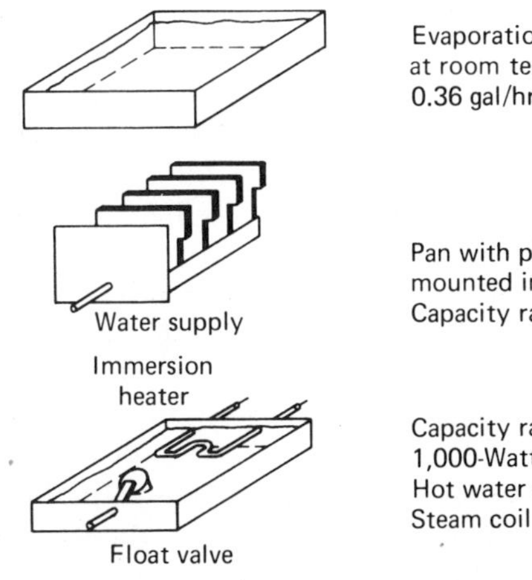

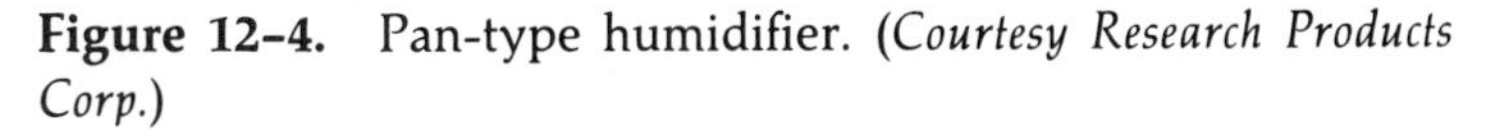

Figure 12-4. Pan-type humidifier. (*Courtesy Research Products Corp.*)

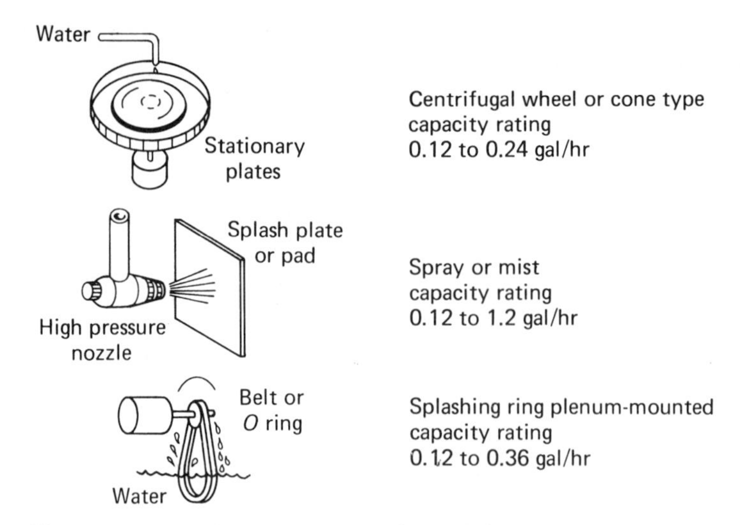

Figure 12-5. Atomizing-type humidifier. (*Courtesy Research Products Corp.*)

The second type of humidifier is the atomizing type (Figure 12-5). This device atomizes the water by throwing it from the surface at a rapidly revolving disc. It is generally a portable or console unit, although it can be installed so that water particles will be directed into a ducted central system.

The third type is the wetted element type humidifier. In its simplest form, it operates in the manner of an evaporative cooler. Here air is either pushed or pulled through a wetted pad material or filter and evaporative cooling takes place (Figure 12-6). By increasing the air flow or by supplying additional heat, the evaporation rate of the humidifier can be increased. The heat source for evaporation can be free from an increase in water temperature or an increase in air temperature.

The air for evaporation can be taken from the heated air of the furnace plenum and directed through the humidifier by the humidifier fan, or it can be drawn through the wetted element by the air pressure differential of the furnace blower system.

Furnace-mounted humidifiers, usually of the wetted type, can be constructed so that they produce 1.2 or more gallons of moisture per hour. Because of their capacity, this type usually has a humidistat or control that will actuate a relay or water valve and start a fan that operates until the control is satisfied. Normally more water is supplied to the unit than is evaporated and this flushing action washes a large portion of the hardness salts from the evaporative element to a floor drain to eliminate them from the humidifying system.

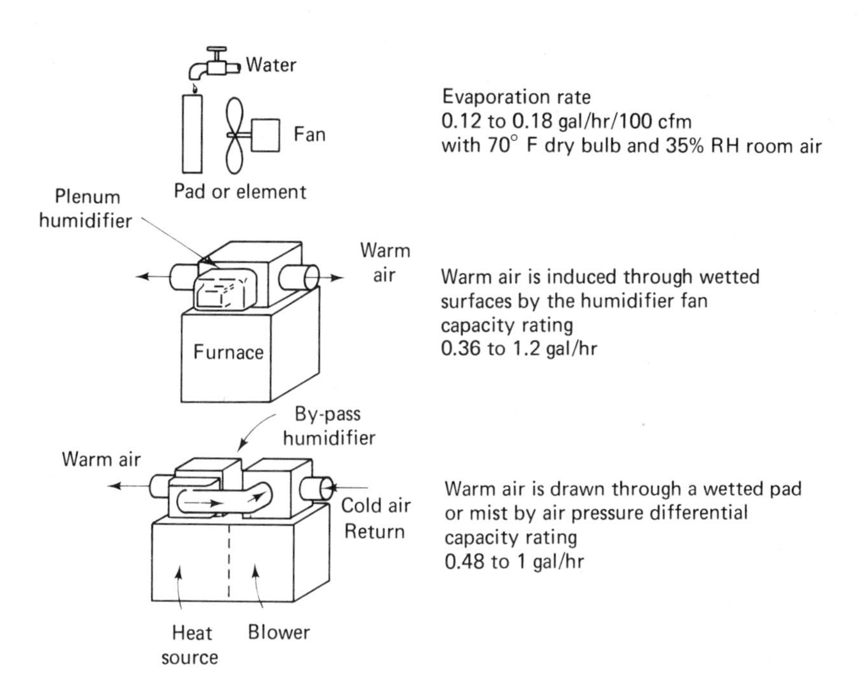

Figure 12-6. Wetted-element type humidifier. *(Courtesy Research Products Corp.)*

HUMIDIFIER SIZING

Sizing for humidification is similar to sizing a system for heating and cooling. The humidifier capacity required will be determined by various factors: (1) the volume of the area being humidified; (2) the air change rate (infiltration or ventilation); (3) the inside and outside design conditions; and (4) other sources of humidity. Each factor is discussed as follows:

1. The volume of the home being humidified: The volume can be determined from a floor plan or from measurements taken within the home. If the basement is heated and ventilated, its volume should be included.

2. Air change rate: The amount of infiltration was probably calculated when computing the heating and cooling load.

The following method may also be used:

An average house will normally have about one air change per hour. A tight house may have as little as one-half air change per hour and a loose house may have as high as two air changes per hour. For definition purposes, an average house is assumed to have insulation in the walls and ceilings, vapor barriers, loose storm doors and windows, and may or may not have a fireplace. If it has a fireplace, however, it will be dampered. The tight house will be well insulated, have vapor barriers, tight storm doors and windows with weather stripping, and its fireplace will be dampered. The loose house will probably be one constructed before 1930, have little or no insulation, no storm doors or windows, no weather stripping, no vapor barriers, and quite often will have a fireplace without an effective damper.

3. Design Conditions: The principal factors involved are (a) the desired indoor temperature and relative humidity; and (b) the prevailing outdoor temperature and relative humidity.

Comfort is usually the prime requirement in residential humidification. The human body is comfortable within a broad range of relative humidity. Humidities of 35% or more are desirable from the health standpoint. However, in the winter these high humidities can present a condensation problem. There is, however, a compromise between the humidities preferable from a health standpoint and the humidities desirable from a construction standpoint (see Table 12-7). Because of the condensation problem, the humidity maintained within a home should be lowered as the outdoor temperature drops.

However, if the house is designed to withstand higher humidities or if for medical reasons a higher humidity must be maintained, then higher values can be used.

CALCULATIONS

When calculating the amount of humidity required, the following formula can be used:

$$H = \frac{VR(W_i \times W_o)}{13.5 \times 8.3}$$

where

H = gallons of moisture per hour required to maintain indoor design conditions

V = number of changes of air per hr

W_i = pound of moisture per pound of dry air at the desired indoor conditions (from the psychrometric chart)

Table 12-7
Recommended Indoor Humidity Levels

OUTDOOR TEMP., °F	RECOMMENDED HUMIDITY, %
20° and above	35%
10°	30%
0°	25%
−10°	20%
−20°	15%

(Courtesy Research Products Corp.)

W_o = pound of moisture per pound of dry air at outdoor conditions (from the psychrometric chart).

The value of 13.5 ft^3 of air per pound of air is an average that is suitable for calculations where extreme accuracy is not required. The value of 8.3 is the number of pounds of moisture in a gallon.

HUMIDIFIER OPERATION

Aprilaire Humidifiers

The operating principle of all Aprilaire humidifiers is basically the same, and all use nature's own process (the introduction of humidity in the form of water vapor), with the refinements necessary to provide positive, accurate control.

The humidistat, preferably located in the living area, is conveniently set at the desired humidity level and activates the unit whenever the humidity falls below the setting. Water is supplied to the distribution pan (Figure 12-7) from where it flows evenly across the water panel evaporator. Thirsty, dry air is forced through the wetted panel and the now humidified air carrying water as a vapor is distributed throughout the living area of the home.

Illustrated in Figure 12-8 is the amount of mineral residue from the evaporation of just 1 gallon of 20 hardness water (that is high hardness). About 360 gallons of water must be evaporated monthly in a typical 13,000 ft^3 home (1,625 ft^2) to maintain proper relative humidity.

In some humidifiers, this imposing amount of mineral deposit build-up can cause a malfunction requiring service. And, in some, the minerals can be distributed throughout the home in the form of white dust.

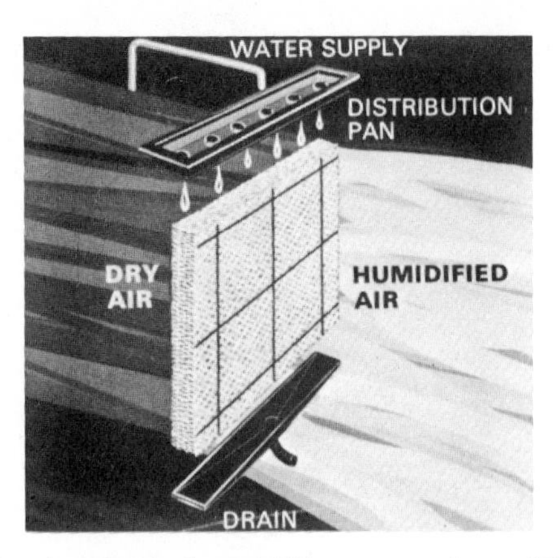

Figure 12–7. Aprilaire humidifier operation. (*Courtesy Research Products Corp.*)

Control

A precision-made, accurately calibrated humidistat is required for maximum performance (Figure 12–9). For maximum convenience, it is usually located in the living area. All that is required is to set the dial (and reset as necessary) for the desired humidity.

Important Aprilaire Humidifier Principles

1. Drain (Figure 12–10). The water flushes a significant amount of the minerals completely out of the system.

2. Distribution Pan (Figure 12–11). The scientifically designed pattern of circular inlets permits water to flow, by gravity, uniformly across the water panel evaporator for more efficient, higher capacity humidification.

Figure 12–8. Mineral residue. (*Courtesy Research Products Corp.*)

Figure 12-9. Humidistat. (*Courtesy Research Products Corp.*)

Figure 12-10. Aprilaire humidifier drain. (*Courtesy Research Products Corp.*)

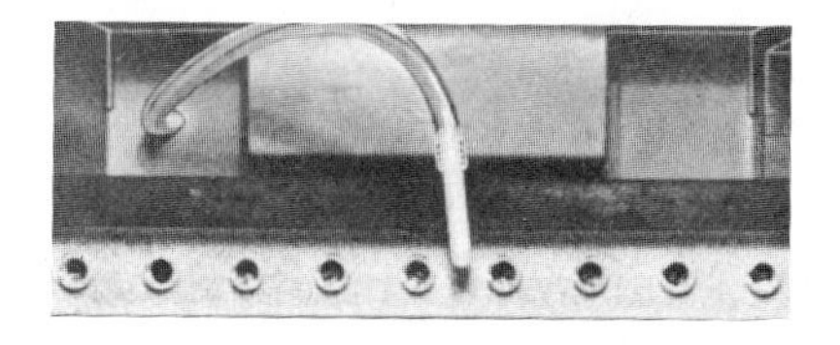

Figure 12-11. Aprilaire distribution pan. (*Courtesy Research Products Corp.*)

3. Water Reservoir and Liner (Figure 12-12). The partitioned reservoir controls scale, separates fresh and circulated water. A removable liner simplifies cleaning on some models.

4. Water Panel Evaporator and Scale Control Insert (Figure 12-13). The water panel is absorbent, with high saturation ability, to insure uniform evaporation, optimum efficiency, and high capacity. It traps trouble-causing minerals. The scale control insert localizes excess mineral deposits—another aid in reducing maintenance.

5. Solenoid Valve (Figure 12-14). This valve is electrically controlled for positive water supply and shutoff. A teflon orifice assures a positive, metered flow of water.

A complete line of Aprilaire humidifiers offers a choice of models to fulfill the requirements of installations from homes to commercial establishments.

Figure 12-12. Aprilaire partitioned reservoir. (*Courtesy Research Products Corp.*)

Figure 12-13. Aprilaire water panel evaporator and scale control insert. (*Courtesy Research Products Corp.*)

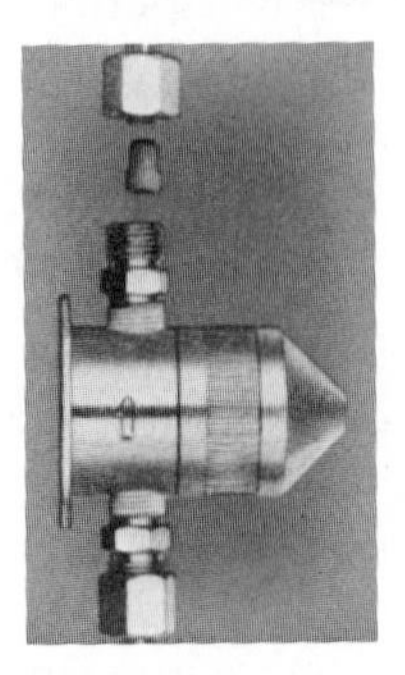

Figure 12-14. Aprilaire solenoid valve. (*Courtesy Research Products Corp.*)

The models 110 and 112 (Figure 12-15) are forced air furnace attached units that incorporate a fan and a drain, and represent the ultimate in controlled humidification for residential applications.

The model 330 (Figure 12-16) is a self-contained unit designed especially for homes heated by hot water or steam systems, and equally

Figure 12-15. Aprilaire models 110 and 112 humidifiers. *(Courtesy Research Products Corp.)*

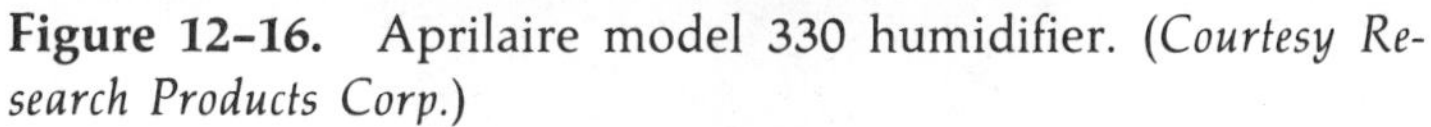

Figure 12-16. Aprilaire model 330 humidifier. *(Courtesy Research Products Corp.)*

suitable for electric or any other type of heating system. It can be installed in the basement, utility room, or heated crawl space.

The Aprilaire models 440 and 445 (Figure 12-17) and the Chippewa models 220 and 224 (Figure 12-18) are bypass-type humidifiers that can be installed on either the supply or the return air plenum of a forced air furnace. They are ideal for installations where plenum space is limited.

Figure 12-17. Aprilaire models 440 and 445 humidifiers. *(Courtesy Research Products Corp.)*

All of these are equipped with a drain except the model 445, which has a water circulating system and is designed for installations where drain facilities are not available, or where high water hardness is not a problem.

Lau Humidifiers

The Lau Vapor-Air residential humidifiers operate on the proven vapor-wheel principle (Figure 12-19). By taking dry air and forcing it

Figure 12-18. Aprilaire Chippewa humidifier. (*Courtesy Research Products Corp.*)

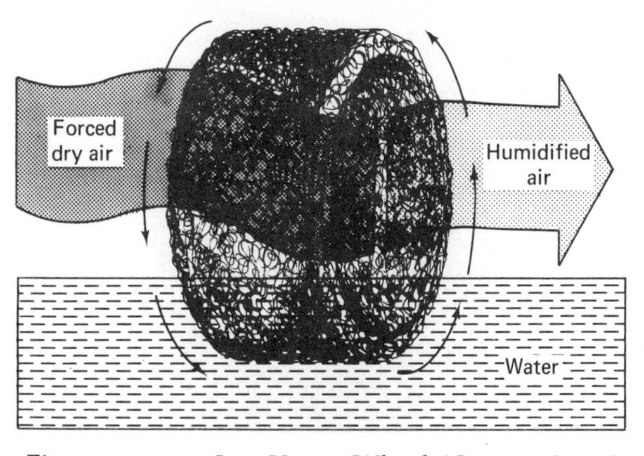

Figure 12–19. Lau Vapor-Wheel (*Courtesy Lau, Inc.*)

through a saturated, rotating media on the vapor-wheel, the dry air becomes comfortably humidified. The moisture-laden air produced by the humidifier is then carried throughout the home (Figure 12–20).

The Lau Vapor-Wheel principle is an extremely economical method of operation. A Lau Vapor-Air Humidifier costs no more to operate per day than a 60-watt light bulb.

Applications: The Vapor-Air I is a quiet, completely self-contained, "lime dust"-free automatic humidifier that operates independently of any heating system (Figure 12–21).

The Lau Vapor-Air I has been developed specifically for commercial and industrial applications in which the maintenance of a proper

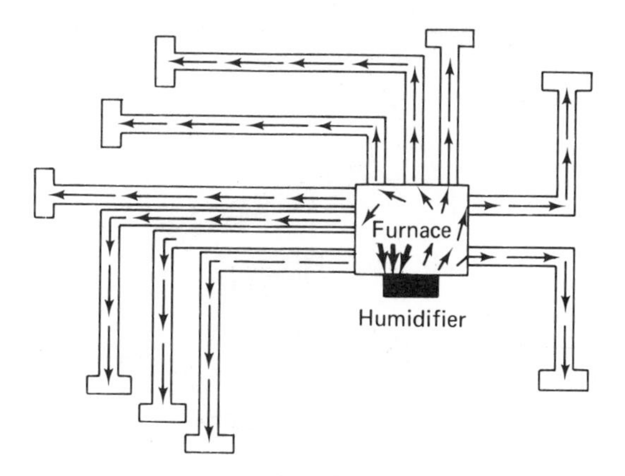

Figure 12–20. Moisture distribution throughout a home. (*Courtesy Research Products Corp.*)

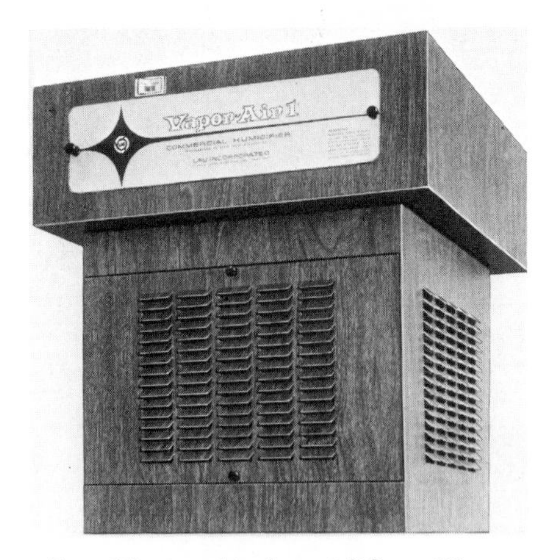

Figure 12-21. Lau Vapor-Air humidifier. (*Courtesy Lau, Inc.*)

level of humidity is necessary to the comfort and well-being of the customers and employees and to protect furnishings, supplies and products. The unit is a humidistatically controlled, wetted-surface type humidifier with a capacity of up to 50 gallons of water per day at a 72° F room temperature.

When the unit is operating, dry air is drawn into the unit through the top by a propeller (Figure 12-22). The air then passes through a moistened media pad and is forced out through the louvers as properly humidified air.

The Lau Vapor-Air 2 is a highly versatile, easy to install, automatic central system humidifier (Figure 12-23). It is designed for use on all res-

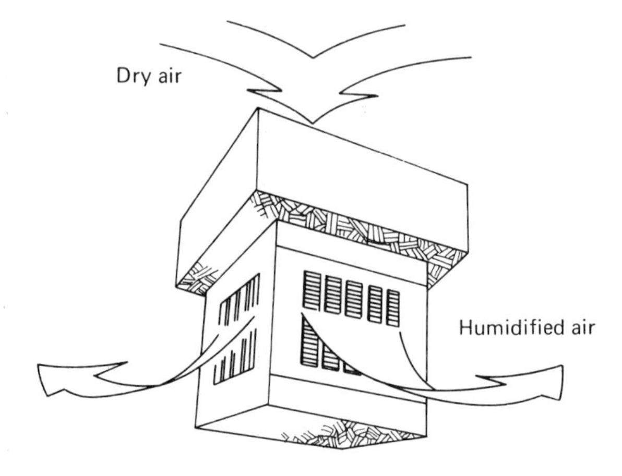

Figure 12-22. Lau Vapor-Air 1 humidifier air flow. (*Courtesy Lau, Inc.*)

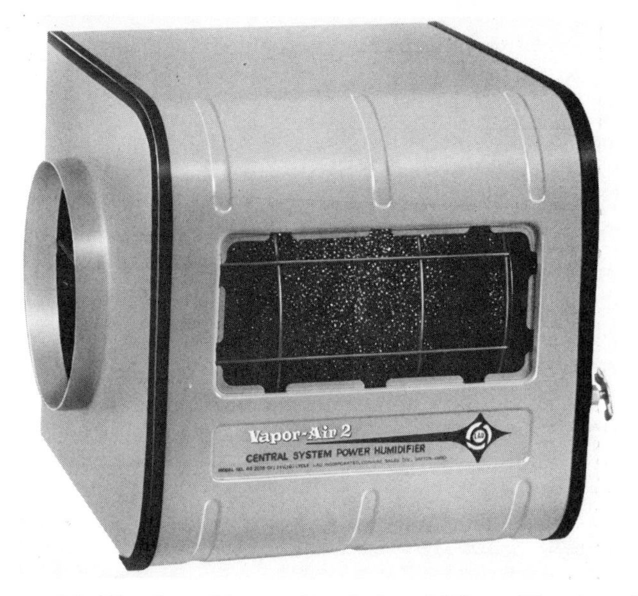

Figure 12-23. Lau Vapor-Air 2 humidifier. (*Courtesy Lau, Inc.*)

idential and commercial forced warm air heating systems. It is capable of delivering up to 38 gallons of water per 24 hours of operation.

During operation, air is drawn into the unit through the flexible ducting by the static pressure differential between the warm air and cold air ducts (Figure 12-24). The unit is designed so that the air drawn into

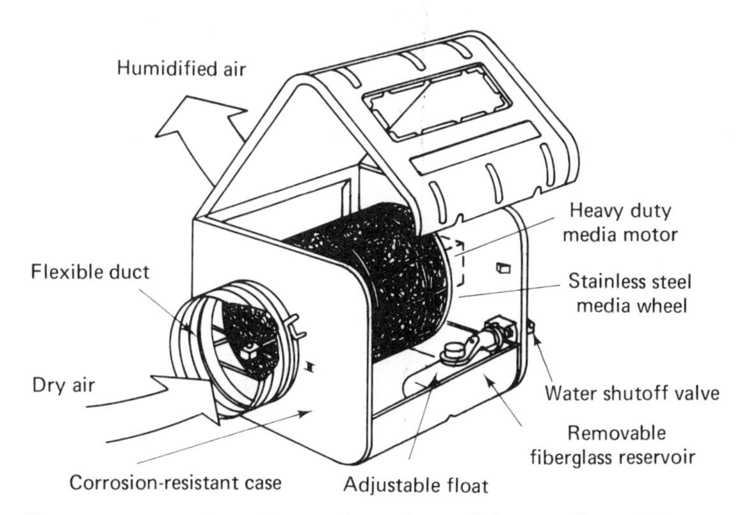

Figure 12-24. Lau Vapor-Air 2 humidifier air flow. (*Courtesy Lau, Inc.*)

the unit passes through the saturated media pad, where it becomes properly humidified, and is forced into the heating system.

The Vapor-Air 2 can be installed on any forced warm air heating system. It may be mounted on either the warm air or cold air plenum. The unit is completely reversible so that, depending on the installation, the flexible duct may be mounted on either the right or left side of the humidifier. It is extremely easy to service. The lid raises for maximum access, the media wheel is easily removable and the reservoir slides out (Figure 12-24). A water shutoff valve on the side of the unit eliminates the need to turn off the water at the saddle valve.

The Vapor-Air 3 is a trim, appliance-styled, completely automatic central system power humidifier designed for average to large homes requiring up to 20 gallons of moisture per day (Figure 12-25).

The specifications for the Vapor-Air 3 are shown in Table 12-8.

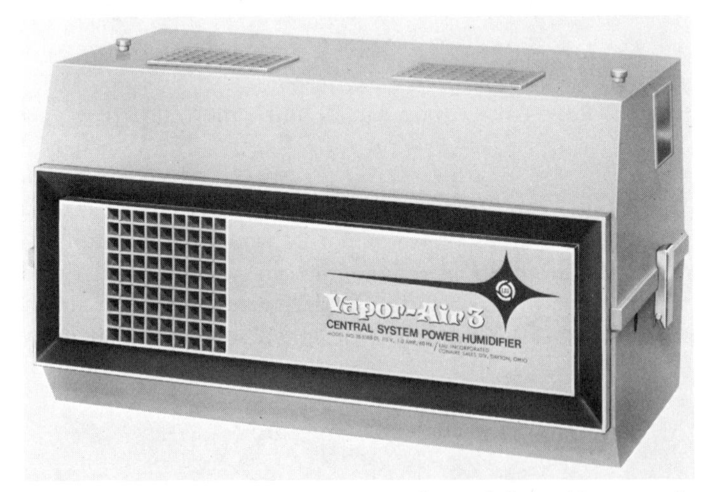

Figure 12-25. Lau Vapor-Air 3 humidifier. (*Courtesy Lau, Inc.*)

Table 12-8
Vapor-Air 3 Specifications

Duct Temp Fahrenheit	Gal/24 Hr	Lb/Hr
180°	19.50	6.8
160°	16.95	5.9
*140°**	*14.75**	*5.1**
120°	11.75	4.1
100°	9.29	3.2
80°	6.88	2.4

(*Courtesy Lau, Inc.*)

For a central system, bypass-type humidifier, the Vapor-Air 4 is designed for smaller homes with forced warm air heating systems (Figure 12–26). This unit requires no electrical connections and has a capacity of up to 14 gallons per day.

In operation, the Vapor-Air 4 uses the Lau Vapor-Wheel principle coupled with the Lau Dyna-Drive Assembly (Figure 12–27). A multiblade propeller is driven off the pressure differential air velocity that takes place between the warm and cold air plenum in a forced air heating system. The air-driven propeller drives the reduction gear that rotates the media wheel at 2 rpm with an 85 cfm bypass at .2 static pressure for the designated capacity.

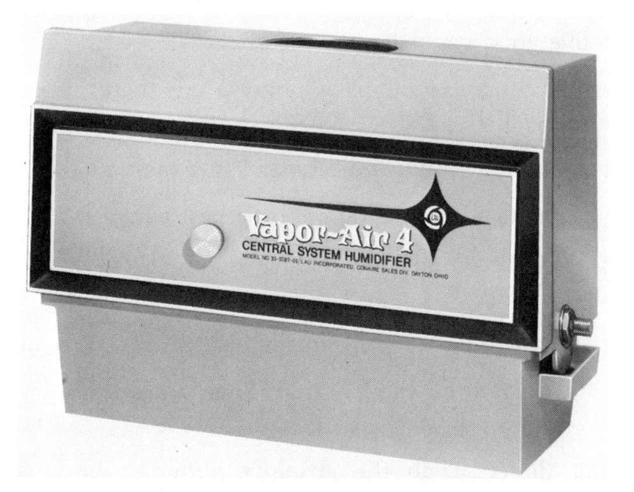

Figure 12–26. Lau Vapor-Air 4 humidifier. (*Courtesy Lau, Inc.*)

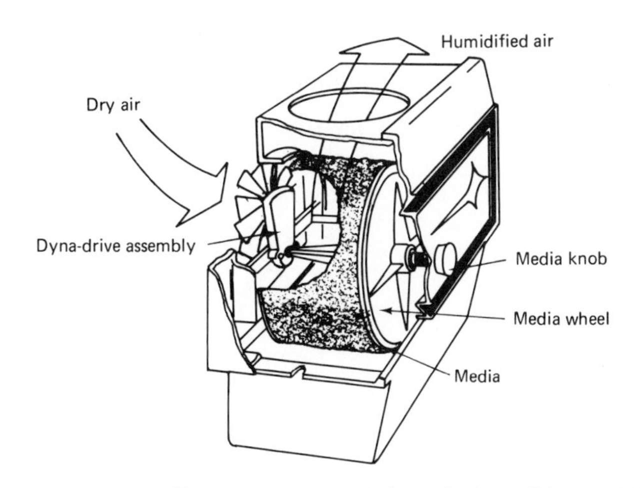

Figure 12–27. Operating principle of Lau Vapor-Air 4. (*Courtesy Lau, Inc.*)

REVIEW QUESTIONS

1. Define humidity.
2. What is relative humidity?
3. Give three reasons why we need humidity.
4. Will warm air or cold air hold the most moisture?
5. Why does a carpet wear quicker when low humidity is prevalent?
6. Will the relative humidity be the same in all parts of a house?
7. Why is a vapor barrier installed in outside walls?
8. Name five good vapor barrier materials.
9. If there is condensation on windows, will there be condensation inside the walls?
10. What is the first health indication of low humidity?
11. What percentage relative humidity is best suited for health?
12. What is the recommended relative humidity for an outside temperature of $-10°$ F?
13. What are two types of instruments used to measure humidity?
14. In what direction do the dry bulb lines run on the psychrometric chart?
15. In what direction do the absolute humidity lines run on the psychrometric chart?
16. Which are the relative humidity lines on the psychrometric chart?
17. What are the three types of humidifiers?
18. Name the three factors determining the size of a humidifier.

13

Basic Heating Controls

The objectives of this chapter are:

- To acquaint you with the function of basic heating controls.
- To provide you with the information necessary for the operation of basic heating controls.
- To provide you with the information necessary for competent servicing of basic heating controls.

The residence should be heated to a comfortable temperature, with little or no variation in temperature between the floor and the ceiling.

This enormous duty falls on the control system, in conjunction with a properly designed and installed heating system. These three factors go hand in hand. Without the proper controls, however, an otherwise perfect system would be ineffective.

THE THERMOSTAT

In general, low-voltage room thermostats should be used for the best temperature control (Figure 13–1). The low-voltage thermostats respond much faster to temperature changes than the greater-mass line-voltage devices. This has been proven on residential heating systems.

From a cost standpoint, the less expensive installation of low-voltage wiring more than offsets the extra cost of the transformer. Also, homeowners are accustomed to the safety of low-voltage thermostats.

Figure 13–1. Low-voltage heating thermostat. (*Courtesy Honeywell, Inc.*)

The room thermostat is provided with a heat anticipator connected in series with the rest of the control circuit (Figure 13–2). These anticipators are made of a resistance-type material that produces heat in accordance with the current drawn through them (Figure 13–3). Heat anticipators are adjustable and are normally set to correspond with the current rating of the main gas valve. The purpose of these devices is to make the room temperature more even.

In operation, when the thermostat is calling for heat, the anticipator is also producing heat to the thermostat. This heating action causes the thermostat to become satisfied before the room actually reaches the set point of the thermostat. Thus, the thermostat stops the flame and the room temperature will not overshoot, or go too high.

The heat anticipator is adjusted to match the amperage draw of the temperature control circuit. This adjustment is made by setting the heat anticipator pointer to the correct number on the scale. The correct amperage draw of the circuit is determined by actual measurement with an ammeter.

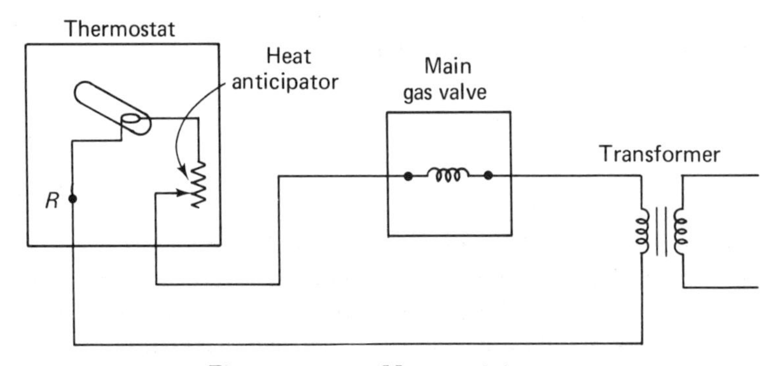

Figure 13–2. Heat anticipator.

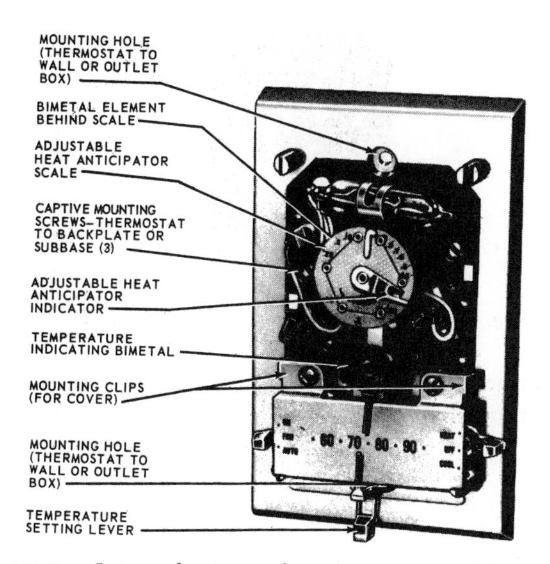

Figure 13–3. Internal view of a thermostat. (*Courtesy Honeywell, Inc.*)

THE TRANSFORMER

The transformer is a device used to reduce line voltage to a usable control voltage, usually 24 V. Transformers must be sized to provide sufficient power to operate the control circuit. Most are oversized sufficiently to provide enough power to operate an air conditioning control circuit also. This rating is usually 40 VA.

FAN CONTROL

The fan control is a temperature-actuated control that, when heated, will close a set of contacts to start the indoor fan motor (Figure 13–4). The sensing element of the fan control is positioned inside the heat exchanger where the temperature is the highest.

This control is actuated by a bimetal element that opens or closes the contacts on temperature change. The fan control is usually set to start the fan at about 150° F and to stop it at about 100° F.

In operation, the burner provides heat to the heat exchanger for a few seconds to warm the furnace before the fan is started. This operation is to prevent blowing cold air into the room on furnace start up.

Figure 13-4. Fan control. (*Courtesy Honeywell, Inc.*)

When the thermostat is satisfied, the main burner stops providing heat, but the fan continues to operate until the temperature in the furnace has been reduced, thus removing any excess heat in the furnace.

LIMIT CONTROL

The limit control is also a heat-actuated switch with a bimetal sensing element positioned inside the heat exchanger (Figure 13-5).

This is a safety control that is wired into the primary side of the transformer. If the temperature inside the furnace reaches approximately 200° F, the power will be shut off to the transformer, which also stops all power in the temperature control circuit.

MAIN GAS VALVE

The main gas valve is the device that acts on demand from the thermostat to either admit gas to the main burners or to stop the gas supply (Figure 13-6). This valve has many functions. It has a gas pressure regulator, a pilot safety, a main gas cock, a pilot gas cock, and the main gas solenoid all in a single unit—the combination gas control.

Figure 13-5. Limit control. (*Courtesy Honeywell, Inc.*)

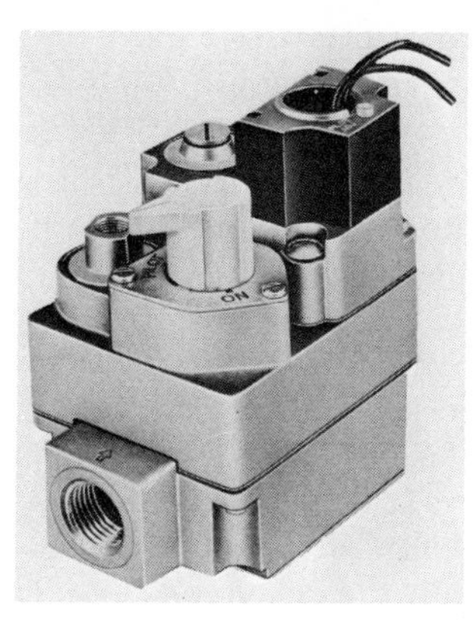

Figure 13-6. A main gas valve. (*Courtesy White-Rodgers Division, Emerson Electric Co.*)

Sequence of Operation

As the thermostat calls for heat, energizing the solenoid coil, the valve lever opens the cycling valve (Figure 13-7). The inlet gas now

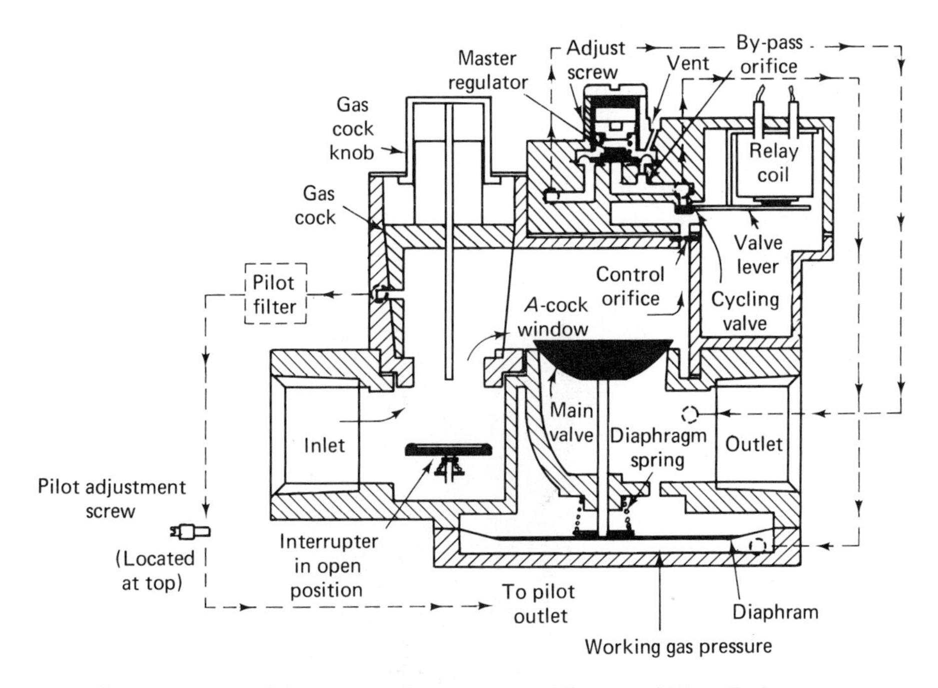

Figure 13-7. Main gas valve cutaway (*Courtesy White-Rodgers Division, Emerson Electric Co.*)

flows through the control orifice past the cycling valve. At this point, gas flow is in two directions, as follows:

1. Part of the flow is to the back of the diaphragm by means of internal passageways. The resulting increase in pressure pushes the main valve to the open position, compressing the diaphragm springs lightly.

2. Part of the flow is through the seat of the master regulator into the valve outlet by means of internal passageways. This causes the master regulator to begin its function.

The gas valve remains in this position and the master regulator continues to regulate until the relay coil is de-energized, at which time the cycling valve seals off.

As the cycling valve closes, the regulator spring causes the seat of the master regulator to close off. The function of the bypass orifice is to permit gas in the passageways to escape into the outlet of the valve, thereby causing the main gas valve to close.

PILOT SAFETY

There are two types of pilot safety controls: non-100% safe and 100% safe. These two names refer to the amount of gas cut off when the pilot light is unsafe. The 100% safe is incorporated in the combination main gas valve. The non-100% device (Figure 13-8) incorporates a set of contacts that open the control circuit during an unsafe condition.

These units are used in conjunction with a thermocouple to keep the control contacts closed, or the valve open, during normal operation.

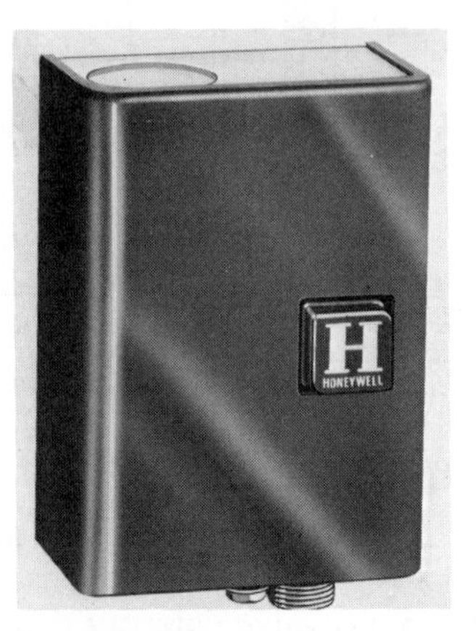

Figure 13-8. Non-100% safe pilot safety. (*Courtesy Honeywell, Inc.*)

If at any time the control "drops out" (the contacts open), the reset button will have to be manually reset before operation of the furnace can be resumed.

THE THERMOCOUPLE

The thermocouple is a device that uses the difference in metals to provide electron flow. The hot junction of the thermocouple is put in the pilot flame where the dissimilar metals are heated (Figure 13-9). When heat is applied to the welded junction, a small voltage is produced. This small voltage, measured in millivolts (mV), is the power used to operate the pilot safety control. The output of a thermocouple is approximately 30 mV. This simple device can cause many problems if the connections are not kept clean and tight.

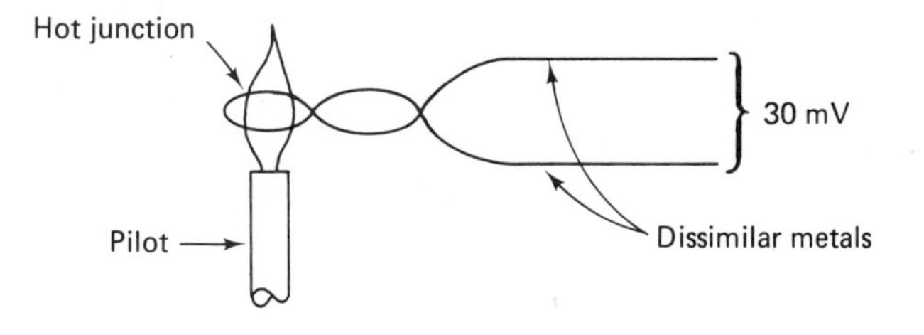

Figure 13-9. Thermocouple operation.

THE FIRE-STAT

The fire-stat is a safety device mounted in the fan compartment to stop the fan if the return air temperature reaches about 160° F (Figure 13-10). It is a bimetal-actuated, normally closed switch that must be manually reset before the fan can operate.

The reason for stopping the fan when the high return air temperatures exist is to prevent agitation of any open flame in the house, thus helping to prevent the spreading of any fire that may be present.

FURNACE WIRING

There are three different circuits and three different voltages in a modern furnace 24 V control system. The following diagrams will illustrate each of these:

Figure 13-10. Fire-stat. (*Courtesy Honeywell, Inc.*)

1. The fan or circulator circuit (Figure 13-11).
2. The temperature control circuit (Figure 13-12).
3. The pilot safety circuit, 30 mV (Figure 13-13).

When all three of these circuits are connected together (Figure 13-14), we have a modern furnace 24 V control system.

The control of electric furnaces is much the same as that just described; however, there are no pilot safety devices and the main gas valve is replaced with relays that actuate to complete the electrical circuit to the heating elements.

ELECTRIC FURNACE CONTROLS

Many of the controls used on gas furnaces are also used on electric furnaces. Some controls are slightly modified while others are a completely separate type of control. There does not seem to be any standard

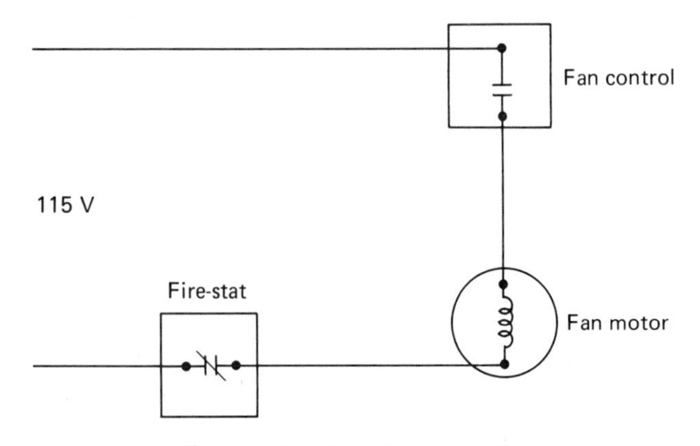

Figure 13-11. Fan circuit.

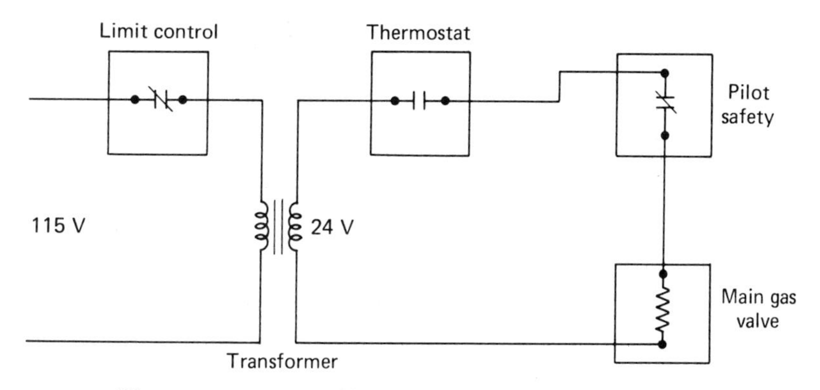

Figure 13-12. 24 V temperature control circuit.

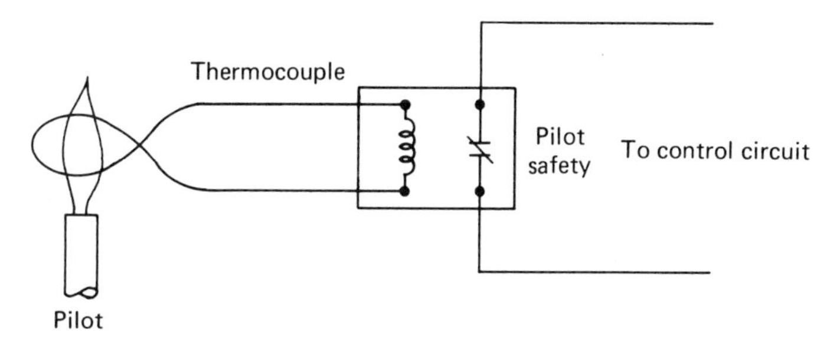

Figure 13-13. Pilot safety circuit.

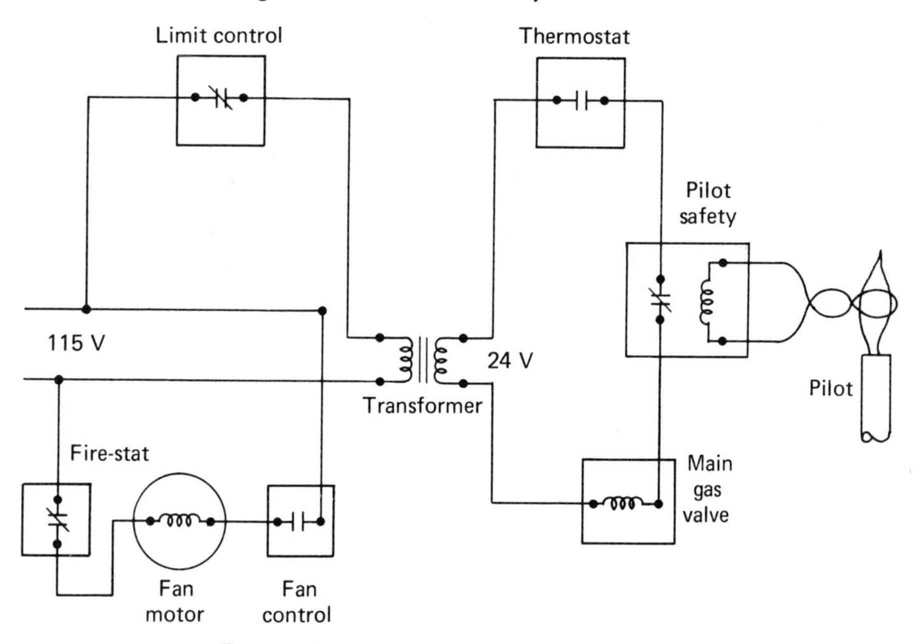

Figure 13-14. 24 V control circuit.

method of wiring the control systems on electric furnaces. The major thing, however, is to know how the control functions, then to wire it into a circuit is relatively simple.

Time Delay Relay

The time delay relay is a device used for controlling electrically operated devices at a predetermined time interval after the heater is energized (Figure 13-15). The heater timer is connected to the 24 volt temperature control circuit which is energized by the thermostat (Figure 13-16). These devices have an inherent time delay before the contacts close or open. Each relay has a given time delay for each function. Each relay will normally control a single load, such as a heating element or a fan motor (Figure 13-17). These relays may be used with a single thermostat or a multistage thermostat.

Sequencer

A more elaborate method of controlling electric heating equipment is by use of a sequencer. These multistage devices may have 10 to 12 stages depending on the size of equipment and the comfort demand (Figure 13-18). When they are used with the proper components, they will

Figure 13-15. Time delay relay. (*Courtesy Cam-Stat Incorporated*)

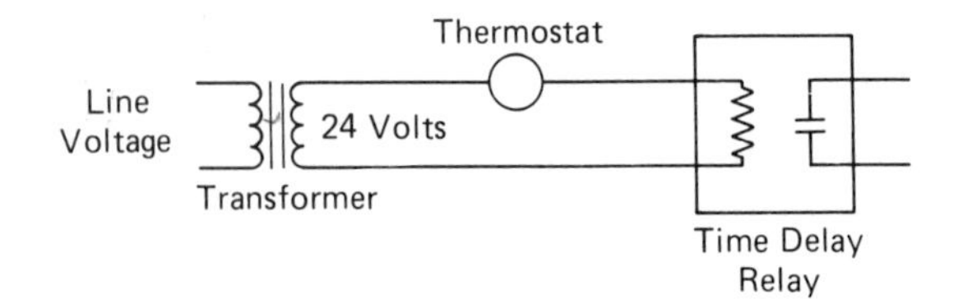

Figure 13-16. Time delay relay wiring diagram (24 Volt).

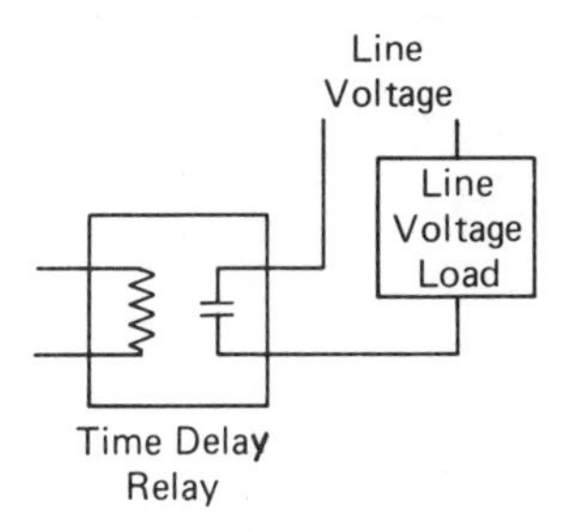

Figure 13-17. Time delay relay wiring diagram (line voltage).

provide proportional control of multistage equipment. Normally, they are used with a modulating-type thermostat and automatically turn on or off the number of heating elements necessary to satisfy the space requirements. However, some of the newer sequencers use a thermal power element, which completely eliminates the need for motors and gears. Accurate temperature control is achieved through the use of a slide wire or solid state sensors.

Limit Controls

Some electric furnace manufacturers use special types of limit control on their furnace. These are SPDT switches used to break one circuit and make another under high temperature conditions. When the temperature inside the furnace reaches approximately 200° F, the limit control will open the electrical circuit to the primary side of the transformer and complete an electrical circuit to the fan motor. (Figure 13-19). This is done to remove heat from the furnace and help prevent damage to the heating elements.

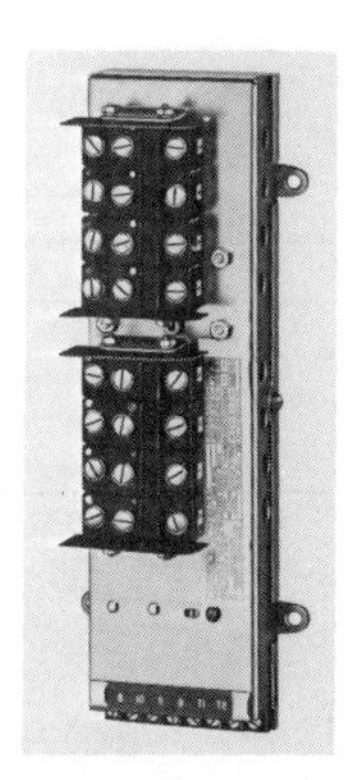

Figure 13-18. Electric heating sequencer. (*Courtesy Penn Controls Division*)

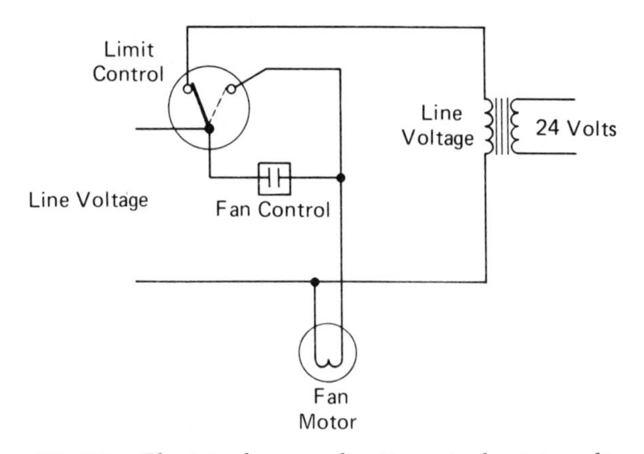

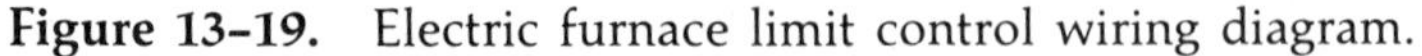

Figure 13–19. Electric furnace limit control wiring diagram.

Element Over-Current Protector

These devices are installed in line with the heating element and are designed to open the electrical circuit to the element in case of an over-current condition (Figure 13–20). They are current sensitive and are usually very accurate. Overcurrent protectors may be automatic reset, manual reset, or replaceable type devices.

Element Over-Temperature Protector

Over-temperature protectors are placed in line with the heating element and open the electrical circuit in case of an over temperature condition of the heating element (Figure 13–21). These devices are tem-

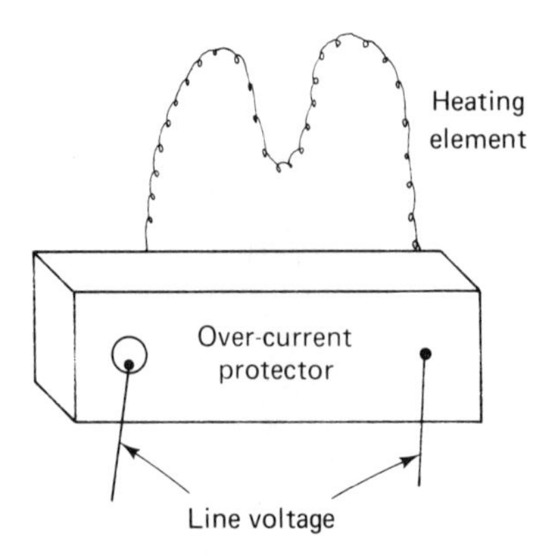

Figure 13–20. Element over-current protector wiring diagram.

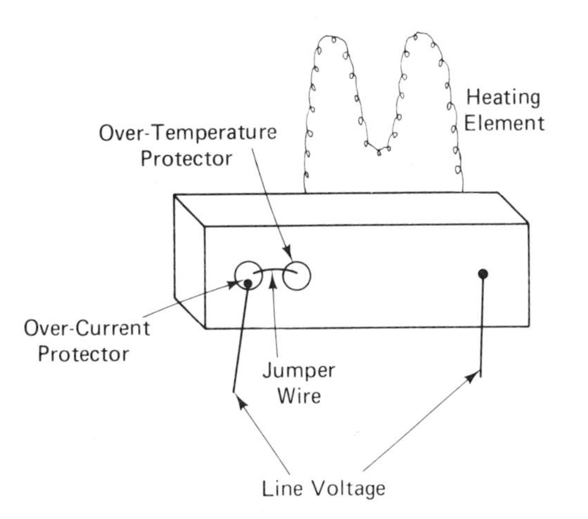

Figure 13-21. Element over-temperature protector wiring diagram.

perature sensitive and usually require replacement if such a condition occurs. They are sometimes referred to as thermal fuses.

The variety of functions performed by a heating system is limited only by the use of controls. The more a service technician knows about controls, the easier will be his job in servicing such equipment. Reference to a comprehensive book on controls, such as B. C. Langley's *Electric Controls for Refrigeration and Air Conditioning* (Prentice-Hall, Inc., Englewood Cliffs, N.J., 1974) is recommended.

REVIEW QUESTIONS

1. Why is a low-voltage thermostat desired over a line-voltage thermostat?
2. Which is more economical to install, the low-voltage or the line-voltage thermostat?
3. How does a heat anticipator operate?
4. To what is the heat anticipator adjusted?
5. What is the purpose of the transformer?
6. How are transformers rated?
7. What is the insert made of on the fan control?
8. Are the fan control contacts normally open or normally closed?

9. In what circuit is the fan control?
10. In what circuit is the limit control?
11. At what temperature does the limit control function?
12. Is the limit control contact normally open or normally closed?
13. What causes the main gas valve to open or close?
14. In what circuit is the main gas valve?
15. Name the six components incorporated in the combination gas control.
16. Name the two types of pilot safety devices.
17. What provides the electricity for the pilot safety device?
18. Of what is the thermocouple made?
19. How many circuits are in a modern furnace?
20. How many voltages are in a modern furnace control circuit?
21. What is the purpose of a time delay relay in an electric furnace?
22. Where is the over-current protector installed?
23. What type of special limit controls are sometimes used on electric furnaces?
24. What are over-temperature protectors sometimes called?
25. What type of thermostat is normally used with electric furnace sequencers?

14

Solar Energy

The objectives of this chapter are:

- To acquaint you with the basic theory of solar energy.
- To acquaint you with the advantages of using solar energy for heating purposes.
- To acquaint you with the disadvantages of using solar energy for heating purposes.
- To acquaint you with basic solar system components.
- To acquaint you with the operating cycle of a solar system.
- To acquaint you with the different types of solar systems.
- To acquaint you with operating check-out procedure.

Solar energy is the most popular of the many alternate energy sources being discussed today. Volumes have been written and much said about solar energy. However, many people do not understand its possibilities and limitations.

Each year approximately 20% of the nation's energy use is for heating and cooling homes. Solar energy is an alternative that can reduce our dependence on scarce fossil fuels with their ever-increasing price.

The energy that the earth receives from the sun is called solar energy. The sun has provided, either directly or indirectly, almost all other sources of energy for the earth since its beginning. Each day the sun deposits an average of 1,400 Btu per square foot of area to the United States; therefore, an area the size of a 1,000 square foot home receives approximately 511 million Btu per year. However, solar energy can vary from season to season, from 2,000 Btu to as low as 500 Btu per square foot per day during a period from June to December. Factors such as cloud cover and geographical location affect the total amount of solar energy received. During the night solar energy drops to 0 Btu per square foot.

Sunshine is made up of a wide variety of electromagnetic waves similar in many ways to radio and television waves. It has three main components: (1) invisible heat waves, (2) visible light rays of various colors, and (3) invisible ultraviolet rays. Most of the ultraviolet portion of the sun is absorbed by the atmosphere of the earth.

Solar energy travels through space from the sun to the earth at a speed of 186,000 miles per second. For all practical purposes, this source of energy is inexhaustible.

For millions of years, sunlight has been captured by photosynthesis in plants. Through the slow heating action, pressure, and aging, plants have been turned into coal, petroleum, and natural gas. These fossil fuels presently provide more than 95% of the energy used by the developed nations.

The human race, and all plant and animal life on earth, has always been dependent on energy from the sun for their existence. A more important fact is that the rays from the sun provide the necessary heat to maintain the required temperature for the survival of humans, plants, and animals.

Using solar energy, however, has two disadvantages. First, this energy from the sun is diffuse; *i.e.*, it is spread out very thinly. It must therefore be collected by some means because only a small amount of it arrives in one place. Second, the energy received is intermittent because the sun shines only during the day and it is often obscured by clouds. Thus, the energy received must be stored until it is needed.

Solar measurements in various locations over the United States have shown that over an entire year (including night and day, summer and winter, cloudy and clear weather) an average of approximately 13% of the original energy from the sun reaches ground level. The actual amount that reaches any particular place at any given time will range from a much higher to a much lower value than the average (Figure 14–1). This 13% average is equivalent to approximately 16.4 watts or 58.5 Btu per square foot per hour.

If some arrangement can be made to collect this diffuse energy over a relatively large area, a tremendous amount can be made available. For example, the average rate per year at which solar energy falls on just one acre is equal to approximately 710 kilowatts or 950 horsepower.

To deal with the intermittent characteristic of solar energy at ground level, expensive techniques are required for storing it in large

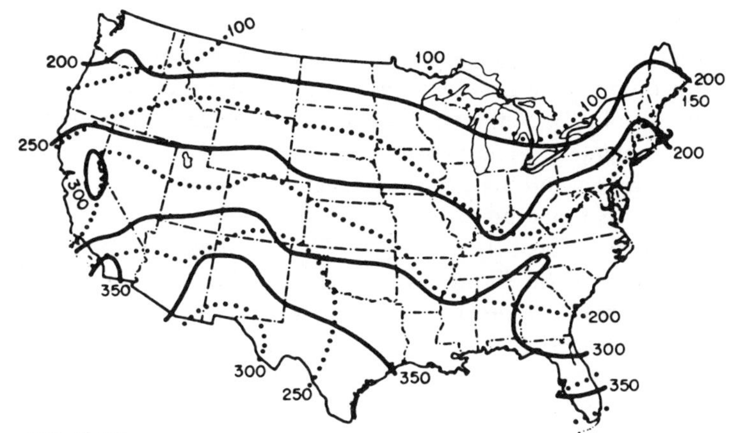

JANUARY

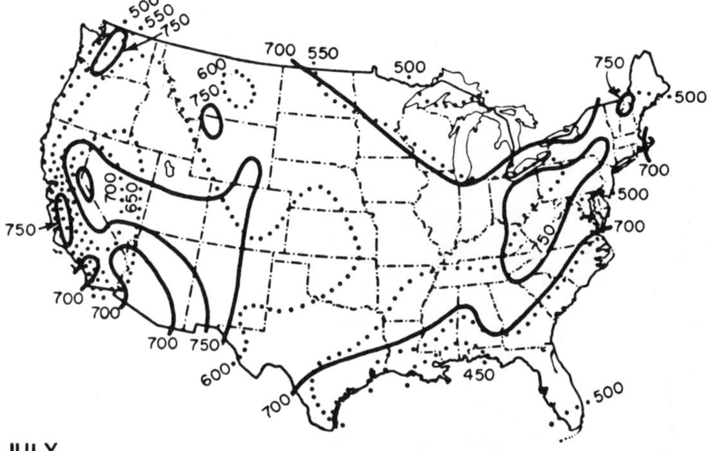

JULY

Lines of equal total daily solar energy at the ground on cloudless days (solid lines) and on days of average cloudiness (dotted lines) in January and July. Units are calories per square centimeter per day. The annual average for the United States is 377, corresponding to 13% of the original energy of the sun. Expressed in American units, this is 58 Btu per square foot per hour.

Figure 14–1. Average solar energy at ground level in calories per square centimeter per day.

quantities. The cost of storage is a large portion of the operating costs of a solar energy system; consequently tremendous effort is being directed toward energy storage research and development.

THE GREENHOUSE PRINCIPLE

Separate from the more obvious uses of energy from the sun, such as drying food and warming people, the most common direct use of solar energy is in a greenhouse. Designed to provide a controlled climate for growing plants, a greenhouse is made basically of glass.

Even though nothing could be much simpler than a greenhouse, it is a very effective method of converting and trapping radiation from the sun in the form of heat (Figure 14–2). This is possible because ordinary glass can transmit the shorter wavelength, visible parts of the sunlight while at the same time preventing the passage of the longer wavelength, invisible heat waves.

The greenhouse principle is based on this specific property of common glass. A large part of the visible energy of the sunlight enters the greenhouse through the glass. When these short wavelength heat waves are absorbed by the plants, ground, and fixtures inside the greenhouse, they are changed to the longer wavelength heat waves which are retained by the glass and cause a rise in temperature inside the greenhouse. This principle applies on cloudy days and sunny days alike and accounts for the effectiveness of this type of structure.

Figure 14–2. Greenhouse heated by solar energy.

BASIC SOLAR SYSTEM COMPONENTS

Although solar radiation is available throughout the universe, it is a diffuse form of energy which requires large collection areas. To get the most from solar energy, a system must be assembled to collect the low concentration of solar energy efficiently and store the heat energy to use at night and on cloudy days.

Instrumentation

Designing a solar installation requires precise information on the amount of solar energy available at the proposed location. This data can be obtained from a pyronometer, a device which measures insolation (energy from the sun) in Btu per square foot per hour (Figure 14-3).

A heat flow sensor can be used to measure the instantaneous rate of heat flow, in Btu per square foot per hour, into the conducting medium of the collector (Figure 14-4). Heat flow readings can be compared to insolation data to show how efficiently the collector is operating; that is, how much of the available energy is being carried away by the collector system.

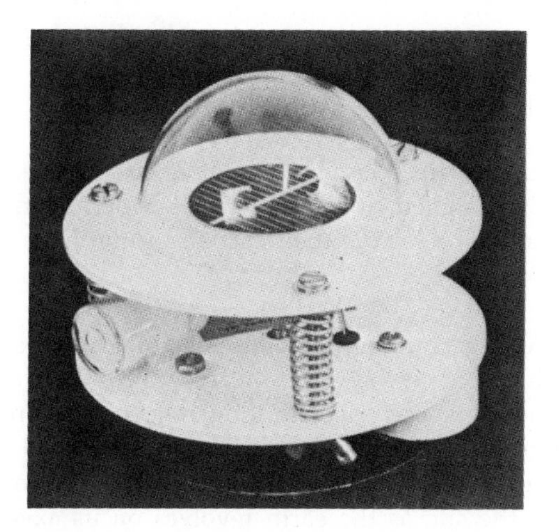

Figure 14-3. Pyronometer. (*Courtesy Rho Sigma Inc.*)

Collectors

To collect energy from the sun for household and hot water heating, the greenhouse principle has been applied to the simple solar heat collector (Figure 14-5). A large part of the energy from the sun passes through the glass cover plate; however, after it is absorbed by the black background material, it cannot escape and therefore heats the water circulating through the tubes. The heated water is then piped to either storage tanks or radiators.

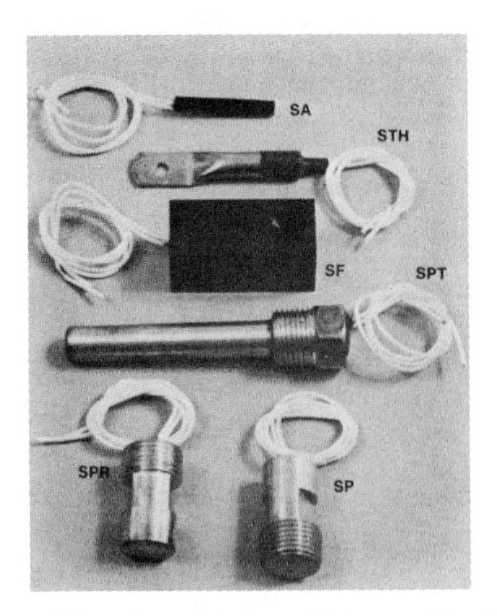

Figure 14–4. Heat flow sensors. (*Courtesy Rho Sigma Inc.*)

Collector Construction

The collector is the key component of a solar heating system. A solar collector is designed to trap the heat inside and transfer it to the heat transfer fluid. The heat is then carried by the heat transfer fluid to the storage tank.

Design: A large number of different collector designs have been developed, each one having specific advantages for different applications. Several manufacturers use concentrating reflectors to intensify the energy input (Figure 14–6). These collectors are dependent upon direct sunlight and must be installed with motor drives and rotational mechanisms to follow the sun as the earth revolves on its axis.

There are flat-plate collectors which will generate usable heat even from diffused light, such as during cloudy sky conditions. This type of collector does not need to be aimed at the sun constantly (Figure 14–7). Normally, the flat plate collector is oriented to receive the maximum insolation at midday during the season for which the system is designed. This noncritical orientation feature makes the flat plate collector the most desirable for residential and commercial heating applications.

Cover Glass: The properties of glass are ideal as a cover glass for providing the greenhouse effect, because glass readily passes the shorter wavelengths of light which are received from the sun (.3 to 2 microns in length) but is almost opaque to the reradiated heat wavelengths (2 to 10 microns in length) (Figure 14–8).

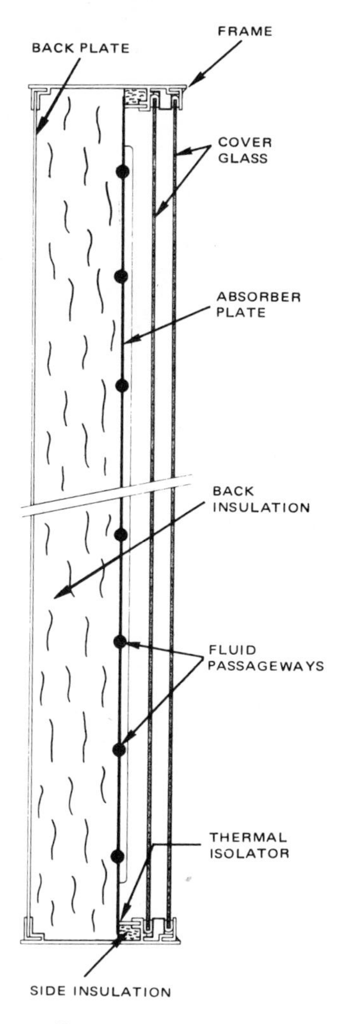

Figure 14–5. Solar collector cutaway. (*Courtesy LOF Solar Energy Systems*)

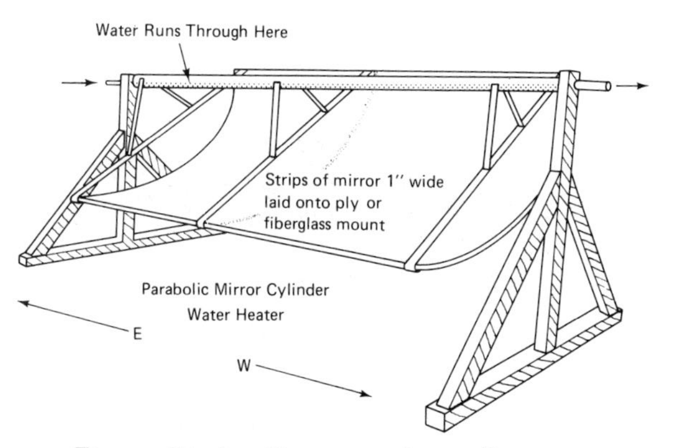

Figure 14–6. Concentrating collector.

Figure 14-7. Flat plate collector. (*Courtesy Solar Research Div. Refrigeration Research, Inc.*)

Absorber Plate: The absorber plate is the heart of the collector panel. Its black surface literally soaks up the sun rays and converts them to heat. Most collectors contain an all-copper absorber plate because of the effective heat transfer characteristics of copper. Copper will transfer two times the amount of heat that aluminum will transfer and eight times the amount that steel will transfer. Copper is also superior for long-term reliability and good performance.

Heat moves through a conductor in an amount proportional to the temperature difference between one side and the other. As the temperature of the absorber plate increases, solar heat is conducted from the absorber plate through a copper fluid-carrying piping arrangement. The heated fluid, moving from the collector to the heat storage area, lowers the temperature of the absorber plate and causes more heat to be conducted.

This continual process, the absorption of heat and its removal by the heat transfer fluid, is the basic function of the absorber plate. Some collectors use an embossed absorber plate to provide a large soldered

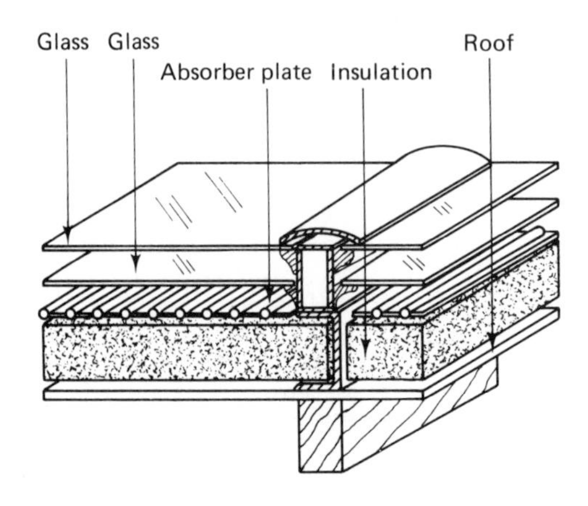

Figure 14-8. Location of collector cover glass.

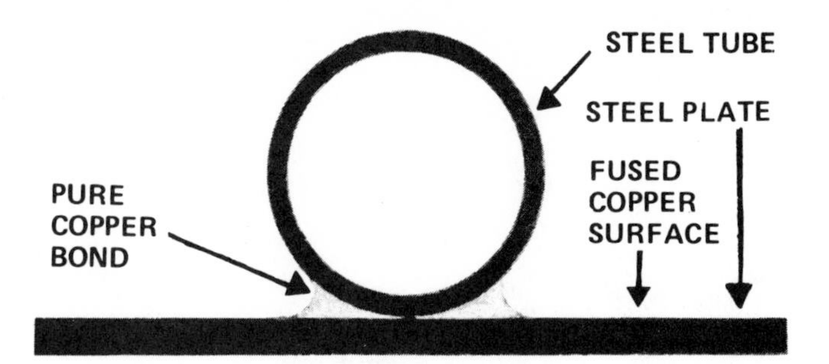

Figure 14–9. Method of bonding absorber plate to tube. (*Courtesy Solar Research Division Refrigeration Research Inc.*)

area with the copper fluid passageways. The bond between the absorber plate and the tubing ensures maximum heat transfer (Figure 14–9).

Insulation: A heat resistant type of fiberglass insulation is used behind the absorber plate to prevent the heat from penetrating to the back of the collector. A special side insulation material reduces the flow of heat from the edges of the absorber plate.

A solar heating system cannot produce more heat than the collector system can obtain from the sun. Therefore, collector performance and reliability are crucial to the performance and reliability of the entire system.

Heat Storage

Solar heating systems require adequate storage for the heat that is collected during sunny days for use at night and in overcast weather. Insulated storage tanks can be placed above or below the ground and may be used to store heat in water, rocks, salt, or special solutions designed for this purpose. The size of the reserve storage tank will depend on the demands of local weather conditions.

Storage Tank Construction: The heat transfer fluid is pumped from the collector to the tank where the heat is stored. Because there are several designs of storage tanks, the best one will depend on the local availability and cost of materials. Some storage tanks use the same fluid for storage and heat transfer in the collector (Figure 14–10). When this design is used, additional heat exchangers must be used to heat the domestic water and the water used for heating purposes.

A variation of this method is to circulate the heat transfer fluid through the heating coils and provide an additional heat exchanger for the domestic hot water (Figure 14–11). With this design, however, the

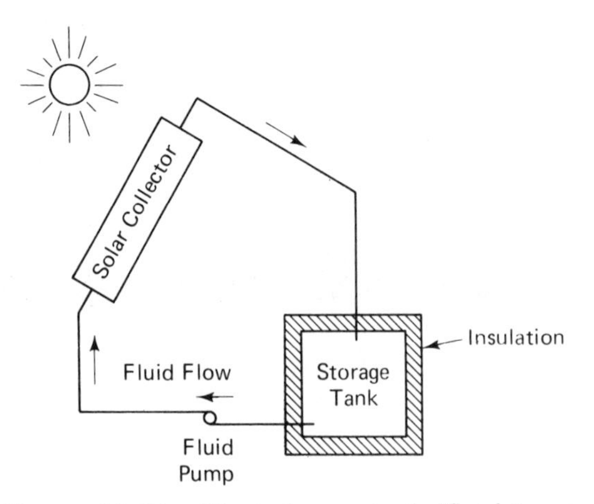

Figure 14–10. Heat storage tank (fluid type).

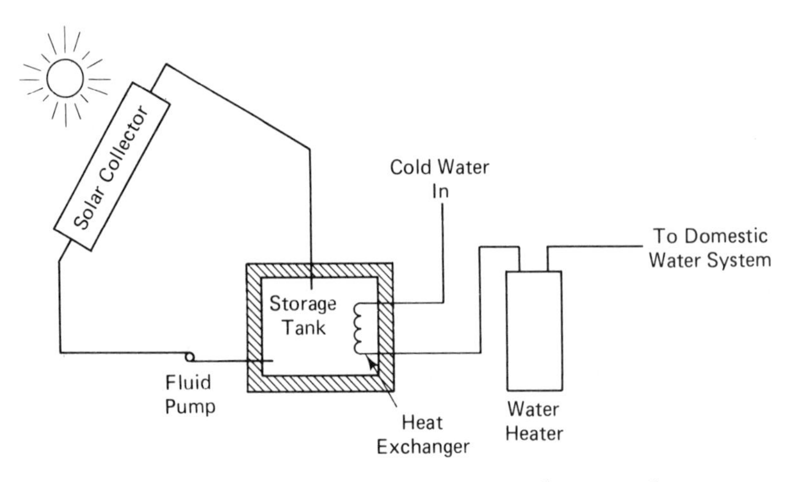

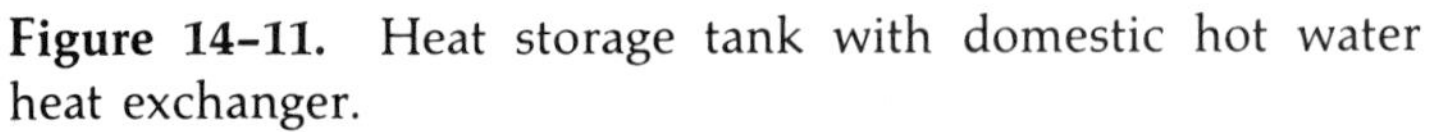

Figure 14–11. Heat storage tank with domestic hot water heat exchanger.

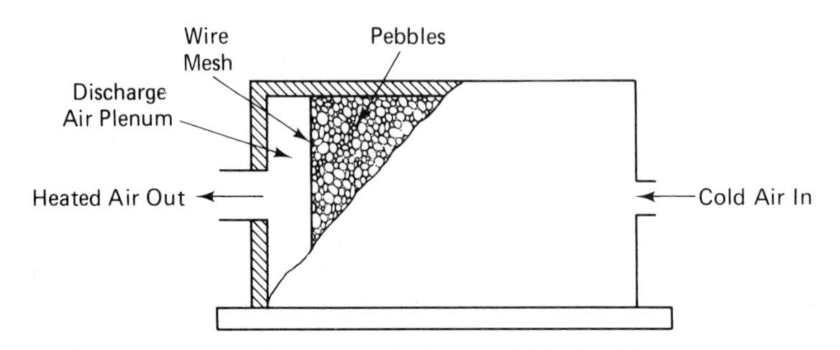

Figure 14–12. Horizontal flow pebble-bed heat storage.

initial cost of the fluid is more expensive than with the other designs. Also, if a leak should occur, replacing the fluid is more costly than when water is used.

Another design uses a heat storage medium, a heat transfer fluid, and additional heat exchangers for the domestic hot water and heating water (Figure 14–12). This design, which can be used with water, salt, or a stone bed as the storage medium, is probably the most common one for heat storage tanks.

Heat Transfer Fluid: Heat transfer fluid is the medium used to transfer the heat from the collector to the heat storage tank or the point of use. This medium should be noncorrosive, very stable, safe, and low in cost. Some manufacturers have developed a fluid especially for use in solar systems. One such fluid, Sun-Temp solar collector fluid, was developed by Solar Research Division of Refrigeration Research, Inc. It has a low specific heat which allows heat to be picked up on many cloudy days. Under certain conditions the use of this fluid makes possible earlier pickup of usable heat in the morning, throughout the day, and longer in the afternoon. Sun-Temp has a freezing point of −40° F, a boiling point above 500° F, and a specific heat of .56.

Water is sometimes used as the heat transfer fluid. It should be mixed with antifreeze or an automatic drain-down system to protect the system from freezing during cold weather and when the solar collector system is shut down, such as when the storage medium has reached the desired temperature and fluid flow through the collectors is stopped.

Storage Capacity and Medium: A storage capacity of one to two gallons of water or one-half to one cubic foot of pebbles per square foot of collector area for domestic hot water or space heating systems is generally recommended. Therefore, the size of storage will depend upon the size of the required collectors.

Water for heat storage is normally used when a liquid is the heat transfer fluid. A heat exchanger transfers the heat from the transfer fluid to the water in the storage tank. Certain precautions must be considered when water is used.

1. *Freezing:* To prevent freezing of the water in the storage tank, the tank should be located within the building or be insulated well enough to prevent freezing. Some type of arrangement should allow drainage of the storage tank if the system is shut down during subfreezing weather.

2. *Corrosion:* The materials used in the construction and installation of solar systems should not be affected by atmospheric conditions or interact with the other components of the system.

The use of dissimilar metals should be avoided or at least held to a minimum. When dissimilar metals are together in a system, dielectric couplings should be used. Dielectric couplings are generally made with an insulating gasket, organic coatings, or both.

When atmospheric conditions are of a corrosive nature, some type of protective coating is needed. The type of environment and material will determine what type of coating is best suited for the application. A reliable coating manufacturer should be consulted.

The use of sacrificial galvanic anodes in the storage tank is sometimes helpful in reducing corrosion in water storage tanks. Water heating specialists can help to determine exactly what to use.

A protective coating should be applied to all steel liquid storage tanks and lines placed underground. This coating must meet the specific conditions in which the tank will exist.

3. *Boilout:* Boilout is the evaporation of water which occurs when the storage water is heated to the boiling point and solar heating continues. A properly sized pressure storage tank should be used or the tank should be vented to the atmosphere and fitted with the proper size relief valves. Some provision for checking the water level inside the tank is needed.

4. *Damage Due to Leakage:* Materials used in the construction of heat storage tanks must be suitable for that purpose. Storage tanks of high temperature type fiberglass are acceptable if they can withstand temperatures above 212° F. Provisions must be made for drainage.

Pebble beds for heat storage generally use a vertical flow storage unit. In rooms less than five feet high, however, a horizontal flow pebble bed may be used (Figure 14-12), in which case a suitable baffling arrangement of the inside of the storage tank is necessary to obtain proper air flow within the tank.

Pebble beds may be placed inside structurally sound wooden boxes, concrete block walls, or concrete bins. When wooden frames are used, they may be built at the construction site where access is limited or when the other construction is completed. Whatever material is used, care should be taken to insure that all joints of the container are air tight and provision made for air plenums at each end.

Crushed rock or pebbles of a fairly uniform size and washed free of sand are suitable as a heat storage medium. For uniform distribution, the pebbles should be placed in the storage bin, by hand or chute, not dumped in directly from a truck.

Pumps: Pumps circulate the fluid through the collectors, through the space heating coils, and sometimes through the domestic hot water system. These pumps are electrically operated and sized to deliver the desired flow, which varies depending upon the system design and usage. The pumps used to circulate the heat transfer fluid are generally low velocity, centrifugal pumps.

Check Valves: Check valves are installed between the heat transfer fluid pump and the collector array. They prevent a reversed flow of the fluid through the system, which would tend to upset the operation of the control system.

Relief Valves: Relief valves are placed at the discharge outlet of the collector array. They are used to relieve any excess pressure which could build up in the heat transfer fluid during abnormal conditions such as pump failure and the like.

SYSTEM CONTROLS

When heat is needed, a conventional room thermostat signals an automatic control system to deliver heat from the storage tank, or during prolonged periods of cloudy weather, from an auxiliary heater. When the storage tank temperature exceeds collector temperature, the collector pump shuts off and allows the heat transfer fluid to drain from the collector bank back into the tank to avoid freezing.

Space Thermostat

The space thermostats used on solar heating systems are of the same types and construction used on any central heating system, and they function in the same manner. Some systems, however, use a two-stage thermostat which also operates the auxiliary heat when the solar system does not supply sufficient heat to maintain the desired temperature inside the structure.

Temperature Sensors

This control measures the transfer fluid temperature at the discharge side of the collector, the temperature of the domestic hot water storage, and the temperature of the storage tank. A domestic water auxiliary sensor activates the domestic hot water auxiliary heater (Figure 14–13).

The storage tank sensor should be installed in a location so that the heat transfer fluid temperature sensor will activate a device, such as a pump, which will add the maximum amount of solar energy to the storage tank. The sensor used for a pebble bed storage should be installed in the pebble area about 6 inches from the cold end of the storage area.

The sensor which measures the temperature of the heat transfer fluid at the collector outlet should be installed so it can sense the average fluid temperature and sense it whether or not the fluid is being circulated by the pump (Figure 14–14).

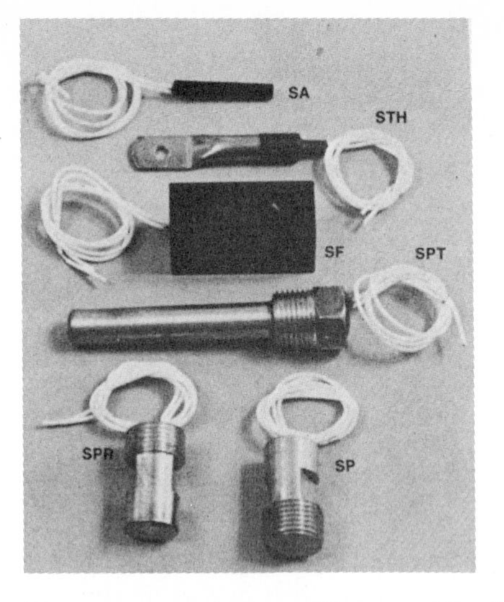

Figure 14–13. Temperature sensors. (*Courtesy Rho Sigma Inc.*)

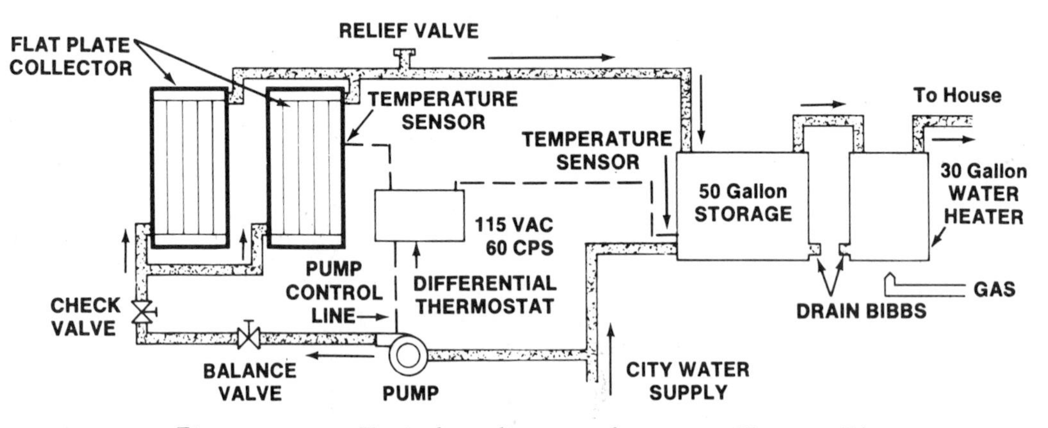

Figure 14–14. Typical application of sensors. (*Courtesy Rho Sigma Inc.*)

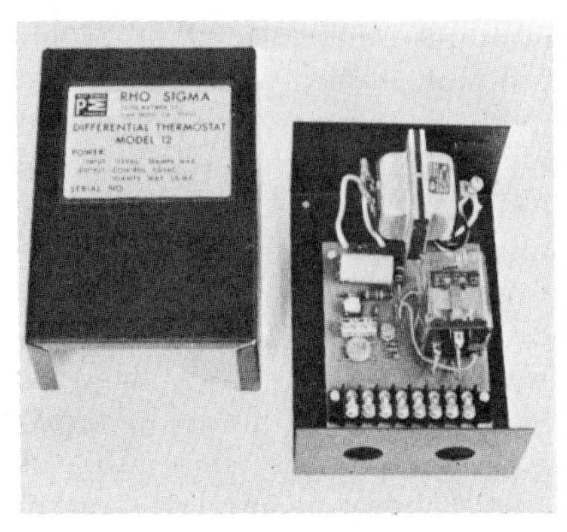

Figure 14–15. Differential thermostat. (*Courtesy Rho Sigma Inc.*)

Control Panel

A central control panel is generally used to house the circuit connections and the relays which provide the control functions. These control panels should be constructed and installed according to state and local electrical codes.

Differential Thermostat

This unit provides the control functions for solar heating and domestic hot water systems by measuring the leaving water temperature at the collector outlet and comparing this with the temperature of the water in the storage tank (Figure 14–15). When the collector outlet fluid temperature is greater by a given difference than the water in the storage tank temperature, the circulation pump is turned on. When the collector fluid temperature approaches the storage tank temperature, the pump is turned off. This operation assures that the maximum heat energy is stored and retained in the tank during all weather and temperature conditions.

TYPICAL SOLAR HEATING SYSTEM OPERATING CYCLE

The following is a sequence of events which occur in a normal daily solar heating/domestic hot water heating cycle (Figure 14–16). Initially, the pump is off. As the sun rises, the collector temperature rises sharply higher than the storage tank temperature. This is shown as ΔT_{on} (ΔT = Temperature difference). This ΔT turns the pump on. The cooler tank water carries off the heat in the collector and causes the temperature to drop to the point marked 2. If the ΔT at point 2 is greater than the ΔT_{off} then the pump will continue to operate.

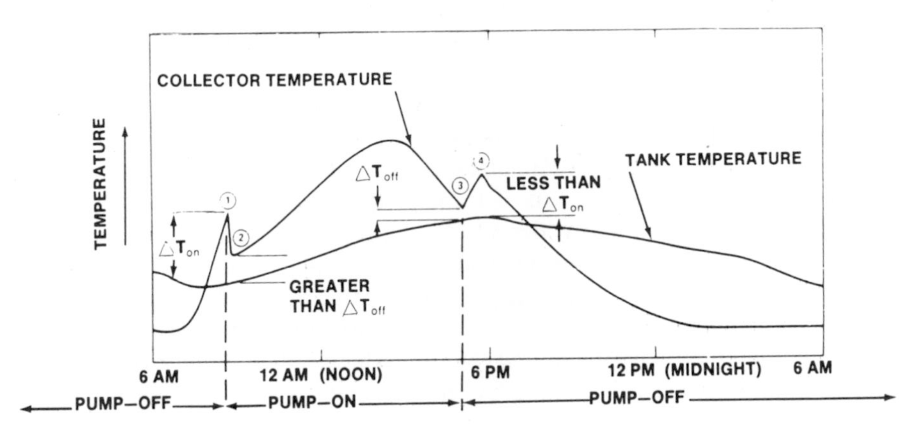

Figure 14–16. Typical solar system operating cycle. (*Courtesy Rho Sigma Inc.*)

As the evening cooler ambient temperatures approach, the collector temperature drops below ΔT_{off} threshold at the point marked 3. The pump is turned off and, due to the stagnation condition (i.e., the sun is heating the collector, but the pump is not operating), the collector temperature rises as shown at point 4. If ΔT_{on} is set sufficiently greater than ΔT_{off}, the pump will remain off. The adjustments of the threshold and hysteresis settings in the differential thermostat are preset at the factory.

The hysteresis circuit is a special feature of the differential thermostat which prevents the pump from being unnecessarily turned on and off immediately following an initial pump turn-on or turn-off event. Without this circuit feature, the sharp peaks of temperature rise shown at points 1 and 4 would cause the pump to repeatedly switch on and off, causing unnecessary wear on the pump, pump motor, and relay contacts which control pump operation, as well as reduced efficiency in transferring heat energy from the collector to the storage tank.

TYPES OF SOLAR SYSTEMS

There are basically two types of solar systems: the active and the passive. Active systems are divided into liquid and air systems. They use pumps and pipes, or fans and ducts, to transfer the heat from the collectors to the storage and from storage to the living space of the building.

Some passive systems use a wall (or walls) of the building or a separate stationary wall as both the collector and storage medium. In the passive building, movable wall panels, or flaps, are often used to direct the heat throughout the living space (Figure 14–17). Another approach is to collect and store heat in water bags on the roof (Figure 14–18).

Active Systems

Active solar systems can be divided into two types: the active liquid collector and the active air collector. The type chosen will depend on the particular design of the system.

The active liquid collector, liquid storage air distribution system is probably the most popular. There are many liquid collector-liquid storage systems possible. A flat-plate collector is located on a south-facing roof or a vertical wall. A heat transfer fluid (which may be water with antifreeze and anticorrosion additives) is pumped through a heat exchanger coil in the storage tank and back to the collector (Figure 14–19). Liquid from the storage tank is pumped through a second heat exchanger coil where it heats air which is then blown into the living space. The storage is provided by a large tank which holds 500 to 1,000 gallons. This tank is usually located in the basement or buried underground.

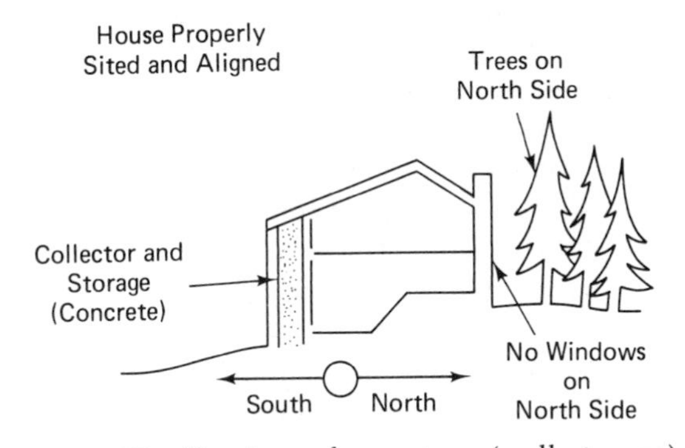

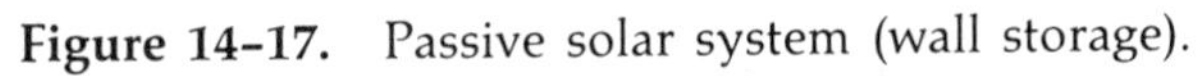

Figure 14–17. Passive solar system (wall storage).

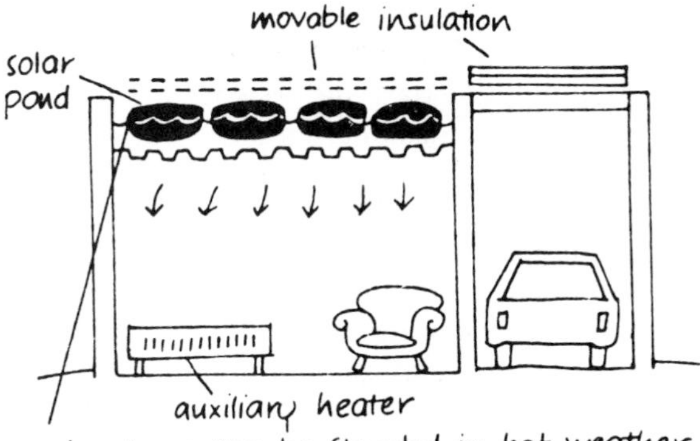

Figure 14–18. Passive solar system (movable flaps).

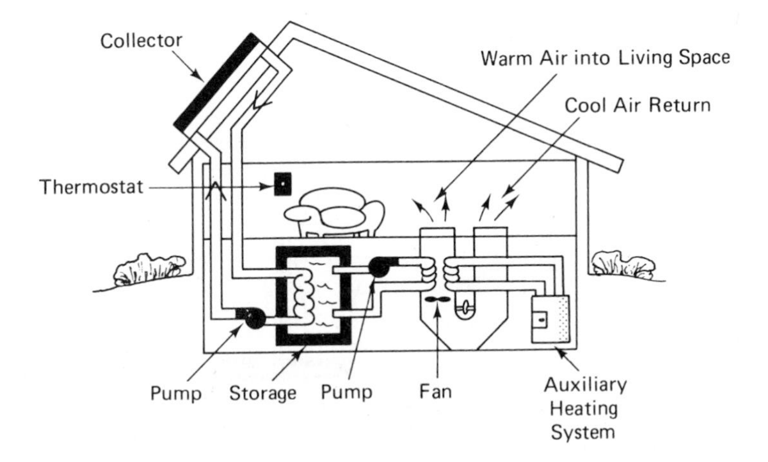

Figure 14–19. Active solar system (liquid collector-liquid storage).

Auxiliary heat from a conventional heating system is needed as a backup system for the solar heating unit. A thermostat controls the temperature in the living area for both the solar and auxiliary systems.

The active air collector rock storage air distribution system is less popular than the liquid type system discussed previously. These systems use a flat-plate collector located on the roof and a bin containing 1 to 3 inch rocks for storage. Hot air flows through the spaces between the rocks and heats them. The bin of rocks must be about 2½ times larger than a water tank of equal heat storage capacity. Air from the cold end of the rock storage bin is blown up to the lower end of the collector, where it is heated, drawn from the hot upper end and returned to the bin (Figure 14-20). When the living space requires heat, dampers in the air ducts are adjusted and hot air is blown from the storage bin into the living space. Heating coils from the auxiliary heating unit are also provided. Depending on the amount of heat in storage, the auxiliary system is used to supply some or all of the heat needed. The same thermostat controls both the solar and the auxiliary system.

An active system can be used in either new or existing buildings and may be useful for an older building if future plans include remodeling or upgrading the present heating system. The addition of insulation, weather-stripping and other energy conserving features should be included. In any case, the building must have a good southern exposure to take full advantage of the light from the sun.

Passive System

A passive solar heating system uses the building structure itself as both a solar collector and a storage medium. Many different designs are possible (Figure 14-21). A large mass, usually water or concrete, collects

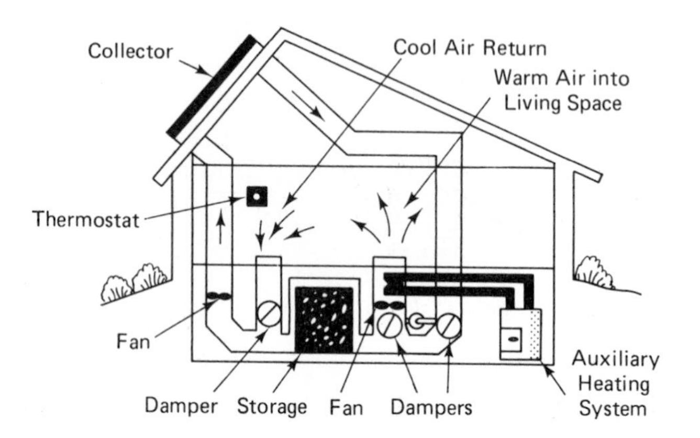

Figure 14-20. Active solar system (air collector-rock storage).

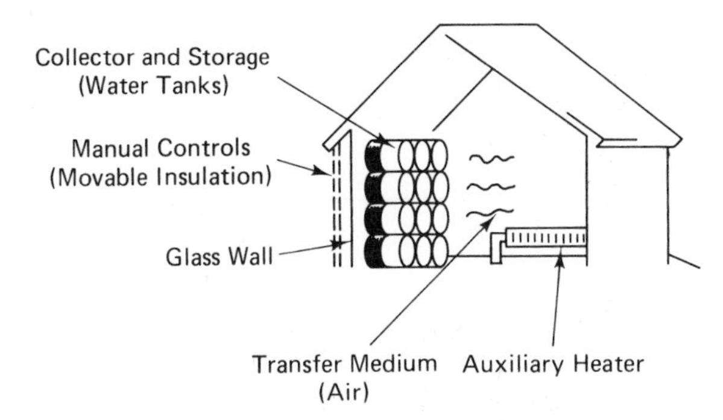

Figure 14-21. Passive solar system (water tanks).

and stores the heat from the sun. Auxiliary heating may sometimes be necessary. The heat reaches the living space by natural air circulation. Movable wall panels, flaps, or shutters are used to control the temperature in the living space.

A passive system should be considered when designing a new building where the main use of solar energy is for space heating. The design of a passive solar building requires careful consideration of siting, north-south orientation, protective landscaping and high quality construction. Consequently, a passive system is seldom feasible for an existing building. However, using movable curtains and awnings on large picture windows can be very effective in collecting and retaining heat.

Domestic Hot Water System

Almost all domestic hot water systems use liquid collectors, although it is possible to use an air system. The physical principles are the same as those for space heating; however, a smaller collector array and smaller storage are needed. Although many domestic hot water system designs are possible, the one shown in Figure 14-22 is typical. The heat transfer fluid in the collector is heated by the sun and gives up its heat in the storage tank. The hot water in the storage tank heats water in a second coil, which heats the water used for washing and cooking. This double set of coils, or a special heat exchanger, is required so that the anticorrosive additives and antifreeze solution in the solar collectors cannot contaminate the potable water system.

SOLAR COLLECTOR PERFORMANCE

When designing a solar heating system, relationships between a number of variables must be considered:

1. *The amount of heat required:* This factor can be determined by

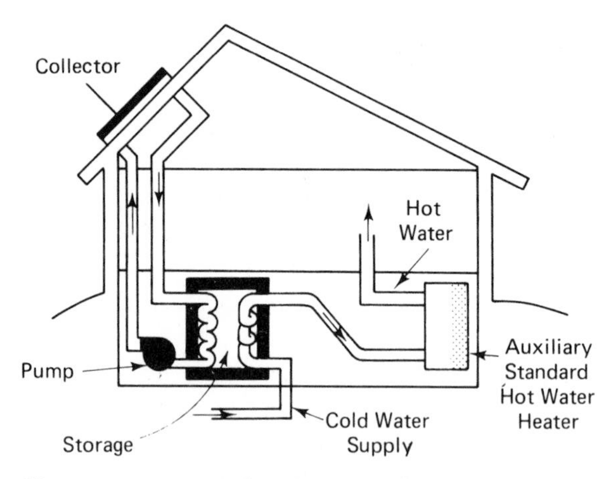

Figure 14-22. Solar domestic hot water system.

standard methods. Local utility companies will usually furnish monthly figures of the Btu's required to heat typical buildings of different sizes in local areas. The size of existing heating plants will give an indication of the expected needs. Generally the solar heating system is not intended to provide all the heating required by the building. The difference in size between a system that will be adequate 80% of the time and one that will handle the entire load even during extreme temperatures may be significant. Oversizing a solar heating system will probably be more expensive than oversizing a fuel-fired system, because nearly all of its cost is in the initial installation.

2. *The daily amount of solar energy expected:* The amount of solar energy that is expected at a specific location also determines the needed capacity of a solar system. The U.S. Weather Bureau publishes maps and charts based on past experience for a given geographical location which cite the number of sunny days each month, the prevailing temperatures, the speed and direction of the wind, and the average daily insolation levels. (Insolation refers to solar energy actually reaching the earth.)

3. *The orientation of the collector:* The orientation of the collector array to the sun has a large effect upon how much energy it collects. The greatest absorption occurs when the panel is directly facing the sun. Since insolation levels are highest at midday, flat-plate collector panels are usually installed facing south, and tilted to face the sun at an angle where it appears at that specific time. The angle, of course, changes with the season, so the installed angle of the collectors becomes a compromise, depending upon the kind of work that the system is designed to do. The following rule-of-thumb (Figure 14-23) will generally provide good results (the angles are measured from the horizontal):

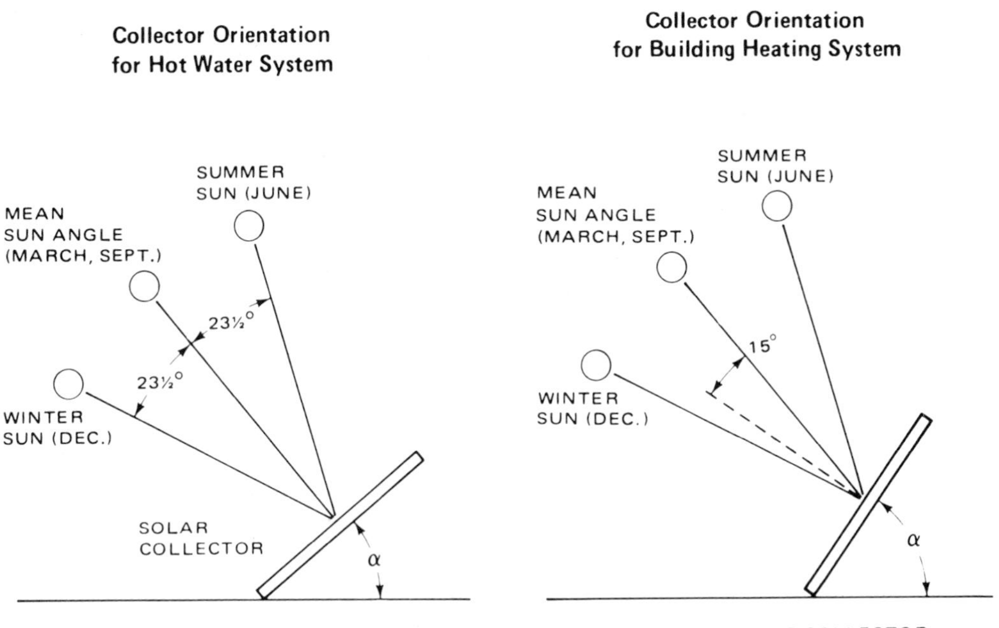

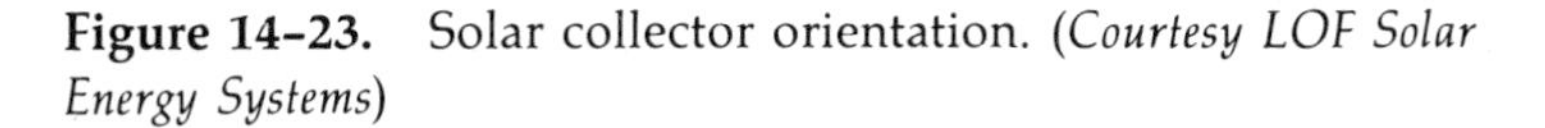

Figure 14–23. Solar collector orientation. (*Courtesy LOF Solar Energy Systems*)

Domestic Hot Water only—collector angle the same as the latitude.

Building Heating only—collector angle is the latitude plus 15°.

For systems designed for multiple functions, the collector angle should be calculated by the design engineer, considering insulation values and system loading throughout the entire year.

4. *The flow rate of the heat transfer fluid:* The flow rate of the heat transfer fluid through the collector affects its absorption rate. The recommended flow rate for most collector panels is 0.5 gallons per minute; performance data is based on this flow rate. Slower flow rates result in higher fluid temperatures, but the absorber plate temperature is also higher and its ability to absorb more energy is thereby reduced. Higher flow rates reduce the temperature of the fluid being returned for use.

5. *The temperature of the incoming fluid:* The temperature of the fluid coming in to the collector will also affect the amount of energy absorbed because heat flow through the absorber plate is a function of the temperature differential across the plate. The higher the incoming fluid temperature, the slower the absorption rate.

6. *Heat loss from the collector:* Heat loss from the collector is inevitable, especially in cold weather. Regardless of the make of collector, the efficiency of a flat-plate collector decreases as the difference in temperature between the absorber plate and the ambient air increases. Obviously, with large variations in seasonal outside temperatures, variations in wind velocity and direction, variations in fluid temperatures and, therefore, variable collector efficiency, calculations for designing an appropriately sized collector array can become complex. Computer assisted analysis may be helpful if energy requirements are critical.

Approximating the Output from a Working Collector

To approximate the output from a collector, the following method may be used:

Btu per hour output = GPM flow × 500 (output fluid temperature − input fluid temperature)

For example: If the output fluid temperature = 130° F and the input fluid temperature = 110° F, there is a difference of 20° F. The flow through a single panel = 0.5 GPM

then: 0.5 × 500 × 20 = 5,000 Btu per hour

If antifreeze is being used in the system, a correction factor is necessary (Table 14–1). The specific heat of ethylene glycol solutions is less than that of water. The heat content of the solution can be calculated for a working system by using the above formula with the proper correction factor:

Table 14–1
Correction Factors for Antifreeze Solutions

% by Weight Ethylene Glycol	Fluid Temperature ° F	Correction Factor (F)
0	100	.995
0	200	.975
25	100	.985
25	200	.970
50	100	.858
50	200	.891

(Courtesy LOF Solar Energy Systems)

Btu per hour = GPM × 500 × (fluid temperature out) − (fluid temperature in) × Factor (*F*).

DETERMINING COLLECTOR AREA

Calculations for determining the number of panels needed in a specific application must include the capacity required for a peak demand. A heating system capacity is calculated as follows:

1. From published data, find the average number of sunny days in the coldest month of operation for the location. On each sunny day, the system must produce enough heat to last through to the next sunny day. For example, if the location has 20 sunny days during the month of January, the system must produce on each sunny day enough heat for 1½ days of consumption, to provide most of the heating requirements.

2. Determine the number of Btu's per day required from the heating system during that month. Multiply the Btu requirement by the number of days for which each sunny day must supply heat. For example, if during January the average heat required is 500,000 Btu per day, the system must supply 750,000 Btu per sunny day.

3. Determine the average daytime temperature and the system operation temperature needed. From a collector performance chart for the particular collector under consideration, find the efficiency of the collector (based on the average hourly solar flux for a typical sunny day, in the plane of the collector).

4. From published data, determine the average daily insolation for the month at that location. Multiply this by the efficiency of the collector to find the collector output per square foot per day. For example, this might come to 1,100 Btu/day.

5. Divide the total requirement by the panel output to determine the number of square feet of collector needed. For example, if the total heat requirement is 750,000 Btu and the collector output is 116 Btu/hr/sq.ft. for 9 hours per day, 719 square feet of collector would be needed.

REVIEW QUESTIONS

1. What is the number of Btu received by the United States daily from the sun?
2. What factors determine the amount of solar energy received from the sun?
3. What are two disadvantages of using solar energy?
4. What is the basis of the greenhouse principle?

5. What is a pyronometer?
6. What two types of collectors are used?
7. What is the purpose of the collector?
8. What is the heart of the collector?
9. Why is proper bonding of the tube to the absorber plate important?
10. What determines the size of the storage tank?
11. What moves the heat from the collector to the storage tank?
12. What precautions must be taken when water is used for heat storage?
13. What is the purpose of temperature sensors?
14. What is the purpose of the differential thermostat?
15. What are the two general types of solar systems?

15

Glossary

The definitions given here are those applying only to heating. They are defined particularly as used in this book.

ABSOLUTE HUMIDITY The weight of water vapor in grains actually contained in 1 ft^3 of the mixture of air and moisture.

ABSORPTIVITY The ratio of radiant energy by an actual surface at a given temperature to that absorbed by a black body at the same temperature.

ACTIVE SOLAR SYSTEM Solar systems that use the forced movement of a fluid to transfer heat from the collector to the heated space.

AIR An elastic gas. It is a mechanical mixture of oxygen and nitrogen and slight traces of other gases. It may also contain moisture known as humidity. Dry air weighs 0.075 lb/ft^3. One Btu will raise the temperature of 55 ft^3 of air 1° F.

AIR CHANGE The number of times in an hour the air in a room is changed either by mechanical means or by the infiltration of outside air leaking into the room through cracks around doors and windows, etc.

AIR CLEANER A device designed for the purpose of removing air-borne impurities, such as dust, fumes, and smoke. (Air cleaners include air washers and air filters.)

AIR CONDITIONING The simultaneous control of the temperature, humidity, air motion, and air distribution within an enclosure. Where human comfort and health are involved, a reasonable air purity with regard to dust, bacteria, and odors is also included. The primary requirement of a good air conditioning system is a good heating system.

AIR INFILTRATION The leakage of air into a house through cracks and crevices and through doors, windows, and other openings, caused by wind pressure and/or temperature difference.

AIR VALVE *See* Vent Valve.

ATMOSPHERIC PRESSURE The weight of a column of air 1 in.^{2} in cross section and extending from the earth to the upper level of the blanket of air surrounding the earth. This air exerts a pressure of 14.7 psi at sea level, where water will boil at 212° F. High altitudes have a lower atmospheric pressure with correspondingly lower boiling point temperatures.

AUTOMATIC FLUE DAMPER A device used between the vent system and the draft diverter on a heating furnace to prevent heat escaping through the vent system.

BOILOUT The evaporation of water which occurs when the storage water is heated to the boiling point and solar heating continues.

BOILER A closed vessel in which steam is generated or in which water is heated by fire.

BOILER HEATING SURFACE The area of the heat transmitting surfaces in contact with the water (or steam) in the boiler on one side and the fire or hot gases on the other.

BRITISH THERMAL UNIT (Btu) The quantity of heat required to raise the temperature of 1 pound of water 1° F. This definition is somewhat approximate but sufficiently accurate for any work discussed in this book.

BUCKET TRAP (Inverted) A float trap with an open float. The float or bucket is open at the bottom. When the air or steam in the bucket has been replaced by condensation, the bucket loses its buoyancy and when it sinks it opens a valve to permit the condensate to be pushed into the return.

BUCKET TRAP (Open) The bucket (float) is open at the top. Water surrounding the bucket keeps it floating and the pin is pressed against the seat. Condensate from the system drains into the bucket. When enough has drained into it so that the bucket loses its buoyancy, it sinks and pulls the pin off its seat and steam pressure forces the condensate out of the trap.

CENTRAL FAN SYSTEM A mechanical indirect system of heating, ventilating, or air conditioning consisting of a central plant where the air is heated and/or conditioned and then by fans or blowers circulated through a system of distributing ducts.

CHIMNEY EFFECT The tendency in a duct or other vertical air passage for air to rise when heated due to its decrease in density.

CIRCULATING PIPE (Hot water system) The pipe and orifice in a Hoffman Panelmatic Hot Water Control System through which the return water bypasses the boiler until the temperature of the circulating stream is too low, at which time part of it is replaced by the correct quantity of hot boiler water to restore its temperature.

COEFFICIENT OF HEAT TRANSMISSION (Overall) *U* The amount of heat (Btu) transmitted from air to air in 1 hr/ft^2 of the wall, floor, roof, or ceiling for a difference in temperature of 1° F between the air on the inside and outside of the wall, floor, roof, or ceiling.

COLLECTOR A device used to gather the rays from the sun and turn them into heat.

COMFORT LINE The effective temperature at which the largest percentage of adults feel comfortable.

COMFORT ZONE (Average) The range of effective temperatures at which the majority of adults feel comfortable.

CONDENSATE In steam heating, the water formed by cooling steam as in a radiator. The capacity of traps, pumps, etc., is sometimes expressed in pounds of condensate they will handle per hour. For instance, 1 pound of condensate per hr is equal to approximately 4 ft^2 of steam heating surface (240 Btu/hr/ft^2).

CONDUCTANCE (Thermal) *C* The amount of heat (Btu) transmitted from surface to surface in 1 hour through 1 ft^2 of a material or construction for the thickness or type under construction for a difference in temperature of 1° F between the two surfaces.

CONDUCTION (Thermal) The transmission of heat through and by means of matter.

CONDUCTIVITY (Thermal) *K* The amount of heat (Btu) transmitted in 1 hour through 1 ft^2 of a homogeneous material 1 inch thick for a difference in temperature of 1° F between the two surfaces of the material.

CONDUCTOR (Thermal) A material capable of readily transmitting heat by means of conduction.

CONVECTION The transmission of heat by the circulation (either natural or forced) of a liquid or a gas such as air. If natural, it is caused by the difference in weight of hotter and colder fluid.

CONVECTOR A concealed radiator. An enclosed heating unit located (with enclosure) either within, adjacent to, or exterior to the room or space to be heated, but transferring heat to the room or space mainly by the process of convection. A shielded heating unit is also termed a convector. If the heating unit is located exterior to the room or space to be heated, the heat is transferred through one or more ducts or pipes.

COOLING LEG A length of uninsulated pipe through which the condensate flows to a trap and which has sufficient surface to permit the condensate to dissipate enough heat to prevent flashing when the trap opens. In the case of a thermostatic trap, a cooling leg may be necessary to permit the condensate to drop a sufficient amount in temperature to permit the trap to open.

DEGREE DAY (Standard) A unit which is the difference between 65° F and the daily average temperature when the latter is below 65° F. The degree days in any one day is equal to the number of degrees F that the average temperature for that day is below 65° F.

DEW POINT TEMPERATURE The air temperature corresponding to saturation (100% relative humidity) for a given moisture content. It is the lowest temperature at which air can retain the water vapor it contains.

DIFFERENTIAL THERMOSTAT A device which provides the control functions for solar heating and domestic hot water systems by measuring the leaving water temperature at the collector outlet and comparing this with the temperature of the water in the storage tank.

DIRECT-RETURN SYSTEM (Hot Water) A two-pipe hot water system in which the water, after it has passed through a heating unit, is returned to the boiler along a direct path so that the total distance traveled by the water from each radiator is the shortest feasible. There is, therefore, a considerable difference in the lengths of the several circuits composing the system.

DOWN-FEED ONE-PIPE RISER (Steam) A pipe that carries steam downward to the heating units and into which the condensation from the heating unit drains.

DOWN-FEED SYSTEM (Steam) A steam heating system in which the supply mains are above the level of the heating units which they serve.

DRY-BULB TEMPERATURE The temperature of the air as determined by an ordinary thermometer.

DRY RETURN (Steam) A return pipe in a steam heating system that carries both air and water from condensation.

DRY SATURATED STEAM Saturated steam containing no water in suspension.

EQUIVALENT DIRECT RADIATION (EDR) See Square Foot of Heating Surface.

EXTENDED HEATING SURFACE Heating surface consisting of fins or ribs that receive heat by conduction from the prime surface.

EXTENDED SURFACE HEATING UNIT A heating unit having a relatively large amount of extended surface that may be integral with the core containing the heating medium or assembled over such a core, making good thermal contact by pressure, or by being soldered to the core or by both pressure and soldering. (An extended surface heating unit is usually placed within an enclosure and therefore functions as a convector.)

FAHRENHEIT A thermometer scale at which the freezing point of water is at 32° and its boiling point is 212° above 0. It is generally used in the United States for expressing temperature.

FLASH (Steam) The rapid passing into steam of water at a high temperature when the pressure it is under is reduced so that its temperature is above that of its boiling point for the reduced pressure. For example: If hot condensate is discharged by a trap into a low pressure return or into the atmosphere, a certain percentage of the water will be immediately transformed into steam. It also is called re-evaporation.

FLOAT AND THERMOSTATIC TRAP A float trap with a thermostatic element for permitting the escape of air into the return line.

FLOAT TRAP A steam trap operated by a float. When enough condensate has drained (by gravity) into the trap body, the float is lifted which in turn lifts the pin off its seat and permits the condensate to flow into the return until the float has been sufficiently lowered to close the port. Temperature does not affect the operation of a float trap.

FURNACE That part of a boiler or warm air heating plant in which combustion takes place. Sometimes also the complete heating unit of a warm air heating system.

GREENHOUSE PRINCIPLE A method of converting and trapping radiation from the sun in the form of heat.

GRILL A perforated covering for an air inlet or outlet usually made of wire screen, pressed steel, cast-iron, or other material.

HEAD Unit of pressure usually expressed in ft of water.

HEAT That form of energy into which all other forms may be changed. Heat always flows from a body of higher temperature to a body of lower temperature. *See also*: Latent Heat, Sensible Heat, Specific Heat, Total Heat, Heat of the Liquid.

HEAT OF THE LIQUID The heat (Btu) contained in a liquid due to its temperature. The heat of the liquid for water is 0 at 32° F and increases 1 Btu approximately for every degree rise in temperature.

HEAT TRANSFER FLUID The medium used to transfer heat from the collector to the heat storage tank.

HEATING MEDIUM A substance such as water, steam, or air used to convey heat from the boiler, furnace, or other source of heat to the heating units from which the heat is dissipated.

HEATING SURFACE The exterior surface of a heating unit. *See also* Extended Heating Surface.

HEATING UNIT Radiators, convectors, baseboards, finned tubing, coils embedded in floor, wall, or ceiling, or any device which transmits the heat from the heating system to the room and its occupants.

HOT WATER HEATING SYSTEM A heating system in which water is used as the medium by which heat is carried through pipes from the boiler to the heating units.

HUMIDISTAT An instrument that controls the relative humidity of the air in a room.

HUMIDITY The water vapor mixed with air.

HYSTERESIS CIRCUIT A special circuit of the differential thermostat which prevents the pump from being unnecessarily turned on and off immediately following an initial pump turn-on or turn-off event.

INSULATION The total amount of solar energy reaching a surface per unit of time.

INSULATION (Thermal) A material having a high resistance to heat flow.

LATENT HEAT OF EVAPORATION The heat (Btu/lb) necessary to change 1 pound of liquid into vapor without raising its temperature. In round numbers this is equal to 960 Btu/lb of water.

MECHANICAL EQUIVALENT OF HEAT The mechanical energy equivalent of 1 Btu, which is equal to 778 foot-pounds.

ONE-PIPE SUPPLY RISER (Steam) A pipe which carries steam to a heating unit and which also carries the condensation from the heating unit.

In an up-feed riser, steam travels upwards and the condensate travels downward while in a down-feed both steam and condensate travel down.

ONE-PIPE SYSTEM (Hot Water) A hot water heating system in which one pipe serves both as a supply main and also as a return main. The heating units have separate supply and return pipes but both are connected to the same main.

ONE-PIPE SYSTEM (Steam) A steam heating system consisting of a main circuit in which the steam and condensate flow in the same pipe. There is but one connection to each heating unit which must serve as both the supply and the return.

OVERHEAD SYSTEM Any steam or hot water system in which the supply main is above the heating units. With a steam system the return must be below the heating units; with water, the return may be above the heating units.

PANEL HEATING A method of heating involving the installation of the heating units (pipe coils) within the wall, floor, or ceiling of the room.

PANEL RADIATOR A heating unit placed on, or flush with, a flat wall surface and intended to function essentially as a radiator. Do not confuse with panel heating system.

PASSIVE SOLAR SYSTEM Solar systems which use the walls of the building as the collector and the heat storage medium.

PLENUM CHAMBER An air compartment maintained under pressure and connected to one or more distribution ducts.

PRESSURE Force per unit area such as pound per square inch. *See* Static, Velocity, Total Gauge, and Absolute Pressures. Unless otherwise qualified, it refers to unit static gauge pressure.

PRESSURE REDUCING VALVE A piece of equipment for changing the pressure of a gas or liquid from a higher to a lower one.

PRIME SURFACE A heating surface having the heating medium on one side and air (or extended surface) on the other.

PYRONOMETER A device used to measure solar insulation in Btu per square foot per hour.

RADIANT HEATING A heating system in which the heating is by radiation only. Sometimes applied to Panel Heating Systems.

RADIATION The transmission of heat in a straight line through space.

RADIATOR A heating unit located within the room to be heated and exposed to view. A radiator transfers heat by radiation to objects it can "see" and by conduction to the surrounding air which in turn is circulated by natural convection.

RELATIVE HUMIDITY The amount of moisture in a given quantity of air compared with the maximum amount of moisture the same quantity of air could hold at the same temperature. It is expressed as a percentage.

RETURN MAINS The pipes that return the heating medium from the heating units to the source of heat supply.

SENSIBLE HEAT Heat which only increases the temperature of objects as opposed to latent heat.

SOLAR ENERGY The energy received by the earth from the sun.

SPECIFIC HEAT In the foot-pound-second system, the amount of heat (Btu) required to raise 1 pound of a substance 1° F. In the centimeter-gram-second system, the amount of heat (cal) required to raise 1 pound of a substance 1° C. The specific heat of water is 1.

SPLIT SYSTEM A system in which the heating system utilizes radiators or convectors and ventilation by a separate apparatus.

SQUARE FOOT OF HEATING SURFACE Equivalent direct radiation (EDR). By definition, that amount of heating surface which will give off 240 Btu/hr when filled with a heating medium at 215° F and surrounded by air at 70° F. The equivalent ft^2 of heating surface may have no direct relation to the actual surface area.

STATIC PRESSURE The pressure which tends to burst a pipe. It is used to overcome the frictional resistance to flow through the pipe. Expressed as a unit of pressure, it may be given in either absolute or gauge pressure. It is frequently expressed in ft of water column or (in the case of pipe friction) in mil-in. of water column per foot of pipe.

STEAM Water in the vapor form. The vapor formed when water has been heated to its boiling point, corresponding to the pressure it is under. *See also* Dry Saturated Steam, Wet Saturated Steam, Super Heated Steam.

STEAM HEATING SYSTEM A heating system in which the heating units give up their heat to the room by condensing the steam furnished to them by a boiler or other source.

STRAINER A device such as a screen or filter used to retain solid particles while the liquid passes through.

SUPERHEAT Temperature of a vapor above the boiling temperature of its liquid at that pressure.

THERMOCOUPLE A device which generates electricity, using the principle that if two dissimilar metals are welded together and the junction is heated, a voltage will develop across the open ends.

THERMOMETER A device for measuring temperatures.

THERMOSTAT A device which is responsive to ambient temperatures.

VALVE, REVERSING A valve used to reverse the flow of a refrigerant depending upon whether heating or cooling is desired.

VAPOR A word used to denote a vaporized fluid rather than the word gas.

VAPOR BARRIER A thin plastic or metal foil used in air conditioning structures to prevent water vapor from entering the insulating material.

VAPOR, SATURATED A vapor condition which will result in the condensation of droplets of liquid as the vapor temperature is reduced.

WET BULB A device used in the measurement of relative humidity. Evaporation of moisture lowers the temperature of a wet bulb compared to a dry bulb thermometer in the same area.

Index

W